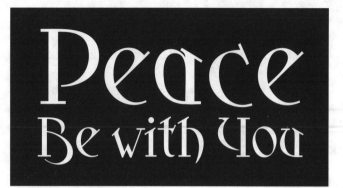

Peace
Be with You

Peace Be with You

Justified Warfare or the Way of Nonviolence

EILEEN EGAN

Maryknoll, New York 10545

The Catholic Foreign Mission Society of America (Maryknoll) recruits and trains people for overseas missionary service. Through Orbis Books, Maryknoll aims to foster the international dialogue that is essential to mission. The books published, however, reflect the opinions of their authors and are not meant to represent the official position of the society. To obtain more information about Maryknoll and Orbis Books, please visit our website at www.maryknoll.org.

Published by Orbis Books, Maryknoll, NY 10545-0308

Manufactured in the United States of America

All biblical citations, unless otherwise indicated, are from the Catholic Study Bible.

Library of Congress Cataloging-in-Publication Data

Egan, Eileen.
Peace be with you : justified warfare or the way of nonviolence / Eileen Egan.
p. cm.
Includes index.
ISBN 1-57075-243-5 (pbk.)
1. Just war doctrines. 2. War—Religious aspects—Catholic Church—History of doctrines. 3. Catholic Church—Doctrines—History. 4. Day, Dorothy, 1897–1980. I. Title.
BX1795.W37E43 1999
261.8′73–dc21 99–17247

With great gratitude
this book is dedicated to
Bernice McCann-Doyle, first reader and critic,
whose immense aid included
keyboarding the text into a computer.
Gratitude is extended
to her generous, peacemaking husband,
Jack Doyle, leader of work for the homeless,
and their children, Sharon and John.

Contents

Preface

This book arose out of a time in the twentieth century when humankind had come to rival the Creator with regard to control of the earth. Creation could now be undone; it could be threatened with destruction.

In case this assertion sounds apocalyptic, we only need to turn to a statement of the bishops of the United States. In their peace pastoral of 1983 they asserted, "We are the first generation since Genesis with the power to virtually destroy God's creation."

It fell to me as a member of the staff of Catholic Relief Services to become part of the lives of peoples plunged into all but inconsolable loss and unbearable misery in the wake of mass violence in Europe. In Asia I experienced the tragic displacement of millions of human beings also wreaked by violence.

One of the things that pierced my heart during these experiences was the realization that followers of Jesus offered justification for all this violence. Along with this was general acceptance of a system whereby governments could order their citizens to starve, maim, kill, and fragment fellow human beings. Justified warfare, with its license to inflict limitless destruction on human life and all that supports it, survived unscathed through the carnage of World War II.

An incident in Europe shortly after the end of World War II remains in my memory. Standing in the ruins of a German city, I saw around me not arms factories or rail junctions but acres of homes, their jagged walls open to the sky. My companion, also from the United States, explained that the bombing attacks on civilian areas were covered by the principle of "double-effect." The aim, the destruction of the war machine, caused a secondary effect, the destruction of those who were presumed to support the machine. "Even families in residential neighborhoods, far from military targets?" I asked. "Of course," was the reply. "There is also the morale factor. Only by destroying the morale of the enemy can a war be brought to an end. We had a just cause for our policies." This was a man of the cloth speaking.

While millions of followers of Jesus were living by the just war teaching, a small man in India, Gandhi, was living by the Sermon on the Mount in a nonviolent response to the oppression and humiliation imposed by an empire and its soldiers. He was teaching the peoples of the West the methods of resistance to evil without becoming part of the evil.

A new moment then arrived for members of the Catholic Church. Indiscriminate warfare was condemned "as a crime against God and man himself" by the bishops of the entire world at the Second Vatican Council. Conscientious objection to military service was validated, as was the witness of nonviolence as alternative service in times of conflict.

The Catholic bishops of the United States praised the nonviolent witness of Gandhi, along with that of Dorothy Day and Martin Luther King Jr., in the 1983 peace pastoral. They acknowledged, however, that "the just war teaching has clearly been in possession for the past 1500 years of Catholic thought." Prior to that fifteen hundred years, the tradition of gospel nonviolence, of living by Jesus' Sermon on the Mount, had moved his followers to love and do good even to their enemies. Transformed by grace they helped transform their society.

The changeover from the nonviolence of the early church to killing in justified warfare occurred as Roman philosophy, in particular that of the great philosopher Cicero, permeated church leaders. Two theological giants, Augustine and Ambrose, borrowed from Cicero in formulating the conditions which could justify organized killing in defense of the community. The Roman and the Galilean clashed on the issue of justified warfare, and the Roman prevailed.

When weaponry in the hands of humankind necessarily renders warfare indiscriminate, bringing suffering to the innocent, to noncombatants in general (and in nuclear warfare, to both sides of a conflict), just war conditions are no longer relevant or applicable.

A changeover in the opposite direction is already in progress, a changeover from justified violence to the nonviolence of Jesus. This turnabout is as important to history as that which occurred fifteen hundred years earlier.

It can be maintained that the thought of many members of the church is no longer held in captivity by just war teaching.

With the closing of the option of resolving conflict, righting wrongs, or achieving justice by the means of violence, the people of God need to pray for grace to accept and live by the implications of nonviolence. Because the resort to violence may now include the undoing of creation, many of those committed to the teachings of Jesus are realizing that there is a profound need to return to those teachings that came from the lips of the Messiah.

Acknowledgments

Beyond those who appear in the pages of this book, there are many members of the peace community who have sustained and inspired me over the years. I name a few co-workers in Pax Christi–USA: Mary Evelyn Jegen, SND, Gordon Zahn, Joseph Fahey, Gerard Vanderhaar, Richard McSorley, SJ, Bishop Thomas Gumbleton, Bishop Walter Sullivan, Antonia Malone, Mary Lou Kownacki, OSB, Marilyn Bertke, OSB, Anne McCarthy, OSB, Nancy Small, and Dave Robinson; representing Pax Christi in our consultative status at the United Nations: Mary Beth Reissen, SSND, Michael Hovey, Joan and Vincent Comiskey, and Edith Ryan, SND. These persons have spoken before various commissions on such issues as child welfare, disarmament, development, and the right to refuse to kill. Working from Pax Christi headquarters in Brussels, Magda and Etienne De Jonghe and the international team have helped those of us at Pax Christi–USA to see ourselves as members of an international community of peacemakers, whether with Valerie Flessati and Bruce Kent of England or Joseph and Rita Camilleri of Australia. Always giving the witness of lives lived in simplicity and the nonviolence of Jesus are the staff of the Catholic Worker, all volunteers. I mention a few at the New York house: Jane Sammon, Frank Donovan, Katharine Temple, Cathy Breen, Patrick and Kathleen Jordan, and Jeannette Noel. Associated with the Catholic Worker in Albany are Mabel Maureen and her husband, Deacon Joseph Gil.

Among other champions of peace, justice, and nonviolence I have been privileged to know are Richard Deats and John Dear, SJ, of the Fellowship of Reconciliation; Mairead Corrigan Maguire of the Peace People of Ireland; Devendra Kumar Gupta and his wife, Prabha, of the Gandhian movement in India; and Niall O'Brien, missioner and defender of justice for the poor in the Philippines.

For all these and for many more, I thank God, who has allowed me to know them, to learn from them, and to communicate with them in the journeying toward, and the fashioning of, the peaceable kingdom.

My deep gratitude is owed to my editor, Robert Ellsberg, who manifested his signal gifts as an editor in his younger years at the *Catholic Worker* and who has been firm and helpful in his suggestions for cutting and redesigning *Peace Be with You.*

Part One

THE WAY OF JUSTIFIED WARFARE

Chapter 1

Will You Restore the Kingdom of Israel?

Conversation on the road to Emmaus:

We were hoping that he was the one who would set Israel free. Besides all this, today, the third day since these things happened, some women of our group have brought us some astonishing news. They were at the tomb before dawn and failed to find his body. —LUKE 24:21–23; JERUSALEM BIBLE

The twentieth century, bloodiest of all centuries, when the heart of the atom was invaded in the search for fire to unleash against humankind, was the century in which war became total. Of all the events of an eventful century, this actuality was the most fateful for the future of human beings and their earthly home.

The century saw two wars destructive enough to be termed world wars. In World War I, called the Great War, more than ten million persons were estimated to have died, of whom 5 percent were civilians. Another twenty million were calculated to have been wounded.

In World War II, often referred to as the Good War,[1] the most horrendous orgy of bloodletting in human history, fifty-two million persons were estimated to have perished, with 48 percent considered to be civilian noncombatants.

Both wars came to birth in Europe, ancient womb of Western Christianity, before involving, to a greater or lesser degree, the peoples of the world. The followers of Jesus were conscripted by the millions for the dread task of killing their kind, the religion of Jesus having become irrelevant to the mutual slaughter of modern warfare.

Never was the rhetoric of war's redemptive violence preached with more intensity, one could call it religious intensity, than in World War II by the Allied powers. Violence was the means to rid humanity of palpable evils. The cause, the destruction of the National Socialist regime of Germany, was so surely just, and the evil to be expunged so unspeakably dreadful, that heights of violence never before achieved in warfare were not questioned.

Moving about our world bloodied by human slaughter, and called upon to describe human affliction for an agency of mercy, led me to search more deeply than I had hitherto for reasons people give to justify the violence of war against their kind. More urgently, it impelled me to concentrate on how the followers of Jesus could in good conscience take part in the infliction of violence, in particular as part of the social organization of large numbers of human beings for the legalized killing and destruction known as war.

The Messiah and the Kingdom

The longing for the Messiah, the Anointed One, was deep in the soul of the disciples of Jesus, as it had been in the people of Israel. Like their fellow Jews, the disciples lived in the conviction of the imminent realization of Israel's restoration to freedom. Their Messiah was to redeem Israel, not only spiritually but in the temporal sense. Their lives were lived in the ardent hope that the Messiah would liberate the chosen people of God from their suffering under a pagan power. The expectation was that after his coming they would be freed from the humiliation of seeing Jerusalem under the heel of idolatrous occupiers.

A biblical scholar, John L. McKenzie, describes this yearning: "The restored Kingdom would be an establishment in which Yahweh would reign supreme over all the nations, and his chosen people Israel, under their chosen king David, would share in the universal reign of Yahweh.... I believe that it can be said that the Messiah of popular Jewish hopes was a national, patriotic and political hero. This is to say that the expected messianic reign was inextricably linked with the political fulfillment of the people of God as supreme over all other peoples."[2]

Christians, on the other hand, are reminded as in the peace pastoral of the American bishops that "there is no notion of a warrior God who will lead the people in an historical victory over its enemies in the New Testament."[3] The war discussed in the New Testament is depicted in the Book of Revelation as the eschatological struggle between God and Satan. It is a war in which the lamb is victorious (Rev. 17:14).

Faithful followers of the covenant were awaiting the day of the redeemer when Jerusalem, now occupied, would no longer be dishonored. It would be the glorious capital of the messianic kingdom to which the nations would resort in fulfillment of the Hebrew Scriptures.

The prophets, including Isaiah, envisioned the kingdom. "In days to come the mountain of the Lord's house shall be established as the highest mountain and raised above the hill. All waters shall stream toward it; Many people shall come and say: Come, let us climb the Lord's mountain, to the house of the God of Jacob, that he may instruct us in the ways, and we may walk in his paths. For from Zion shall go forth instruction and the word of the Lord from Jerusalem" (Isa. 2:2–3).

The expectation of such a redeemer marked the people of Israel as distinct from all other peoples. It was the reason that they maintained their separate identity while living under the Persian and Greek empires, and during periods of exile away from Jerusalem. Jerusalem, and above all the Temple, was the heart of their covenant with Yahweh. The men of Israel were willing to kill and die to free Jerusalem, and they struggled fiercely against those who suppressed the Jewish religion. One example is seen in the armed resistance to Antiochus IV Epiphanes (the latter name means "god manifest") in 164 B.C.E. Antiochus attempted to enforce apostasy on the Jews. It was he who had caused an altar of Zeus to be placed in the sanctuary of the Jewish Temple. Mattathias, the priest, and his five sons had resisted this abomination, since on the Temple altar the Jews sacrificed to the one God, Creator of the universe. The Jewish people resisted the order that sacrifice to the deities of Greece be offered in the towns and villages of Judea. Jewish resistance to apostasy became a war led by Judas, the son of Mattathias, "and they carried on Israel's war joyfully" (1 Macc. 3:1).

Though they were a few against many, Judas encouraged his soldiers, announcing that "victory in war does not depend on the size of the army but on the strength that comes from heaven" (1 Macc. 3:19). The Jews had no doubt that their cause was God's cause.

The bravery in battle of the men who came to be known as the Maccabees (the name comes from their leader, Judas Maccabeus [the hammer]) was matched by the heroic resistance of the Jewish people. They were willing to suffer even death for the purity of their religious practice. When circumcision was forbidden under pain of death, mothers whose infants were circumcised were put to death with their children suspended from their necks (1 Macc. 2:60–63).

Antiochus, as part of his aggressive campaign of Hellenization, ordered Jews to eat foods that were considered unclean and therefore unlawful. A poignant example of resistance is that of a mother and her seven sons. All were imprisoned for their refusal to eat the meat of pigs. Despite torture by whips and scourges, they maintained their faithfulness to the law. In exhortation to her sons, the mother told them that it was not she who had given them the breath of life. "It is the Creator of the universe who shapes each man's beginnings," she told them. In words that reflect a conviction of the afterlife, she comforted them with the hope that "he in his mercy will give you back both breath and life, because you now disregard yourselves for the sake of his law."

Antiochus made a special effort with offers of riches to bring about the apostasy of the youngest son. The young man responded, "I will not obey the king's command. I obey the command of the law given to our forefathers through Moses." That day he joined his mother and brothers in death (2 Macc. 7:20–21).

The holy Maccabees are commemorated in the liturgy of the Catholic Church on the first of August every year.[4] There are readings from the books of the Maccabees during the liturgy for Catholic congregations at various times during the year. Especially commemorated is St. Eleazar, an aged man and chief of the scribes who chose martyrdom rather than transgress the law. The persecutors attempted by bribery and threats to force him to comply with the edict of Antiochus to eat unclean meat. When they failed, some of those in charge of the ritual meal offered to allow him to secretly substitute his own meat so as to circumvent the law. In this way he could escape torture and death.

The holy man refused the subterfuge, pointing out that by such an act he, a ninety-year-old man, would lead young people astray. "Even if, for the time being, I avoid the punishment of men, I shall never, whether alive or dead, escape the hands of the Almighty. Therefore, by manfully giving up my life now, I will prove myself worthy of my old age, and I will leave to the young a noble example of how to die willingly and generously for the revered and holy laws" (2 Macc. 6:26–28).

The martyrdom of Eleazar, and of so many of those persecuted by Antiochus, was a vivid memory for the Jews of Jesus' time and depicted the cost their ancestors had paid for fidelity.[5]

In 164 B.C.E., the armed struggle of the Maccabees was crowned by victory and won for the people of Israel control of Jerusalem. In celebration of the victory, Judas Maccabeus took his whole army to Mount Zion. The desecrated altar of the Temple was destroyed, and the Temple was rededicated with singing and the playing of harps, flutes, and cymbals. The lamps on the lampstands newly illuminated the Temple, and sacrifices were restored to the renewed altar of holocausts. Judas and his brothers decreed that the anniversary of the rededication of the altar should be marked with joy for eight days every year (1 Macc. 4:36; 37:52–59). This festival, Channukah, the festival of lights, has been celebrated by Jewish people throughout the world since that time.

A war carried on joyfully and blessed by heaven because it was fought for the preservation of the covenant was deep in the consciousness of the House of Israel. Would not the strength of heaven empower the people of Israel to throw off subjugation by Rome?

Is This the Time?

The proclamation of the kingdom came from an unlikely source, a man garbed in rough garments and subsisting on food found in the barren desert of Judea. He was John, announcing to his fellow Jews, "Repent, for the kingdom of heaven is at hand" (Matt. 3:2). When the people of Jerusalem and all Judea came to him, he had them immerse themselves in the Jordan River as they acknowledged their sins. Repentance was the entryway to the kingdom announced by John the Baptizer, as he proclaimed the mighty one to come who would baptize with the Holy Spirit and fire.

When Jesus joined the throng asking for baptism, John tried to prevent him, but Jesus insisted. It was then that the heavens were opened for Jesus; the Spirit descended on him like a dove; and he was proclaimed by a voice from above as the son of God (Mark 1:1–11). The Messiah would lead the earthly realm of God's people.

The coming of the Messiah was to signal the lifting of the yoke of subjugation over the House of Israel. The disciples chosen by Jesus were as deeply imbued with the yearning for freedom and for the messianic kingdom as were their suffering people. They had lived from childhood with the living hope that the Messiah would liberate Israel. They accepted John's message of repentance, contained in the word *shub,* a turning away from evil ways, a turning toward God. They rejoiced that there was among them one who performed "signs" never before seen by humankind. As they followed him in the towns and villages of Israel, they looked on as he cleansed in the twinkling of an eye lepers whose flesh was rotting. They were present as he gave sight to the blind and fed five thousand from a few small loaves and a couple of fish. He had calmed the storming waters of the Sea of Galilee. But all these were as nothing when compared with his power to bring back to life those who had entered the kingdom of death.

Lazarus walked among them. The power of God had been exercised mightily before their eyes and in their hearing.

Who was this Jesus who was performing these "signs" by the power of the Spirit of God, which descended upon him at his baptizing by John? He taught them to pray, using the word "Abba," the familiar name for father. He told them that in their search for the kingdom, they should put all things aside and rely on Abba for their needs. He began to reveal to his disciples that they would have to make another *shub,* another turn, away from their growing hope for an earthly kingdom. They were to expect scourgings and arrests and times when they would face hostile interrogations. They were to rely on the Spirit of the Father "speaking through you" (Matt. 10:20).

The apostles were given signs by Jesus so that they could be strengthened for the trials to come. Three of the apostles were with him when he was suddenly transformed into a dazzling white figure in the act of conversing with Moses and Elijah. At that point, the terrified three were overshadowed by a cloud from which issued a voice: "This is my beloved Son with whom I am well pleased; Listen to him" (Matt. 17:5). They needed the reassurance that Jesus gave them, and their fears subsided.

The greatest "sign" which Jesus gave his apostles for their future life was his last meal with them. On the eve of his death on the cross, he instituted the covenant meal by which they, and all his followers until the end of time, could become one with him. Breaking the bread, as his body would be broken in the coming day, he told them to eat it. Using words of unutterable power and love, he said, "This is my body." Passing the cup he told them to drink it with the words, "Drink from it all of you for this is my blood of the covenant which is to be shed on behalf of many for the forgiveness of sins" (Matt. 26:26–28). This was his legacy for his followers, a legacy demonstrated every time they meet at the Table of the Lord. It would be the indissoluble bond uniting them to the sacrifice consummated with the crucifixion and resurrection of the Messiah.

Those who had shared the body and blood of divinity itself and who had experienced the power of the inbreaking of the eternal into their daily lives wondered if this might not be the time when the earthly kingdom would be restored. After the crucifixion, before word had come about Jesus' resurrection, the disciples met him on the road to Emmaus. Unaware of his identity, they explained to Jesus, "We were hoping that he was the one who would set Israel free" (Luke 24:21). They needed grace to discern the reality of the freedom that Jesus brought to them.

Will You Restore the Kingdom?

When the first community of Christians was gathered in Jerusalem after the resurrection, their mission was still not clear to them. They put the question to Jesus: "Lord, are you at this time going to restore the kingdom of Israel?" (Acts

1:6). This is one of the most unforgettable and poignant questions in the entire New Testament.

In his reply to the question regarding the restoring of the kingdom, Jesus told the disciples: "It is not for you to know the times or seasons that the Father has established by his own authority." That reply might have confounded them except for the promise that immediately followed: "But you will receive power when the Holy Spirit comes upon you, and you will be my witnesses in Jerusalem, throughout Judea and Samaria, and indeed to the ends of the earth" (Acts 1:7–8).

As the disciples were standing with him, a cloud took Jesus from their sight: "While they were looking intently at the sky as he was going, suddenly, two men dressed in white garments stood beside them. They said, 'Men of Galilee, why are you standing there looking at the sky? This Jesus who has been taken up from you into heaven will return the same way as you have seen him going into heaven'" (Acts 1:10–11).

The disciples had not long to wait. It was ten days to the feast of the fiftieth day, Pentecost, marking the firstfruits of the harvest: "When the time of Pentecost was fulfilled, they were all in one place together. And suddenly there came from the sky a noise like a strong driving wind, and it filled the entire house in which they were. Then there appeared to them tongues as of fire, which parted and came to rest on each one of them. And they were all filled with the Holy Spirit and began to speak in different tongues, as the Spirit enabled them to proclaim" (Acts 2:1–4).

With the descent of the Spirit and the shining tongues of fire, the hearts of the apostles became tremblingly aware of the new power within them. The dream of a restored Israel was shattered before a vision of a reign of hearts open to the Spirit of God. An unstoppable power emanated from them. Standing with the eleven, Peter raised his voice and began speaking fearlessly to the Jews living in Jerusalem, Jews "from every nation under heaven." In amazement they heard his message as if it had been spoken in their own tongue (Acts 2:5–6).

Choosing passages from the Hebrew Scriptures, Peter presented the prophecies that concerned Jesus, citing the words of the prophet Joel and King David. He made it clear that these prophesies were fulfilled in the crucifixion and resurrection of Jesus. He ended with the words, "Therefore let all the House of Israel know for certain that God has made him both Lord and Christ — this Jesus whom you crucified" (2:36). His hearers were "pierced to the heart, crying, 'Brethren, What shall we do?'" (Acts 2:37). About three thousand were baptized on that day (Acts 2:37–42).

Jesus had not deserted his followers. They could call on the Spirit and find nourishment at the Table of the Lord through the living bread and saving cup. No human power could vanquish those so empowered, and so nourished.

It was settled for the twelve and for the other disciples. The matter was closed. The restoration of the territorial political Kingdom of Israel had not been the mission of the Messiah. The community of disciples were to learn that the kingdom was within each one of them, within each human soul. It was a kingdom which could not be defended by weapons.

For many in the House of Israel the Messiah had not come in the person of Jesus. When they saw before them Jesus, the man of sorrows, and Barabbas,[6] the convicted murderer, they did not hesitate in replying to Pilate's question, "Which of the two do you want me to release to you?" Barabbas was their choice (Matt. 27:16). It must be added that they were urged on to make this choice by the chief priests and elders. In Matthew, Barabbas is described as "notorious" (Matt. 27:16). Mark explains that Barabbas had been convicted with a band of rebels who in the course of the rebellion had committed murder (Mark 15:7).

To many members of the Jewish community, seething under Roman rule, Barabbas would be viewed as a hero, a freedom fighter, according to John M. Oesterreicher.[7] He was a man who had risked his life to redeem Israel from alien domination. It was possible that he had been part of one of the smaller, failed uprisings, of which there had been many. The Romans had experienced a series of such revolts since Judea had been added to their empire by Pompey (Gnaeus Pompeius Magnus) in 63 B.C.E.

One such revolt arose in Sepphoris, capital of the province of Galilee in 4 B.C.E. under one who was called Judas the Galilean. Sepphoris was approximately five miles from Nazareth. Judas supplied weapons to his comrades-in-arms by raiding the arsenal of the Roman client-king Herod. The Jewish historian Josephus described the horrendous response of the Roman military leader, Varus, who crushed the rebellion. The town of Sepphoris was ordered burned to the ground and the survivors enslaved. As a deterrent to further insurrections, Varus then "sent a part of his army into the country, against those who had been the authors of this commotion, and as they caught great numbers of them, those that appeared to have been the least concerned in these tumults he put into custody, but such as were the most guilty he crucified; these were in number about two thousand."[8]

Such repression, with rebels on crucifixes lining the roadsides, was part of the very air breathed by members of the Jewish community, including the early followers of Jesus. According to some calculations, the destruction of Sepphoris and the mass crucifixion took place a few years before the birth of Jesus. It was undoubtedly a searing memory for the people of Nazareth, since some of the insurrectionists might well have come from Nazareth itself. According to other historians, if Jesus' birth is figured as occurring in 6 B.C.E., Jesus himself might have viewed the dread spectacle of agonizing human figures, tribute to the might and cruelty of Rome. In any case, the tragedy of Sepphoris must have been part of the conversation around Jesus in his boyhood. If the date of 4 B.C.E. is arrived at for the revolt, Jesus might well have walked the five miles to Sepphoris and seen the spots where the crosses had been planted in the earth.

Was there reason for ordinary Jews to sympathize with Judas the Galilean, with the victims of Sepphoris, and with Barabbas?

In *The New Encounter between Christians and Jews,* John M. Oesterreicher put it this way: the freedom fighters "held firmly that God's kingdom — rather His reign — could not come as long as the Roman idolaters and their idols polluted the land. Thus, the freedom fighters saw it as imperative that the Romans be expelled by force, if necessary, in blood-drenched combat." Jesus, who claimed to be the Messiah, should have proved his Messiahship by undertaking the task of expelling the Romans and restoring the kingdom of Israel. Oesterreicher continued:

> Here was the one who could have led them and all Israel to victory, yet he refused to act. He could have brought freedom from the idolaters that ruled the land and new life in the sight of the Lord, true Ruler of Israel and of the universe; yet he did not take the needed step. Like Samson he could have laid the Palace in ruins; he could even have made the Empire collapse....
>
> Jesus had become a scandal to them [the freedom fighters]. They were furious, nay fury took possession of them. No wonder they demanded that he be crucified. In a way, they relished the irony: Crucifixion, Rome's punishment for rebels and seditionists, was to become the penalty of him who refused to lead the revolt....
>
> Their line of thinking may have been like this: "Jesus bolted our cause, no, the cause of God; he thus deserves to be punished as a deserter."[9]

Two Crosses

The abyss between those who expected a mighty Messiah who would challenge Rome and those who had a vision of a Messiah who would challenge people's hearts was never more dramatically symbolized than in the juxtaposition of Barabbas and Jesus.

The cross on which Jesus was raised was a scandal to the freedom fighters who believed in direct action, if necessary by the sword. It became, if not a scandal, then certainly a dilemma to his followers, since on the cross Jesus presented a new way to drive out evil. This way, almost incomprehensible to his followers, consisted of the willing acceptance of suffering by the innocent. In order to make the sacrifice of the cross comprehensible there had to be an opening of the heart and a turning away from old ways.

The new vision transformed the concept of the kingdom from an earthly one to a reign of the Spirit. Only those whose lives were spiritually transformed could pursue the kingdom by "weapons" never before taught to humankind. This arduous but necessary transformation had hardly been glimpsed by those in Jesus' company, including Peter, who drew his sword to defend Jesus from those who came to arrest him. The words of Jesus to Peter, "Put away your sword," have come down the ages as a call to the followers of Jesus to relinquish the sword even in the pursuit of good ends. Could there have been a better end than defending Jesus from unjust arrest?

The violence of Peter's sword, in striking off the ear of Malchus, was undone by the miracle of restoring the ear. What was the significance of the words of Jesus to Peter, "Put away your sword," and of the "sign" by which he restored the ear? Was the former merely a one-time command of the Master applicable to a particular time and a particular place? Or was it more than that? Was it Jesus commanding his followers to do likewise by putting swords out of their lives?

What weapons does Jesus give his followers aside from the weapons of the Spirit? As Peter was left weaponless, so were those others who followed Jesus. It was the weaponlessness of utter nonviolence which separated the members of the Jewish community who followed Jesus from those who rejected him. The word *shub,* in its root of "turn," gave Jesus' followers a new path that changed lives. Their "weapon" was to be the cross, the cross of suffering love.

The two crosses influenced the history of the world: that of Barabbas, the freedom fighter who risked his life for a cause and paid the penalty, and that of Jesus, the Lamb of God who, out of love, paid the penalty to redeem the human family from sin.

Ancient Wound: New Encounter

Any treatment of peace and the church could hardly avoid the tragic wound that opened between the House of Israel and the Jewish believers in Jesus.

In the twentieth century a light appeared in a new encounter between Christians and Jews. It emanated from the world's Catholic bishops meeting in the Second Vatican Council in Rome and was inspired by such leaders as John M. Oesterreicher, whom I was privileged to know as a friend. After deliberation by hierarchs who had come together from every corner of the globe, a declaration was promulgated on October 28, 1965, in Rome. Its overall subject was the relationship of the Catholic Church to non-Christian religions. Of crucial importance was the section devoted to "the spiritual bond linking the people of the New Covenant with Abraham's stock." The declaration, called *Nostra Aetate* (In Our Time), informed the world that "Christ underwent his passion and death freely out of infinite love because of the sins of all humanity so that all may obtain salvation."[10]

The document points out that

> while the Jewish authorities and those who sided with them pressed for the death of Christ (John 19:6), still, what happened in His passion cannot be attributed indiscriminately to all Jews then alive, nor can it be attributed to the Jews of today....
>
> The church, moreover, rejects all persecution against all human beings. For this reason and for the sake of her common patrimony with the Jews, she decries outbursts of hatred, persecution, and manifestations of contempt directed against the Jews at whatever time in history and by whomsoever.

Slightly earlier in the declaration, the bishops state:

> The Church of Christ acknowledges that, according to the mystery of God's saving design, the beginnings of her faith and her election are already found among the patriarchs, Moses, and the prophets. She professes

> all who believe in Christ, Abraham's sons according to faith (Gal. 3:7), are included in the patriarch's call, and likewise that the salvation of the church was mystically foreshadowed by the chosen people's exodus from the land of bondage.
>
> The church, therefore, cannot forget that she received the revelation of the Old Testament through the people with whom God in his inexpressible mercy deigned to establish the ancient covenant. Nor can she forget that she draws sustenance from that good olive tree onto which had been grafted the wild olive branches of the Gentiles (Rom. 11:17–24). Indeed, the church believes that by His cross, Christ, our peace, reconciled Jew and Gentile, making them both one in Himself (Eph. 2:14–16). (*Nostra Aetate* 4)

Would that such teaching had been imprinted more deeply on the minds and hearts of Christians by the leaders of the church in earlier ages. Rather than lament, however, we must give thanks. In the fulfillment of time, the entire leadership of the church, not a part of it, not a few bishops here or there, reasserted the church's roots in ancient Israel. These roots uninterruptedly sustain the church's life, its teaching, and its worship.

The Break between the Covenants

The fateful break between the followers of Jesus — the new covenant — and those who maintained their ties to the old covenant took its root in the question of violence versus nonviolence.

Jesus had announced: "This is the time of fulfillment. The kingdom of God is at hand. Repent and believe in the gospel" (Mark 1:15). Those who could not accept Jesus could have seen it this way: having announced that he was the promised one, the Anointed One of the Most High, he had left them as he found them, bound and subject to a pagan empire. Jesus had gone about proclaiming that he had brought the "good news" (*besorah tovah*) to the human family, but, they might ask, "Where are the signs of the good news?" They saw no sign of the *besorah tovah* around them. Thus, even if they had begun by listening to Jesus, they "returned to their former way of life and no longer accompanied him" (John 6:66).

While the difference between the communities sprang from the abyss between violence versus nonviolence, there was another gulf that the followers of Jesus had to confront, whether his message was for the chosen people of God or for humankind. The issue of universality entered when the question arose regarding the obligation of Jewish Christians to observe Jewish laws. Among the questions was that of the imposition of circumcision on Gentile men. Paul answered this by taking Titus, a Gentile, with him to Jerusalem and announcing that there would be no compulsory circumcision (Gal. 2:3–5).

Another issue concerned the necessity of Gentiles to obey Jewish laws on food. Should the Gentiles be bound by Jewish laws which included the refusal to eat foods considered unclean? Jews were mindful of how Eleazar and others had suffered death for their obedience to the law against eating the meat of swine. It took a special revelation from on high to clarify this matter for Peter. He was accorded a vision in which four-footed animals, wild beasts, crawling creatures, and birds of the air appeared to him as on a great sheet. Peter shrank from the command to "kill and eat." A voice from heaven reassured him that what God had cleansed could no longer be considered unholy (Acts 11:5–9).

Even eating with Gentiles was frowned upon by those whom Paul referred to as "the party of the circumcision." Paul rebuked Peter for withdrawing from taking meals with Gentiles, a hypocrisy in which he was being joined by other followers of Jesus (Gal. 2:11–16).

It was Peter, however, who received the most crucial revelation when he was brought to a Gentile home in Caesarea. After he began to speak to the people there, the Holy Spirit fell on them. Believing in Jesus Christ, the Gentiles received the same gift as the disciples. "Who was I," asked Peter, "that I should stand in God's way?" (Acts 11:17).

The fiery, fervent Paul included synagogues and groups of Gentiles in preaching the message of Jesus. He and the other disciples frequented the Temple in Jerusalem, many of whose priests joined with the followers of Jesus. Scattered far and wide after the death of Stephen, the disciples went first to the communities of Jews with the word of the risen Christ. Though seen initially as a sect of Judaism, the infant church moved beyond its roots to enter the lives of the many people gathered within the empire of Rome.

It was the Pharisee Paul, as apostle to the Gentiles, and seen as the great renegade by his own kinsmen of the flesh, who drove home the heart of the teaching. "For all of you who were baptized into Christ," Paul announced, "have clothed yourself with Christ. There is neither Jew nor Greek, there is neither slave nor freeman, there is neither male nor female; for you are all one in Christ Jesus. And if you belong to Christ, then you are Abraham's descendant, heirs according to the promise" (Gal. 3:27–29).

Exclusivity gave way to universality.

National Liberation

Though the Jewish believers in Jesus and the Jewish community struggled with their differences, the pivotal break between the two communities came as a result of two wars of national liberation. In both wars undertaken by Jews against the oppression of Rome, the followers of Jesus abstained. In modern terminology, they would be called conscientious objectors to war's violence. The Christian abstention from violence was seen by the Jewish community as treason to the Jewish people, or possibly as a secret alliance with Rome, or as both.

The first war of national liberation was the revolt in the province of Judea in 66 C.E. It was ignited by such provocations as the demand by the Romans for large amounts of gold from the treasury of the Jerusalem Temple. In this uprising the Jewish forces were able to drive the Roman troops out of Jerusalem.

The members of the Jerusalem church, Eusebius of Caesarea points out, did not take part in the violent uprising for the liberation of Israel. Though they had every reason to oppose Rome, the Christians did not participate in the bloodshed necessary to overthrow its rule. Eusebius notes, "To Pella those who believed in Christ migrated from Jerusalem."[11] Some scholars maintain that the migration to Pella across the Jordan occurred after James the first bishop of Jerusalem was killed in 62 C.E.

The movement to Pella might also have been a response to the words of Jesus: "When you see Jerusalem surrounded by armies, you will know that its desolation is near. Then let those who are in Judea flee to the mountains" (Luke 21:20–21).

Between 66 and 68 C.E., the war spread to hundreds of towns and villages in Judea. The Jewish forces held off the Roman troops under Vespasian (Titus Flavius Vespasianus) and declared an independent nation.

One of the commanders of the Jewish forces in the war was Josephus (Joseph ben Matthias). He described himself as the "son of Matthias of Jerusalem" and as "a member of a priestly family," explaining that with Jews, to be of sacerdotal dignity "is an indication of the splendour of the family."[12] Josephus was to write *History of the Jewish War.* He made clear his opposition to extreme Jewish nationalism.

From Josephus's *History of the Jewish War* we learn of the stubborn resistance and atrocious sufferings of the Jews, soldiers and noncombatants. Convinced that his people could only meet defeat and death, he urged a negotiated peace. His advice was disregarded. Titus, the son of Vespasian, ordered his soldiers to build a wall around Jerusalem so that the Jews would then, either in despair or wasted by famine, surrender to the Romans. The holy city under siege was a place of agony. Josephus wrote:

> All hope of escaping was now cut off from the Jews, together with their liberty of going out of the city. Then did the famine widen its progress, and devoured the people by whole houses and families; the upper rooms were filled with women and children that were dying of famine; and the lanes of the city were full of the dead bodies of the aged; the children also and the young men wandered about the market places like shadows, all swelled with famine, and fell down dead wheresoever their misery seized them. As for burying them, those that were sick themselves were not able to do it; and those that were hearty and well were deterred from doing it by the great multitude of dead bodies, and by the uncertainty how soon they should die themselves; for many died as they were burying others, and many went to their coffins before their fatal hour; but the famine confounded all natural passions; for those who were just going to die, looked upon those who were going to their rest before them with dry eyes and opened mouths.
>
> A deep silence also, and a deadly night, had seized upon the city; while yet the robbers were still more terrible than their miseries were themselves; for they broke open those houses which were no other than graves of dead bodies and plundered them of what they had; and carrying off the coverings of their bodies, went out laughing and tried the points of their swords on their dead bodies; and, in order to prove what mettle they were made

> of, they thrust some of those through that still lay alive on the ground; but for those that entreated them to lend their right hand, and their sword to dispatch them, they were too proud to grant their requests, and left them to be consumed by the famine. Now every one of these died with their eyes fixed upon the Temple.[13]

In thousands of words of horrendous and heart-scalding detail Josephus describes the progress of the war and the fall of Jerusalem, always stressing the need for peace and defending the decisions of Titus. An unforgettable part of the narrative of Josephus recounts how Titus discussed with the leaders of his legions and tribunes what should be done about the Temple, which was occupied by a rearguard of the Jewish army. Some believed that it would be best to act according to the rules of war and demolish it since the Jews would not stop their rebellions while the Temple was still standing. Others thought that if the Jews did not surrender and used the Temple as a military citadel, the evil of burning it would be on the heads of the Jews. Titus wanted the Temple preserved if only the Jews would cease their resistance. He did not want to take revenge upon a building, an inanimate thing that if preserved could be a reminder of Rome's lenient government.[14]

Nevertheless, the Temple was set on fire through the unauthorized actions of soldiers. Josephus relates:

> While the holy house was on fire, everything was plundered that came to hand, and ten thousand of those that were caught were slain; nor was there a commiseration [for] any age, or any reverence of gravity; but children and old men and profane persons and priests were all slain in the same manner; so that this war went round all sorts of men and brought them to destruction, those that made supplication for their lives as well as those that defended themselves by fighting. The flame was also carried a long way and made an echo, together with the groans of those slain.[15]

In the end Titus ordered the holy places and the Temple to be razed to the ground, leaving only a few lofty towers that had escaped the conflagration.

To Titus belonged the victory in the four-year-long Jewish War, which ended in 70 C.E. It was he who brought to Rome the spoils of the Temple and made them part of the procession of victory. Josephus recounts that of all the spoils, the ones that stood out

> were those captured in the Temple at Jerusalem. These consisted of a golden table, many talents in weight, and a lamp stand likewise made of gold, but constructed on a different pattern from those we use in ordinary life. Affixed to a pedestal was a central shaft from which extended slender branches arranged trident fashion, a wrought iron lamp being attached to the extremity of each branch; of these there were seven, indicating the honor paid to that number by the Jews. After these, and last of the spoils, was carried a copy of the Jewish Law.[16]

These treasures can still be seen, depicted in stone, in the arch of Titus that has stood through the centuries in Rome.

Judea to Palestina

The decisive breaking point between the House of Israel and Jewish believers in Jesus came in the second Jewish revolt against Rome. Violence, always seething against the imperial power, erupted when the Roman conquerors decided to erect a temple to Jupiter on the site of the Temple Mount. The defilement of the holy place moved Simon bar Koziba to lead an uprising in 132 C.E. This rebellion became a full-scale war. Eusebius recounts that the revolt was considered so threatening that the governor of Judea received military reinforcements directly from the Roman emperor.

The Jews fought with intensity for their liberation and the liberation of their holiest site, the Temple, symbol of God among his people. For three years the insurrectionists succeeded in withstanding the power of Rome's legions. Rabbi Akiba ben Joseph in the Talmud credited the leader with the title "bar Kokhba" (Kokhba signifies "star"). He was declared the "son of light," the longed-for warrior-messiah.

As in the Jewish War of 70 C.E., Jewish believers in the message of Jesus did not participate in the violence. In the end the revolt was mercilessly crushed. Bar Kokhba and the other leaders were executed. The Romans took vengeance on the Jewish people by banning them from the city of Jerusalem and from the part of the Judean province around the city.

Eusebius relates in *The History of the Church:*

> From that time on (c 135 C.E.) the entire race had been forbidden to set foot anywhere in the neighborhood of Jerusalem under the terms and ordinances of the law of Hadrian which ensured that not even from a distance might Jews have a view of their ancestral soil. Aristo of Pella tells the whole story. When in this way the city was closed to the Jewish race and suffered the total destruction of its former inhabitants, it was colonized by an alien race, and the Roman city which subsequently arose changed its name so that now, in honor of the emperor then reigning, Aelius Hadrianus, it is known as Aelia. Furthermore, as the church in the city was now composed of Gentiles, the first after the Bishops of the circumcision to be put in charge of the Christians was Mark.[17]

The temple of Jupiter was constructed on the site of the Jerusalem Temple. The full name given to Jerusalem was Aelia Capitolina, Capitolina referring to Jupiter. The tragedy of the Jewish people in being banned from their holy city, in seeing idolatry replacing God's Temple, and in being sent into diaspora is beyond imagining.[18]

The name given to the blood-drowned province of Judea was Palestina. For the remainder of the century, Jerusalem remained a place of ruined buildings. According to most accounts, the Temple was not restored for worship.

The followers of Jesus saw their abstention from bloodshed as central to fidelity to his message. The kingdom that he preached to them was not an earthly kingdom to be defended by earthly means. They strove to spread this kingdom of God, at the center of which was love. In the first four centuries after the resurrection of Jesus, ordinary human beings — fishermen, men and women who farmed the fields around the villages, weavers, dyers, and tentmakers — took him at his word. They became a spectacle to the world by the ways in which they gave witness to his word.

There came a time when love was interpreted in a vastly different way. As the American bishops explain it in their peace pastoral, "Faced with the fact of an attack on the innocent, the presumption that we do no harm even to our enemy, yielded to the command of love understood as the need to restrain an enemy who would injure the innocent." In this concept is the genesis of the just war doctrine.

The following chapters will describe how the changed interpretation of the "command of love," as involving the use of force, and if need be violence, came to be accepted by the followers of Jesus. It became grafted onto the teaching of love as it came from the lips of the Messiah.

Chapter 2

The New Way

That Providence which once gave us the law now has given us the gospel.
—Origen

"The warrior-king who annihilates the enemies of Israel is often more prominent than the shepherd of righteousness and judgment,"[1] says the biblical scholar John L. McKenzie in reflecting on Hebrew Scriptures. Yet the gentle shepherd and suffering servant prefiguring Jesus are also persistent strands of Hebrew prophecy.

When Jesus read the words from the scroll of Isaiah, given eight centuries before he came in human form, his hearers in the Nazareth synagogue found them familiar. "The spirit of the Lord is upon me, because the Lord has anointed me. He has sent me to bring glad tidings to the poor. He has sent me to proclaim liberty to captives and recovery of sight to the blind, to let the oppressed go free and proclaim a year acceptable to the Lord" (Isa. 61:32; Luke 4:18–19).

When Jesus asserted that on that day the scripture passage had been fulfilled in the hearing of his listeners, the congregants did not demur. It was only when he went on to say that Elijah and Elisha had performed their miracles for people outside the House of Israel that their hearts were inflamed and they were ready to throw him over a nearby cliff.

The Servant Song of Second Isaiah prefigures in heart-stopping concreteness the suffering that would accompany Jesus' Messiah-ministry:

> There was in him no stately bearing to make us look at him,
> nor appearance that would attract us to him.
> He was spurned and avoided by men,
> a man of suffering, accustomed to infirmity,
> one of those from whom men hide their faces,
> spurned, and we held him in no esteem.
> Yet it was our infirmities that he bore,
> our suffering that he endured,
> while we thought of him as stricken,
> as one stricken by God and afflicted.
> But he was pierced for our offenses,
> crushed for our sin. (Isa. 53:2–5)

The prophetic oracle states that the guilt of all would be laid upon one who went like "a lamb to the slaughter" and who would submit "without opening his mouth" (Isa. 53:7). The Servant Song foretells the bringing of justice by the

Lord's servant gently, so that the bruised reed would not be broken, nor the smoldering wick quenched (Isa. 42:1–3). Though he accepted insults, giving his back to be beaten and his beard to be plucked, Jesus triumphs, for through him the will of the Lord was accomplished (Isa. 53:10).

Jesus at the Last Supper states in grief that the one who dips his hand in the dish with him is his betrayer (Matt. 26:23). In this, he is bearing out the prophecy contained in Psalm 41, "Even my friend who had my trust and partook of my bread, has raised his hand against me" (Ps. 41:10).

To Jeremiah a "new covenant" was promised, written on the heart (Jer. 31:31–34). When Jesus before Pilate was faced with pouring out his blood in an agonizing execution, he refused to call to his defense the legions of angels who would come at his bidding (Matt. 26:53). In dying as an unresistant lamb he had fulfilled what had been written about him in the Prophets and the Psalms (Luke 24:44).

The way of reconciling the human family to God was the willing acceptance, out of love, of the worst that humankind could mete out to him. Betrayal, humiliation, laceration of the body by scourging, the bearing of the cross, and finally the raising of his pierced body on the cross — he was called upon to endure them all. The cross of Jesus was given to his followers as the sign and symbol of the ultimate in loving, the ultimate in forgiving. By his blood he took away the sin of the world. From Calvary came the glorious truth that his blood had been shed to redeem not only the children of Israel but all the children of God, of "every tribe and tongue and people and nation" (Rev. 5:9).

Exalted Command: Utopian or Irrelevant?

To his disciples, Jesus addressed an exalted command, to love as he had loved. "As I have loved you, so also you should love one another" (John 13:34). This was to be the mark of his followers, the mark by which they would be recognized.

Frail human beings were commanded to emulate the limitless love, selflessness, and endurance of sufferings of their Lord. Such love must be possible, or he would not have given them this "new commandment." How on earth could it be accomplished? Certainly not by any power on earth, but only by the power of God.

Many theologians have found ways to dismiss this command and the lifestyle that flows from it, as either utopian or even irrelevant in the practical world. Yet it has never been rescinded, and those who attempt even to approximate it are convinced that, in the end, evil can only be overcome by good. To reach that end, they are taught to call on the Holy Spirit for enabling grace.

When Jesus was no longer with them in the flesh, the believers met at what Paul called the Supper of the Lord (1 Cor. 11:20), the great mystery of the sacramental meal. The new covenant was also called the breaking of the bread (Acts 2:42). Later, though not in the New Testament, it was called the Eucharist, the thanksgiving for the Lord's coming and giving of himself.

When the Christians gathered together, they not only memorialized Christ's atoning death but actually partook of the body and blood of the risen and glorified Lord in the form of bread and wine. They were reminded that they were expected to offer their lives for others as Jesus had. Suffering for the followers of the Messiah took on a new and creative significance. When willingly endured, it was co-redemptive with the Lord's sufferings. No such suffering was ever lost or useless when it was joined with that of Jesus nailed to the cross. The new covenant, sealed by his blood, was the bond that united men and women across all divisions of the human family. The unbloody memorial of the death of Jesus served as a proclamation of his saving death and of the hope of reunion with him when he comes again. In that sense, the Eucharist prefigures the messianic banquet when God's will is accomplished in a transfigured world.

It became clear that those for whom the Lord's Supper was the central act of their lives were transformed and enabled by grace to live according to a new way. The new way was a way of love for those who were of "one heart and soul" (Acts 4:32).

The New Moses

"The way" had been traced out for them by their Lord in teachings that they took into their hearts. In the Beatitudes the new way is contrasted with the old law. In a series of antitheses, Jesus asserted the old ways and then amended them with the expression, "But I say to you." This was repeated six times in Matthew. It was not enough to obey the command against killing; it was necessary to refrain even from anger. The most startling antithesis was the setting aside of the old law of retaliation. Instead of "an eye for an eye" and "a tooth for a tooth," and "hate your enemy," Jesus gave his new and opposing teaching, "But I say to you, love your enemies and pray for those who persecute you" (Matt. 5:21–48). These form part of what came to be called the Sermon on the Mount.

It has been pointed out by McKenzie that the antitheses show Jesus as the new Moses, promising a new law of equal authority with the old. According to McKenzie, the fact that the location of the discourse "is on a mountain . . . is a deliberate invocation of the first revelation of the law upon a mountain."[2]

Surrounded by Violence

The Sermon on the Mount, including the Beatitudes, presents a way of righteousness that must revolutionize daily living and overturn merely human values (Matt. 5:1–48; 6:1–29; Luke 6:17–49).

It is hard for modern Christians, accustomed to bland recitations of the Beatitudes, to appreciate the shock with which they must have assailed the ears of those belonging to the old covenant. Beatitudes abound in the Hebrew Scriptures as a description of the blessedness or happiness that accrues from virtue or good fortune. The word *baruch* was attached to the blessedness of the enjoyment of good fortune.

The most shocking of the eight beatitudes in Matthew (four in Luke) was of course that related to persecution, the acceptance of which is reiterated in the antitheses. Almost as shocking was the one that concerned the land, which to the Israelites was the holy land of Israel. The meek are to be accounted blessed for they shall inherit the land. They are not told how or when, but they must wait their inheritance with meekness. What about their fierce desire to control the land of their fathers, the land promised to them but now controlled by idolaters?

As for counting poverty, hunger for righteousness, and sorrowing as blessed, that was a promise to be fulfilled eventually. Only the heavenly Father can fulfill the promise of a place in the reign of God, the satisfaction of spiritual hunger, and consolation for the sorrowing. Everything rebounded to God and his mercy, and that mercy could only be obtained by first showing mercy. In this spiritually overturned world, Jesus' followers had much to ponder. Each beatitude presented a paradox, a holy paradox, which would make sense only when eyes were opened to the new vision of the reign of God that Jesus was bringing to them.

The Sermon on the Mount in Matthew has to be considered in studying the lives of Jesus' first followers as they confronted the world of violence around them.

Roman Thanksgiving, Christian Thanksgiving

Violence was accepted and honored in an empire built on campaigns of war. In Rome, the successful wars were celebrated by victory parades in which captives in chains were put on display and often executed.[3] To celebrate such victories, the populace was invited to the amphitheater for the "games." At the "games" that were part of the thanksgiving celebration for the victory in the Dacian Wars (Dacia corresponds to modern Romania), the emperor Trajan gave a series of gladiatorial combats. It was in 107 C.E., and five thousand pairs of gladiators took part. In the main, these men were war prisoners, condemned criminals, or slaves. For the excitement of the spectators, the gladiators watered the ground of the arena with their blood. Often, persecuted Christians were thrown into the arena with wild animals to satisfy the frenzied bloodthirstiness of the crowd.

Public executions by crucifixion, preceded by scourging, were seen as a deterrent to crime and were applied chiefly to miscreants who were slaves or members of the humble classes.

Christians, viewing human beings in the light of the incarnation, and seeing even opponents as children of God, shrank from the violence around them. The teaching of the Beatitudes told them that the peacemakers, not the war-makers, are children of God. The lives of the disciples of Jesus were a total contradiction to the Roman society in which they found themselves. They lived by such precepts as hungering and thirsting for righteousness rather than for possessions or land, and showing mercy to those in need in order to deserve the mercy of God.

There came a time when what had been done to their Lord, betrayal, suffering, humiliation, and death, became the fate of his followers. Wave after wave of persecution descended on them because they were disciples of Jesus Christ. This strange new sect gave rise to questions and calumnies because they refused to recognize and give homage to the ancestral gods of Rome and had no image of their god to replace them. They were considered by upstanding Roman citizens to be atheists. Among the calumnies that were spread about the celebration of the Eucharist was that of cannibalism, the eating of babies. Another problem to the Romans was the use by the Christians of the word "kingdom." This was perceived as the plan of this new and strange group to replace the rule of Rome.

The nonviolent response of the Christians to the evils heaped upon them was a mystery to those around them, and above all to their persecutors. It was not only a mystery but a madness, a madness inexplicable without the example of the willing death on the cross. Only the hope in the promises of the Beatitudes could help them to withstand the cruelties heaped upon them. Their new faith taught them, "Blessed are you when they insult you, and persecute you, and utter every kind of evil against you because of me. Rejoice and be glad for your reward will be great in heaven" (Matt. 5:11–12).

The Christians followed the first of the martyrs in forgiving their persecutors. Deacon Stephen, suffering under a hail of stones, died with words of forgiveness on his lips. He was the first of a long train of men and women who followed Jesus in transcending agony to utter the forgiving word. Forgiving as the nonviolent response to hurt or hatred became the mark of the followers of Jesus. Their way of life clearly excluded retaliation. The early martyrs, who were venerated as saints, achieved the spiritual power to return gentleness and forgiveness for violence and torture. Among them were a special group of saints who died because they refused to perpetrate violence on others, the soldier-martyrs.

Before describing the soldier-martyrs it is well to emphasize the witness of two apostolic fathers, Ignatius of Antioch and Polycarp (the latter's name meaning literally "much fruit") of Smyrna. Both had known the apostles of Jesus. Both met the cruel violence of martyrdom with love and forbearance. Their stories serve to reveal the amazing growth of the new way in Asia Minor by about 100 C.E.

After his conquest in the Dacian campaigns, Trajan, out of gratitude to the gods of Rome, decreed a persecution against those who refused to pay homage to the Roman deities. One of those arrested was Ignatius, the bishop of Antioch. Sentenced to be consumed by wild beasts, he had to be taken to Rome for the execution of the sentence. He was guarded by ten soldiers, whom he described as wild leopards. En route, his captors broke the journey at Smyrna, where Bishop Polycarp received him, kissing his chains. Deputations of the church communities from Ephesus, Magnesia, and Tralles came to be with the condemned man. To the churches in each of their cities, Ignatius addressed a letter. At Troas, when the guards had to await a vessel to take them to Neapolis on the opposite coast, there was time to write three more letters. Two went to the congregations of Philadelphia and Smyrna, and a third was a personal letter to Smyrna's bishop, Polycarp.

These letters, authenticated by historical sources, are among the most glorious treasures of the early church, coming as they do from an apostolic father, the third to head the church in Antioch. Becoming bishop in 69 C.E., Ignatius served in one of the largest cities of the Hellenistic-Roman world, ranking behind only Rome and Alexandria. Ignatius reflects not only the gentleness of the Christians before their persecutors but their fearlessness before the most barbarous of martyrdoms. In one of the letters, Ignatius tells the Christians of Ephesus, "Give them a chance to learn from you, or at all events, from the way you act. Meet their animosity with mildness, their high words with humility, and their abuse with your prayers. Let us show by our forbearance that we are their brothers, and try to imitate the Lord by seeing which of us can put up with the most ill-usage or privation or contempt."[4]

Ignatius advises the Ephesians to attend the church gatherings in a state of grace and to "share in the one common breaking of bread — the medicine of immortality, and the sovereign remedy by which we escape death and live in Jesus Christ for evermore."[5] In another letter, this to the Philadelphians, he emphasizes the importance of the Lord's Supper, by then called the Eucharist, the thanksgiving. "Make certain, therefore, that you all observe one common Eucharist, for there is but one cup of union with the blood, and one single altar of sacrifice — even as there is but one bishop, with his clergy, and my fellow servitors the deacons. This will ensure that all your doings are in full accord with the will of God."[6]

Best known of all the utterances of Ignatius is his passionate appeal to the Christians of Rome not to put any obstacle in the way of his achieving martyrdom. He knew that they would make every attempt to save him from the amphitheater. "Pray leave me to be a meal for the beasts, for it is they who can provide my way to God. I am his wheat, ground fine by the lion's teeth to be made purest bread."[7] The letter to the Romans, with its burning ode to the triumph of martyrdom, has become for Christians down the ages a sort of manual for martyrdom. Shining through his willingness to undergo the fangs of the beasts is his certainty of joining the resurrected Lord.

There is no doubt that at journey's end, Ignatius's prayer was answered; he was thrown to the wild beasts. Tradition states that he arrived in Rome on the last day of the Dacian games. The emperor's letter was presented to the Roman prefect, who had the soldiers deliver him to the amphitheater, where his body was devoured.

Polycarp: Blood Contending with Fire

Polycarp, the bishop of Smyrna who had kissed the chains of his fellow bishop Ignatius, was to follow him in martyrdom about fifty years later. An eyewitness account exists of his martyrdom, the earliest authentic account of a death by persecution. In his youth in Ephesus, Polycarp had been a disciple of the last surviving apostle, John. For this, he was especially revered during the fifty years he served as bishop of Smyrna; in particular he was revered by his own disciple St. Irenaeus.

A letter from Irenaeus takes us back to his time. He wrote:

> I can tell the very place in which the blessed Polycarp used to sit when he discoursed, and his goings-out and his comings-in and his manner of life and his personal appearance, and the discourse he held before the people and how he would describe his intercourse with John and with the rest of those who had seen the Lord and how he would relate their words from memory; and how the things he heard them say about the Lord, his miracles and his teachings, things he had heard from the eyewitnesses of the word of life, were proclaimed by Polycarp in complete harmony with scripture.[8]

The Christians of Smyrna must have been transfixed to be in the presence of one who had been close to John, who had been close to the Messiah.

According to the account of Eusebius, persecution of the Christians came to Smyrna by order of Marcus Aurelius. "Sometimes," Eusebius wrote, "they were torn with scourges to the innermost veins and arteries, so that even the secret hidden parts of the body, the entrails and internal organs, were laid bare; sometimes they were forced to lie on pointed shells and sharp spikes. After going through every kind of punishment and torture, they were finally flung to the beasts as food."[9]

The eyewitness account of the death of Polycarp comes from a man named Marcion, who begins with a description of the bravery of Polycarp's fellow martyrs. So caught up were they in the grace of Christ that "they made light of the cruelties of this world and at the cost of a single hour purchased for themselves life-everlasting."[10]

The endurance of the Christians in the arena, and the wild heroism of one of them in dragging a resisting beast toward him, whetted the appetite of the crowd for more bloodletting. "Down with the godless," they shouted. "Fetch Polycarp."

Polycarp did not impatiently offer himself to the torturers, writes Marcion, but like Jesus "patiently waited the hour of his betrayal." He chose to stay in Smyrna, but was prevailed upon to go to a country property near the city. For three days, he and his companions prayed day and night for all the people and for all the churches of the world. During this time Polycarp had a vision of his pillow in flames and told his companions that he would be burned alive. Late in the evening his pursuers found the aged man in bed. Probably to their surprise he ordered the table to be laid for them to take a meal and asked for time to pray. Placed on a donkey he began his journey to the city. The chief of police transferred Polycarp to his carriage and spent the time trying to persuade him to make the sacrifice and say the words "Lord Caesar." When he refused he was put out of the carriage so roughly that he scraped his shin. Despite this, he set off on foot to the stadium.

The proconsul made several attempts to have Polycarp swear by Caesar's fortune. "Swear," he repeated, "and I will set you free: execrate Christ." "'For eighty-six years,'" replied Polycarp, "'I have been his servant and he has never done me wrong. How can I blaspheme my King who saved me?' Unmoved by threats of wild beasts or fire, Polycarp stated simply that he was a Christian."[11] At that point the proconsul ordered the crier to announce three times from the middle of the arena, "Polycarp has announced that he is Christian." The crowd in fury began to shout: "This fellow is the teacher of Asia, the father of the Christians, the destroyer of our gods who teaches numbers of people not to sacrifice or even worship." Their cry to set a lion immediately on Polycarp was refused because the sports had been concluded. They then set up the cry to burn him alive. The crowd took the matter into their own hands, collecting logs and faggots and standing them up for a pyre. Polycarp began calmly taking off his outer garments and loosening his belt, and he bent to remove his shoes. He was unused to doing this. Marcion reports in a touching detail, "Each of the faithful strove at all times to be the first to touch his person." Polycarp was bound, not nailed, to the stake, and his prayer was recounted by Marcion. He said in part, "I bless thee for counting me worthy of this day and hour, that in the number of the martyrs, I may partake of Christ's cup to the resurrection of eternal life of both soul and body in the imperishability that is the gift of the Holy Ghost." When the old bishop had said "Amen," the pyre was lit. A great flame "took the shape of a vaulted room, like a ship's sail filled with the wind, and made a wall around the martyr's body, which was in the middle not like burning flesh but like gold and silver refined in a furnace."[12]

When the body of Polycarp was not consumed by fire, a swordsman was called to give him the "coup de grace." Blood contended with fire and conquered when a stream of the martyr's blood poured out and quenched the fire. Only afterward was the body of the holy one burned by the centurion. The Christians were able to gather up his bones "more precious than stones of great price, more splendid than gold."[13]

Not a Human Kingdom

An early statement of the explicit Christian refusal to use violence was made by a philosopher who studied most of the philosophies circulating in his era — those of Pythagoras, Plato, and others — before accepting Christ. He was about thirty at the time of his conversion, a conversion which arose from his admiration of the courage of the martyrs. He was Justin, whose name is conjoined with the word "Martyr." A native of Palestine, of Greek parentage and language, he was convinced that many people would accept the teachings of Jesus if they were clearly presented to them. Wearing the philosopher's cloak, he engaged in disputations with people of intellect — pagans, Jews, and members of various sects. He eventually reached Rome, where he not only witnessed publicly to his faith in Jesus, but wrote his *Apology for Christians,* addressed to Emperor Antoninus Pius. The word "apology" meant not what it has come to mean, but a defense of the faith. It is considered the first recorded address by a Christian to a head of state.

Like Paul, he asserts his loyalty to the Roman government and points out that Christians took special care to pay their taxes, as instructed to do so by Christ himself. He explains that when Christians speak of the kingdom, they do not mean a human kingdom, which would have been treason, but rather the kingdom of God.[14]

In explaining Christ's counsel to turn the other cheek, he states the Christian view on violence: "It is not right to answer fighting with fighting, nor does God wish us to imitate the wicked; but he has exhorted us to lead all men away from the shame and cupidity of the wicked by patience and gentleness."[15]

Justin explains to the emperor how twelve men, illiterate and without skill in speaking, were able by the power of God to proclaim to every race the teaching of Christ. He describes how Christ's teaching made a difference in their lives. "We who formerly used to murder one another," he wrote, "do not only refrain from making war on our enemies, but also . . . may not lie nor deceive our examiners."[16] Justin compares the military oath which binds the emperor's soldiers to him to the binding commitment of Christians to their Lord.

During his second visit to Rome, Justin founded a school for the study of Christianity. Denounced by an enemy, he was arrested and brought before the Roman prefect Rusticus. He was given a chance to state his faith in Jesus Christ. To the question of Rusticus, "You, then, are a Christian?" he replied, "Yes, I am a Christian." In his refusal to make the required sacrifice to the gods, Justin was joined by six of his pupils, one a woman. The sentence was read by the prefect Rusticus: "Those who will not sacrifice to the gods and obey the Emperor will be scourged and beheaded according to the laws."[17] None in the group weakened before the executioner.

The date was 165 C.E. To Justin we owe the description of the gathering of the early Christians for the sacramental meal:

> On Sunday, the inhabitants of a town or region gather and the gospels or the prophetic writings are read. . . . The one presiding exhorts us to imitate the great things we have heard. Next, we all stand and pray, and when we have finished, bread and wine are brought. The priest offers prayers and thanks as best he can and the assembly cries, "Amen." Distribution is made of the consecrated food, with each one present sharing it and the deacons bringing it to those who are absent.

The day following the Sabbath was chosen for the gathering to recall the day on which the Lord had risen from the dead. Justin describes the collection which was given to the priest so that he could help "orphans, widows, the sick, prisoners, and guests from other places." "In all our offerings," he writes, "we praise the creator of the universe through his son Jesus Christ and through the Holy Spirit."[18]

Hospitality to strangers was a distinguishing mark of the Christian community. Day-to-day community living was grounded in peacemaking. Any person who had a quarrel with another was to make peace and become reconciled before placing his or her gift on the altar.

Origen, born in 185 C.E. in Alexandria, Egypt, of Christian parents, knew the reality of martyrdom at seventeen years of age. His father, Leonides, was a martyr for the faith. Origen himself was prevented from accompanying his father to the place of execution by his mother, who, knowing his intention, hid his clothes.

Like Justin Martyr, he was an apologist, using his immense knowledge and writing skill to defend the teaching of the church. Also like Justin, he conducted a school of Christian theology at which students prepared for living the Christian life and also for giving up their lives.

Origen was so carried away by the teachings of asceticism that he took literally the figurative statement of Matthew about becoming a eunuch for the kingdom of God (Matt. 19:12). He underwent castration. Origen later regretted the rash act, which was considered a grave sin by the church. Eusebius explains that he wanted to "rule out any vile imputations on the part of unbelievers. For in spite of his youth he discussed religious problems before a mixed audience."[19]

As a scholar, Origen was able to broaden the influence of the church among the educated classes. His earlier studies of philosophy gave way to concentration on scripture and the study of Hebrew. His voluminous commentaries on scripture were used by the bishops and by St. Jerome. A monumental contribution was his sixfold edition of translations of the Psalms set in parallel columns along with the original Hebrew text.

Eventually settling in Caesarea, Palestine, he preached daily on aspects of the gospel. According to Eusebius his sermons as well as his other writings were taken down by a team of scribes. It was not surprising that it fell to Origen to reply to the first comprehensive polemic against Christianity, *The True Discourse.* The work of Celsus, a Greek and Platonist, had been a matter of discussion in the Roman world since it had made its appearance in 180 C.E. Celsus asserted that the refusal of Christians to pay the divine honor due to the emperor might lead to dire consequences. If all followed the Christians, Celsus maintained, "nothing would prevent his [the emperor's] being left all alone and deserted while all earthly affairs fell under the sway of the most lawless and uncivilized barbarians." To Celsus, and probably to others, the negative attitude of the Christians to emperor-worship implied a negative attitude toward defense of the empire. The text of Celsus is only known through copious citations in Origen's response *Against Celsus.*

Celsus was acquainted with Hebrew Scriptures and pointed out that in them the deity had commanded wars to be fought and cities to be destroyed by the sword. Origen explains the mission of Christians and makes the distinction between the old covenant, with its divinely sanctioned wars, and the new dispensation brought by Christ.

Origen writes that Jesus counsels us "to cut down our hostile and insolent 'wordy' swords into ploughshares, and to convert into pruning-hooks the spears formerly employed in war. For we no longer take up sword against nation, nor do we learn war anymore, having become children of peace for the sake of Jesus who is our leader."[20] He continues: "We know that it was said to the ancients: An eye for an eye and a tooth for a tooth (Exod. 21:24). But we also read: But I say unto you, if a man strike thee on one cheek, turn to him also the other (Matt. 5:39)."[21]

"That Providence," Origen asserts, "which once gave us the law now has given us the gospel." He emphasizes the universal mission of Christianity, saying, "Because God desired the doctrine of Jesus Christ to benefit all nations, all the counsels of men against the Christians have been brought to naught; and the more emperors, rulers of the nations, and peoples persecuted them, the more their numbers have grown and their strength increased."[22]

Responding to the urging of Celsus to help the emperor in the maintenance of justice and fight for him, Origen states that the Christians do give help to rulers by obeying the injunction of the Apostle Paul to honor those in authority. By their piety, Origen maintains, the Christians give more help than that given by soldiers who slay as many enemies as they can.

"To those enemies of our faith," he states, "who require us to bear arms for the commonwealth and to slay men, we can reply: Do not those who are priests at certain shrines, and those who attend on certain gods, as you account them, keep their hands free from blood, that they may, with hands unstained and free from human blood, offer the appointed sacrifices to your gods: and even where the war is upon you, you never enlist the priests in your army?"[23]

Origen claims that if the priests' action is praiseworthy while others are engaged in battle, then the action of the Christians in praying on behalf of those fighting in a righteous cause should also be recognized.

The reply to Celsus was written toward the end of Origen's life. The Christian who had longed for martyrdom was imprisoned during the persecution by Decius. He was kept in darkness and tortured but was eventually released.

In the words of Eusebius, the judge "strove with might and main at all costs to avoid sentencing him to execution."[24] This was probably because of Origen's fame as a scholar and teacher. Origen died after his release and was therefore not listed as a martyr.

Origen, according to John C. Cadoux, is particularly significant since he proves that into the middle of the third century the "hard sayings" of Jesus regarding nonretaliation and violence were taken literally by Christians. Calling Origen the finest thinker the church had produced in many generations, Cadoux made the point that he based the arguments against army service not on the idolatrous oath required of soldiers, but on the basis of the refusal to shed blood. Cadoux terms Origen's defense of the Christian refusal of army service "a series of arguments that have never since been answered."[25]

Origen's championship of gospel nonviolence and the clear witness of other Christians like Justin Martyr were anchored in the "hard sayings" of Jesus. They supplied the followers of the new way with the groundwork for their response to the demand that they inflict violence on others. Hitherto, the Christians had been called upon to endure suffering nonviolently and unite their suffering to that of their Lord as part of the redemption. When they were called upon to be part of an army to take the lives of others, they refused. As conscientious objectors, they were executed.

Soldier-Martyrs

The function of the army of imperial Rome was diverse. To be part of the military force (*militare*) meant to be the emperor's policeman. Included, besides apprehending criminals, were duties arising from the guarding of the areas occupied by Rome, customs control, and the construction of military encampments and other buildings. Only during martial campaigns was there bloodshed, and the word for this was *bellare,* to wage war. It is of note that during the reign of Septimus Severus (193–211) much civil administration was placed in the hands of military personnel. Since the army offered regular pay and security, there were sufficient numbers of voluntary enlistments. During the early years of the Christian era there was no necessity for obligatory conscription. (There was, however, conscription of one particular class, which will be discussed later.) A man could be a soldier for all of his life without being called upon to slay an enemy in battle. There is some evidence that the Roman army included Christians in its ranks before 200 C.E. It could well have been that they were converted during their service.[26]

The sporadic outbreak of persecution against the Christians, now in Asia Minor, now in North Africa, now in the evangelized areas of southern Europe, now in Rome itself, did not encourage patriotism for the empire. However, Peter and Paul had both counseled obedience to lawful authority, and Rome was the only authority over them all. Not even in the Great Persecution of Diocletian and Galerius in 303–13 did the Christians take part in secret or open rebellion. Their obedience, however, was a limited one, as enunciated by Theopholus, a successor to Ignatius as bishop of Antioch: "I honor the Emperor, not indeed worshipping him, but praying for him. God, living and true, I worship, knowing that the emperor is appointed by him. He is no god but a man appointed by God, not to be worshipped but to judge justly."[27]

Significant guidance for Christian behavior regarding military service was presented by Hippolytus, a prominent person in the church of Rome. According to Canon 13 of the Canons of Hippolytus, a man who has accepted killing as part of his profession should not be admitted to the faith. A Christian should not volunteer as a soldier, but if he is conscripted into military service, he should not shed blood. If, however, he has shed blood, he must not take part in the mysteries (the Eucharist) until he undergoes a complete change of conduct and has repented openly. One version of the Canons expressly forbade magistrates from ordering capital punishment and soldiers from killing, even when commanded to do so.

Maximilian: The Leaden Seal

Christians down the ages are grateful that the precious records of soldiers who were martyred for their faith in Jesus have survived. Included here are three imperishable records of soldier-martyrs. Among these is the shining record of the martyrdom of a young man, Maximilian, who refused conscription into the Roman army. Only the sons of army veterans were subject to obligatory conscription, and Victor, Maximilian's father, was present at his trial. It occurred in Numidia, proconsular Africa (now Algeria). The proconsul Dion ordered that Maximilian be given the military badge and be measured for military garb. It was noted that his height measured five feet, ten inches. The young conscientious objector was constant in his refusal.

The *acta,* or record, from 295 C.E. is most significant in that it does not mention as a reason for refusing army service the oath and the sacrifice required of officers. There has been a tendency among some commentators on the conscientious objection of early Christians to link it with the refusal to commit idolatry. Although the objectors would not assent to idolatry, their absolute refusal to enter the army was based on a much deeper commitment, the incompatibility of army service as such with the message of Jesus. Even when bloodshed was not immediately involved, there was other questionable conduct among the troops inimical to living a Christian life.

Nothing can give the impact of Maximilian's witness more vividly than the stark recital of his trial and death:

> [*The proconsul Dion announced:*] "You must serve or die."
>
> MAXIMILIAN: I will never serve you. You can cut off my head, but I will not be a soldier of this world, for I am a soldier of Christ.
>
> DION: What has put these ideas into your head?
>
> MAXIMILIAN: My conscience and He who has called me.
>
> DION (to Fabius Victor): Put your son right.
>
> VICTOR: He knows what he believes, and he will not change.
>
> DION (to Maximilian): Be a soldier and accept the emperor's badge.
>
> MAXIMILIAN: Not at all. I carry the mark of Christ my God already.
>
> DION: I shall send you to your Christ at once.
>
> MAXIMILIAN: I ask nothing better. Do it quickly, for there is my glory.
>
> DION (to the recruiting officer): Give him his badge.
>
> MAXIMILIAN: I will not take the badge. If you insist, I will deface it. I am a Christian, and I am not allowed to wear that leaden seal around my neck. For I already carry the sacred sign of the Christ, the Son of the living God, whom you know not, the Christ who suffered for our salvation, whom God gave to die for our sins. It is He whom all we Christians serve, it is He whom we follow, for He is the Lord of life, the Author of our salvation.
>
> DION: Join the service and accept the seal or else you will perish miserably.
>
> MAXIMILIAN: I shall not perish: my name is even now before God, and I cannot fight for this world. I tell you. I am a Christian.
>
> DION: There are Christian soldiers serving our rulers Diocletian and Maximilian, Constantius and Galerius.
>
> MAXIMILIAN: That is their business. I also am a Christian, and I cannot serve.
>
> DION: But what harm do soldiers do?

MAXIMILIAN: You know well enough.

DION: If you do not do your service I shall condemn you to death for contempt of the army.

MAXIMILIAN: I shall not die. If I go from this earth my soul will live with Christ my Lord.

DION: Write his name down.... Your impiety makes you refuse military service and you shall be punished accordingly as a warning to others.

[*He then read the sentence:*] "Maximilian has refused the military oath through impiety. He is to be beheaded."

MAXIMILIAN: God lives.

[*Maximilian's age at the time he was beheaded was twenty-one years, three months and eighteen days.*][28]

Marcellus the Centurion

An older man, Marcellus, a centurion, shared the fate of Maximilian three years later. The place was Tingis, now Tangiers, in North Africa.[29]

Marcellus chose a dramatic way to announce his departure from the military life. In the midst of the feasting for the emperor's birthday, Marcellus condemned feasting as pagan and threw down his soldier's belt before the standards of the legion. He declared in a loud voice, "I serve Jesus Christ, the Eternal King. I will no longer serve your emperors, and I scorn to worship your gods of wood and stone which are deaf and dumb idols."

The soldiers were stunned at hearing such things; they laid hold of him and reported the matter to the president, Astasius Fortunatus, who ordered him to be thrown into prison. When the feasting was over, he gave orders, sitting in council, that the man should be brought in:

FORTUNATUS: What did you mean by ungirting yourself contrary to military discipline, and casting away your belt and vine-switch? [The distinctive badge of the centurion].

MARCELLUS: On July 21, in the place of the standards of your legion, when you celebrated the festival of the emperor, I made answer openly and clearly that I was a Christian and that I could not accept this allegiance, but could serve only Jesus Christ, the Son of God the Father Almighty.

FORTUNATUS: I cannot pass over your rash conduct, and therefore I shall report this matter to the Emperors and Caesar. You shall be sent to my lord Aurelius Agricolan, deputy for the praetorian prefects.

On October 30 at Tingis, the centurion Marcellus having been brought into court, it was officially reported: "Fortunatus the president has referred Marcellus, a centurion, to your authority. There is here a letter from him, which at your command I will read." Agricolan said, "Let it be read." The official report was read.

AGRICOLAN: Did you say these things as set out in the president's official report?

MARCELLUS: I did.

AGRICOLAN: Were you serving as a regular centurion?

MARCELLUS: I was.

AGRICOLAN: What madness possessed you to throw away the badges of your allegiance and to speak as you did?

MARCELLUS: There is no madness in those who fear God.

AGRICOLAN: Did you say each of the things contained in the President's report?

MARCELLUS: I did.

AGRICOLAN: Did you cast away your arms?

MARCELLUS: I did. For it was not right for a Christian man, who serves the Lord Christ, to serve in the armies of the world.

AGRICOLAN: The doings of Marcellus are such as must be visited with disciplinary punishment. [*He pronounced sentence:*] Marcellus, who held the rank of a regular centurion, having admitted that he degraded himself by openly throwing off his allegiance, and having moreover used insane speech, as appears in the official report, it is our pleasure that he be put to death by the sword.

When he was being led to execution, Marcellus said, "May God be good to you, Agricolan." He went to his death in 298 C.E.

It is possible and even probable that Marcellus became a Christian during his army service. While slaves and poor people could become attached to the army without taking the oath or offering sacrifice to the gods, members of the officer class, to which Marcellus belonged, were bound by army rules to pay homage to the emperor and the gods. It is unlikely that a believing Christian could serve and advance in an army which called for homage to "deaf and dumb idols."

The decision of Marcellus against military service, like that of Maximilian, was based on something deeper than the idolatrous sacrifice; it related to the fidelity of Christians to their eternal king.

Martin of Tours

Martin of Tours, who refused to shed blood though already a member of the Roman army, escaped the penalty of death only because an armistice was declared. Martin stood up in the battlefield during a lull in the campaign in Gaul and informed his commander that he would not fight. This was equivalent to desertion and was punishable by death. Martin was one of the young men who had been forced to enter the army as the son of a veteran. He was only fifteen at the time of his induction. Despite the fact that his parents followed the old gods, Martin became a Christian catechumen. While he was stationed at Amiens, he performed the works of mercy with which his name has traditionally been linked. He was met at the gate of the city by a beggar shivering for lack of covering. People passed by without giving him alms. Martin, having no alms to give, drew his sword, and taking off his cloak, cut it into two. He gave one half to the beggar. This was the only use he made of his sword.

When he declared on the battlefield his intention not to fight, Martin and his comrades-in-arms were lined up to receive a war bounty from their commander, Julian Caesar. It was common practice in the course of military campaigns for the commander to reward his soldiers with money beyond their salary. Martin refused to accept it and gave his reason: "Hitherto, I have served you as a soldier, let me now serve Christ. Give the bounty to these others who are going to fight, but I am a soldier of Christ and it is not lawful for me to fight."[30]

Julian, astounded, accused Martin of cowardice. On the following day he expected to lead all his men against the enemy. Martin responded that he was prepared to join the battle unarmed and would advance alone against the enemy in the name of Christ. Martin was arrested and confined to prison. Unexpectedly, rather than a battle, an armistice ensued. Martin was freed from prison and given his discharge from the army.

After he received baptism Martin journeyed to Poitiers, where Bishop Hilary accepted him as a disciple. Martin is called one of the "soldier-saints"; more appropriately, he could be termed a "conscientious-objector saint."

The next journey of the newly baptized Martin was to cross Europe to visit his parents in Pannonia (modern Hungary). He helped bring his mother into the Christian faith, but his father remained with the old gods. Bishop Hilary gave Martin a piece of land where his prayerful life drew many disciples. For ten years he and his disciples preached throughout the countryside, awakening the people to the knowledge of Jesus. The people of Tours, in 371 C.E., demanded that Martin become their bishop. Neighboring bishops agreed to their choice. Bishop Martin, installed in Tours, made a yearly visit to the people under his care, journeying on foot, on a donkey, or by boat. His mercifulness extended to all who came his way. When Avitian, an officer of the imperial army, came to Tours with a group of prisoners whom he was planning to execute on the following day, Bishop Martin rushed to Avitian to intercede for them. He arrived at midnight and would not leave until Avitian had agreed to spare the captives.

Bishop Martin was deeply beloved by his people during the more than quarter of a century that he served them. He died in 397 C.E. Many accounts of his life and goodness circulated among the communities of Europe then in the process of accepting Christianity. New churches being erected were dedicated to Martin, and pilgrimages to pray at his tomb at Tours became frequent. An oratory built over what was thought to be the half of his divided cloak was called "capella," or little cloak. This may be the origin of the word "chapel."

The three early conscientious objectors to fighting and killing, Maximilian, Marcellus, and Martin, are known to us through the providential preservation of their *acta*. The accounts of others who suffered for their refusal to maim or kill fellow human beings may never have been committed to writing, or if they were, did not survive those tragic years when Christian writings were summarily destroyed. There were cases when Christians who were ordered by their persecutors to hand over Christian documents, including the *acta* of the persecuted, did so in fear of their lives.

It needs to be emphasized that the fact that a Christian was in the army of imperial Rome did not mean that he would be called upon to shed blood. As mentioned, there was a wide variety of services that he could perform, tasks necessary to the functioning of the extensive empire. Those who were conscientious objectors to army service were not simply objecting to an idolatrous act. They were called upon to shed blood, to be part of *bellare.* At that point, they were prepared to suffer for their commitment to the faith of Jesus Christ, a faith which they saw as incompatible with army service.

Sacrosanct Creature

As a "sacrosanct creature," the human being must never be killed, according to Lactantius (Firmianus), a North African so moved by the courage of the persecuted Christians that he accepted Christ.

The life of Lactantius mirrors the changing situation of the church of Christ in a tumultuous epoch. He was born during the persecution of Decius and lived through succeeding persecutions including the Great Persecution of Diocletian, which began in 303. In that persecution he did not lose his life but lost his post as teacher of rhetoric in Nicomedia (northwest Asia Minor). His life was spared in all likelihood because it was Diocletian himself who had invited Lactantius to come to the Nicomedian School of Rhetoric.

Lactantius lived to experience the peace that came with the recognition of Christianity as a permitted religion. As an old man, eminent as a teacher, writer, and scholar, he was chosen as the tutor of Crispus, the eldest son of the great Emperor Constantine.

One of his best known works, *On the Death of the Martyrs,* allowed him to describe and pay tribute to men and women who gave their lives as witnesses to their faith.

He began his major work, *The Divine Institutes,* as a response to the accusations against Christians, but extended it so that it became an "apology," a defense of the faith of Christians. Along with the other Christian apologists, he deals at length with ceremonies and cults of the imaginary gods. Man, he states, is God's handiwork, while the images of gods and goddesses are man's handiwork.[31] Repeatedly, Lactantius expresses his revulsion toward war. Cadoux points out how in Lactantius, and in Christian thinking of the time, warfare and murder are connected. Cadoux cites the words of Lactantius on the Romans: "They despise indeed the excellence of the athlete because there is no harm in it; but royal excellence, because it is wont to do harm, extensively, they so admire that they think that brave and warlike generals are placed in the assembly of the gods, and that there is no other way to immortality than by leading armies, devastating foreign countries, destroying cities, overthrowing towns, and either slaughtering or enslaving free peoples."[32]

The Christians, according to Lactantius, are "those who are ignorant of wars, who preserve concord with all, who are friends even to their enemies, who love all men as brothers, who know how to curb anger and soften with quiet moderation every madness of the mind."[33]

The Divine Institutes are revered by scholars. The epitome, or summary of the larger work, which he himself prepared, was widely studied. Of all his extensive writings, a particular paragraph has been cited over the centuries, the one referring to the human being as sacrosanct. It speaks to every age whenever human beings are dragooned into armies to kill other human beings, whenever they are put to death or die as the result of the word that orders execution. "It will not be lawful," said Lactantius, "for a just man to serve as a soldier, for justice itself is his military service, nor to accuse anyone of the capital offense, because it makes no difference whether you kill with a sword or with the word, since killing itself is forbidden. And so, in this commandment of God, no exception at all should be made to the rule that it is always unlawful to put to death a man whom God has wished to be a sacrosanct creature."[34]

The Army That Sheds No Blood

"Now . . . the trumpet sounds with a mighty voice," wrote Clement of Alexandria, "calling the soldiers of the world to arms, announcing war; and shall not Christ, who has uttered his summons to peace even to the ends of the earth, summon together his own soldiers of peace? Indeed, O Man, he was called to arms with his blood and his word an army that sheds no blood; to these soldiers he has handed over the kingdom of heaven."[35]

From priests and teachers like Clement came the delineation of the Christians as spiritual warriors to whom bloodshed was abhorrent and antithetical to Christ's teaching. Such a delineation was a constant for the first centuries of the Christian dispensation. Clement, a philosopher, studied the philosophies and cults of his time before he came to Christ. Ordained a priest, he taught at the renowned Alexandrian School of Christian Theology. He saw the stance of Christians as being opposed to that of the world around them.

"The trumpet of Christ is his gospel," said Clement. "He has sounded it in our ears and we have heard it. Let us be armed for peace, putting on the armor of justice, seizing the shield of faith, the helmet of salvation, and sharpening the sword of the spirit which is the word of God."[36]

Christians must be on their guard, they are warned. There could be no slackening in the spiritual warfare. At any time, they might be tested by persecution. In fact, persecution came to Alexandria in 202, and Clement escaped to Caesarea in Cappadocia (Asia Minor). When persecution reached Caesarea, Clement was able to take over the duties of the persecuted bishops. It is not known how he died, but records reveal that he had died by 215.

One of the works from which Clement taught, which he cited as scripture, was *The Didache* (The teaching). In it, hatred, the antecedent to bloodshed, is condemned, and a Christian way to respond to persecutors is set forth. Lost for many centuries, its rediscovery in the nineteenth century gives modern Christians an insight into the thinking of people who learned from this work, considered to be the oldest example of Christian literature outside of the New Testament (dated not later than 150 C.E., possibly 100 C.E.). It might well have been a sort of catechism used in the preparation of candidates for baptism.

Describing "The Way of Life," *The Didache* opens with the command to love the Lord, the Creator, as oneself and one's neighbor, and to follow the golden rule of not doing to anyone else what one would not wish to be done to oneself. It then asserts some of the gospels' hardest sayings and states: "What you may learn from these words is to bless them that curse you, to pray for your enemies, and fast for your persecutors. For where is the merit in loving only those who return your love? Even the heathens do as much as that. But if you love those who hate you, you will have nobody to be your enemy."[37] The concept of fasting on behalf of persecutors is surely a nonviolent response to ire, a sign of the new way.

The Didache stresses the unity of the human family as central to the celebration of the Eucharist. "As this broken bread, once dispersed over the hills, was brought together and became one loaf, so may thy church be brought together from the ends of the earth into thy kingdom."[38]

The Eucharist as Defense

Cyprian, bishop of Carthage, suggested a nonviolent defense against the evil of persecution. Cyprian, like many other Christians in Roman North Africa of the third century, was obsessed with the prevalence of violence in his society. He points to the wars scattered everywhere with "the bloody horror of the camps." He continues: "The world is soaked with mutual blood, and where individuals commit homicide, it is a crime; it is called a virtue when it is done in the name of the state. Impunity is acquired not by reason of innocence but by the magnitude of the cruelty."[39]

Cyprian's description of the attitude of the Christians of his day to enemies is quoted by the American bishops in their pastoral letter on war and peace: *The Challenge of Peace*. They choose the following telling example: "Then do not fight against those who are attacking since it is not granted to the innocent to kill even the aggressor but promptly to deliver up their souls and blood, that, since so much malice and cruelty are rampant in the world, they may more quickly withdraw from the malicious and the cruel."[40]

A teacher of rhetoric, a pleader in the courts, Cyprian found his way to Christ as he reached middle age. Soon ordained a priest, he was named bishop of Carthage in 248 C.E.

The cruelty inflicted on the martyrs is described by Cyprian, possibly to help prepare his flock for what they might soon experience:

> Long and repeated flogging, for all its cruelty, could not beat away the faith of these men, even after their inner organs had been exposed.... Blood flowed to extinguish the fire of persecution and the flames of hell. What a sublime spectacle for the Lord, sealed as it is with the solemn pledge and dedication of his soldiers! "Precious in the sight of the Lord is the death of his holy ones"; precious death, indeed, that was the price of immortality and crowned because of courage.... He who once overcame death for our sake, conquers it ever anew in us.[41]

During one wave of persecution, some Christians ordered to sacrifice to idols escaped punishment or death by purchasing a certificate, a *libellus,* stating that they had made the sacrifice. These persons were known as the *libellaticii.* Others who had actually made the sacrifice and therefore committed apostasy were termed *sacrificati.* The church faced the dilemma of reconciling both groups with the church community. Those who had obtained the certificate, the *libellaticii,* were given long periods of penance and separation from the Eucharist. Only if they became dangerously ill could they receive Communion. The *sacrificati* could only receive the sacrament when they were clearly in danger of death.

A new persecution was threatened. Cyprian told his flock that during peaceable times there existed the possibility of serving long penances, but not in the present time of peril. He told the Christians of Carthage that Christians were in such danger that the living had as much need of Communion as the dying. Those under penances should be allowed to take part in the Lord's Supper. "Otherwise," said Cyprian, "we should leave naked and defenseless those whom we are exhorting and encouraging to fight the Lord's battle: whereas we should support and strengthen them with the Body and Blood of Christ. The object of the Eucharist being to be a defense and security for those who partake of it, we should fortify those for whose safety we are concerned with the armour of the Lord's banquet. How shall they be able to die for Christ if we deny them the blood of Christ? How shall we fit them for drinking the cup of martyrdom, if we do not first admit them to the Chalice of the Lord?"[42]

When the emperor Valerian commanded all bishops, priests, and deacons to participate in worship of the gods of Rome or suffer exile, Cyprian was banished from Carthage. The proconsul demanded the names of his priests, and Cyprian refused.

When the decision was made to put him on trial, the people gathered before the gate of the jailer's house. It was an act of peaceful resistance, since all Christian assemblies had been forbidden. Convicted as an "enemy of the Roman gods and their religion," Cyprian was sentenced to die by being beheaded. The Christians cried out, "Let us be beheaded with him." The people followed him to the place of execution. Cyprian asked his friends for twenty-five pieces of gold, which he gave to the executioner. In a continuation of resistance, the Christians carried his body to the cemetery in a night procession lit by candles and torches.[43]

The besieged Christians, though they refused the weapons of the world, did not consider themselves defenseless when they could view the Eucharist as their defense.

He Feeds Sheep, Not Wolves

"If we are sheep, we conquer," asserted St. John Chrysostom. "Even though ten thousand wolves prowl around us we overcome and prevail."[44]

St. John himself knew what it was to be at war with wolves during his lifetime, and he gave a warning how not to respond: "But if we become wolves, we are lost, for we deprive ourselves of the shepherd's help. He feeds sheep, not wolves, and therefore abandons you if you will not let him prove his power in you."[45]

Born in Antioch in 347 C.E., the great city that was by then about half Christianized, Chrysostom owed his faith to his mother. His father, a pagan, was a commander of the imperial troops. His choice of study was rhetoric, and his teacher, Libanius, was considered the most famous orator of the day. This training was to bear fruit when Flavian of Antioch ordained Chrysostom and appointed him as his preacher.

By that time, John was nearly forty years of age. According to the custom of his time, he had not been baptized until he was over twenty. Seized with a longing for the ascetic life, he joined the hermits in the mountains around Antioch and for a time lived as a solitary in a cave.

There could be no greater contrast to the quiet of his cave than the unceasing tumult at the heart of Antioch. The church itself was divided, with many embracing the Arian heresy. When the emperor Theodosius levied a war tax on the Antiochenes, they rioted. In their fury, they tore down the gold statue of Theodosius and those of his father and two sons. For such crimes, the city itself was to suffer many penalties, including the stopping of free grants of corn to its people. For those who actually committed the crimes, a series of executions began under commissioners dispatched by the emperor. The people's fury turned to fear.

Meanwhile John began a marathon of twenty-one sermons, all preached during the season of Lent of 387. The Antiochenes came to listen, hoping to find in John's words a message to allay their fears. They prayed that the emperor's commissioners would recommend mercy. John pointed out that destroying statues was not a crime, though destroying God's images (that is, human beings) would be. John's sermons helped bring reason to the rioters. Bishop Flavian of Antioch interceded with the emperor. Further executions ceased.

John's eloquence, which earned him the title of "golden mouth" (Chrysostom), was the reason for his being raised to eminence as patriarch of Constantinople. He was, in effect, kidnapped from Antioch and brought to the center of the eastern empire.

While gentle with individuals, John was vehement against the evils of his day, especially those of the military life. "What sins do not they commit every day," he says of soldiers, "insulting, reviling, frantic, making gain of other men's calamities, being like wolves, never clear from offenses?... What disease does not lay siege to their soul? Can one indeed reckon up in words the trespass of their actions?"[46]

John's reply came in a sermon to the agitated people of Constantinople:

> Violent storms encompass me on all sides: yet I am without fear, because I stand upon a rock. Though the sea roars and the waves rise high, they cannot overwhelm the ship of Jesus Christ. I fear not death, which is my gain; nor banishment, for the whole earth is the Lord's; not the loss of goods, [for] I came naked into the world and I can carry nothing out of it.[47]

To prevent a possible riot among his people, he surrendered secretly to the officials, who took him to his place of banishment.

His first exile was short, but on a second imperial order, John was exiled to a place that called for him to undertake a seventy-day journey from Constantinople. In defiance of the decision of the pope that John should be restored to the patriarchate, the emperor in 407 C.E. ordered that his banishment be spent at a place on the Black Sea further still from Constantinople. He died on the journey in that year.

In all of this, John maintained the response of gentleness that he preached to others. He refers to the twelve apostles as examples who were sent out unarmed by Jesus. He says in the same sermon on Matthew:

> Then that they may learn that this system of war is new, and the manner of the array unwanted, he sends them bare and with one coat and unshod, and without staff, and without girdle or scrip.... When sheep get the better of wolves, and being in the midst of wolves, and receiving a thousand bites, so far from being consumed, do even work a change on them, [then] a thing far greater and more marvelous than killing them [has occurred];... and this being only twelve while the world is filled with the wolves.[48]

The message of John Chrysostom, reaching people through hundreds of sermons, was crucial for that period of the early church. A new relationship had developed now that Caesar had embraced Christianity. Caesar's authority was being exercised not against the church, but over the church.

Pursuing further the argument for gentleness on the part of Jesus' followers, John linked it with participation in the Eucharist: "The Lord has fed us with his own sacred flesh.... What excuse shall we have, if eating of the lamb we become wolves? If, led like sheep into pasture, we behave as though we were ravening lions? This mystery requires that we should be innocent not only of violence, but of all enmity, however slight, for it is the mystery of peace."[49]

A *Revolution of Peace*

For Christians in the first days of the new community of the church, it was shiningly, overpoweringly clear that the Lord's Supper was indeed the mystery of peace.

In Jerusalem, where Bishop James and the presbyters he chose presided at the sacramental meal, people who had formerly been enemies joined at the common table, above all, Jews and Samaritans. In Ephesus, where Timothy had been placed as bishop by St. Paul, Christian Greeks sat beside Christian Jews and Christian Romans. In Crete, traders of the Mediterranean world joined with local Cretans and Greeks in a church community led by Titus, upon whom the hands of St. Paul had been laid. Antioch's lively intellectual life attracted a large variety of peoples. Greeks could take part in the sacrifice beside a Roman rhetorician, and he beside a rough-mannered, hirsute Goth, as long as he was not an Arian.

A revolution of peace was occurring in the hearts of men and women in communities from Asia Minor, across Jerusalem to Gaul, Rome, Egypt, and Spain. Those who partook of the body and blood of Jesus could not be the ones to participate in the destruction of the bodies and the shedding of the blood of members of the human family — members made in the image of God himself.

The Christians had a new way of viewing the human family, seeing all in the new light of the incarnation. Christians were reminded of their special dignity as "temples of the Holy Spirit." "Do you not know that your body is a temple of the Holy Spirit within you, whom you have from God, and that you are not your own?" (1 Cor. 6:19).

Chapter 3

The Way of Justified Warfare

It is a higher glory still to slay war itself with the word, than men with the sword, and to procure or maintain peace by peace, not by war. For those who fight, if they are good men, doubtless seek peace; nevertheless it is through blood. Your mission, however, is to prevent the shedding of blood. — St. Augustine

In the fourth century a change occurred in church teaching that had crucial significance for the Christians of the world. It was a watershed that had a decisive influence on human history for the coming fifteen hundred years. The early witnesses, martyrs, bishops, priests, teachers, and laypeople, all agents of transformation through their transformed lives, had achieved a revolution of peace and nonviolence in hearts and minds. There followed a revolution in the other direction. The new revolution brought the followers of Jesus to an acceptance of violence and warfare.

Could the followers of Jesus, who at the Lord's Supper partook of the Savior's body and blood, participate as soldiers in the destruction of the bodies and shedding of the blood of those children of God called "enemies"? Until the fourth century, the answer given by churchmen had been in the negative, but during that century, the answer was changed to "yes." Of course, there were conditions and limitations, but the apocalyptic had happened: the door to violence and warfare had been opened for Christians of future generations.

Many forces and people were involved in the changeover. Among the people involved were the emperor Constantine, along with his friend, Bishop Eusebius of Caesarea, Marcus Tullius Cicero, St. Ambrose of Milan, St. Augustine of Hippo, and Ulfilas, the Goth.

Two Dreams

The dreams of two men dramatized aspects of the changeover. A young general of the Roman army, Constantine, ready at a bridge over the Tiber to lead his legions into Rome, had a vision and a dream. He said he saw — in the heavens, in early afternoon — a cross of light above the sun and with it the words, "In This Sign Conquer." During sleep that night, he dreamed that he saw the same sign, now with Christ, who commanded him to make a likeness of the sign and use it as a safeguard in all engagements.

In the Battle of Milvian Bridge in 312, Constantine led his troops to victory over Maxentius, who had lost favor with the Romans. The Roman senate and the populace welcomed him as a deliverer. Constantine credited the victory to the vision and dream.

Before this battle, Constantine had shown himself a charismatic leader of men. In York, England, on the death of his father, Constantius Chlorus, he had been acclaimed Caesar by the legions in the field. The circumstances of his elevation to leadership give an insight into the future course of his life. Constantius Chlorus, as a member of the tetrarchy that ruled the Roman Empire, was the Caesar to whom fell the governance of Britain and Gaul. The Caesar Augustus, and source of all authority, was Diocletian, who chose for his dominion Egypt, Asia, and Thrace. Galerius and Maximian divided between them the rest of the empire.

Constantine was attached to the court of Diocletian, at whose death he became part of the court of Galerius. It was at this time, in 303, that one of the most savage persecutions of the Christians was launched. Constantine was held at the court as a virtual hostage for the good behavior of Constantius Chlorus.

A man of mercy, Constantius Chlorus took no part in the persecutions. He himself was a worshiper of the invincible sun. Eusebius relates that "he stayed outside the campaign against us and saved God's servants among his subjects from injury and ill-usage, and he neither pulled down church buildings nor caused any mischief whatsoever."[1] Eusebius makes the point that after his accession to power in Rome, Constantine "determined to emulate his father's reverent attitude to our teaching."[2]

One can estimate that an initial reverence for Christianity learned from his father had been deepened in Constantine by the knowledge of atrocious sufferings imposed on innocent Christians, men, women, and even children, and by their bravery in enduring it. It is Hugo Rahner's view that Constantine's vision of the cross inscribed on the sun was a sign of his turning toward the church rather than the cause of it.[3]

It was only when his father asked for his presence in Britain that Constantine was freed from the court and gained freedom of movement. Traveling with all possible speed, he reached his father in time for the victory over the Picts, who were menacing England's northern border. After his father's death he set out for Rome with a small but experienced army. He was determined to claim his full right as a Caesar, a right that had been put in question. His progress was marked by victories in Gaul and at Turin and Verona. The victory over Maxentius at the Milvian Bridge was the culmination of a life marked by success.

"With this battle," wrote Jacob Burckhardt, "the entire west found its master; Africa and the islands also fell to the conqueror.... Constantine, who had previously been known only through border wars, suddenly stood forth in the public eye with the radiance of a hero's glory. Now the problem was to base his new power, wherever possible, on foundations other than mere military strength."[4]

Another dream, this time that of the great St. Jerome, was revelatory of the source from which Christians drew justification for the new acceptance of violence. The source was the great Roman philosopher Marcus Tullius Cicero. On a visit to Antioch in the year 374, St. Jerome became very ill and fell into delirium. In his fever, he had a dream about which he wrote to St. Eustochium, his pupil. He dreamed he died and was brought before the judgment seat of Christ. When asked who he was, he answered, "I am a Christian." "You lie," was the reply. "You are a Ciceronian: for where your treasure is there is your heart also."[5]

The sudden realization of his undue love for the "copious, flowing eloquence" and the thought of Cicero struck Jerome as so true that he reordered his life. He threw himself at the feet of Jesus in fasting and repentance and moved to the desert. Many early Christians found that their love of Cicero and other Roman classical writers and poets interfered with their devotion to the gospels.

Cicero, who met death by execution over a century before Christ's death, was part of the historic change, since two great saints, Ambrose and Augustine, finding nothing in the gospels, turned to Cicero for a rational formulation of conditions that might "justify" warfare. It might be said that Cicero had more influence on church thinking with regard to war than church leaders and theologians.

Ambrose, Roman of the Romans, modeled one of his important works, *On Duties* (*De Officiis*), on Cicero's *On Duties.*[6] Augustine, the provincial young man from North Africa, was honored in having the great Ambrose as his mentor. It was Augustine who went on to be considered "the theologian" of the church, one of the four great doctors of the Western church.

Eusebius, author of *The History of the Church,* was a friend and fervent admirer of Constantine. It is to him that we owe the story of the vision and the dream, and his acceptance of the sign as the "victory-giving cross."

Ulfilas, the Goth, became a convert to Christianity in Constantinople. He translated the Bible into the Gothic language, only some small tracts of which remain. He changed history by being an instrument of the conversion of the Goths to Arianism.

The "Just" War

Aristotle (384–322 B.C.E.) is credited with first using the term "just war." He used it in his *Politics* to describe wars conducted by the Hellenes against non-Hellenes who were considered barbarians. Through wars, the Hellenes by their virtue extended their control over less worthy people. The expansion of control by the Hellenes redounded, according to the Hellenes, to benefits for the governed.[7]

Marcus Tullius Cicero (106–43 B.C.E.), Rome's most mighty orator, was also renowned as a philosopher, writer, politician, and poet. The issue of war was addressed in *On Duties.* In the work, he emphasized the ethical concept of the Stoics regarding the natural law by which man, through reason, can know what is right. He considered war as a last resort, pointing out that there are two methods of settling a dispute, discussion or the use of physical force, meaning violence or war. No war could be just, according to Cicero, unless waged to recover lost goods, repel aggression, forestall enslavement, or honor pledges to allies. Before a war could be initiated, an official demand of satisfaction had to be submitted, a warning given, and a formal declaration made. In accordance with Roman practice, the formal declaration should be made by Roman priestly officials, the *fetiales,* charged with treaties and wars. The just war included punishment of the enemy for crimes. On victory, mercy should be shown to enemies unless they had acted with unnecessary cruelty. When an enemy was besieged, only those who resisted Rome should be punished, while noncombatants ought to be spared. Captured soldiers and civilians should be enslaved rather than killed. The word for peace, *pax,* was related to the making of pacts between hostile parties.

"There are two sorts of injustice," wrote Cicero in *On Duties,* "the one, of those who commit an injury; the other, of those who do not avert the injury from those against whom it is committed, if they have the power to do so."[8]

Ambrose, the great saint of the church, echoed Cicero in a key statement in his *On Duties.* He asserted, "He who does not ward off injury from his comrade, when he is able to, is just as guilty as he who does the injury."[9]

This concept of intervention on behalf of those suffering injustice or aggression was applied to war. This sentence of Ambrose may sound like an assertion of the obvious. In point of fact it had enormous reverberations in history, since it was utilized as the justification for military intervention on behalf of the Christians suffering injustice in the Holy Land, in other words, justification for the Crusades.

Constantine

The Unholy Embrace

Before continuing with the discussion of Ambrose and Augustine, it is necessary to return to Emperor Constantine. One year after his conquest of Rome, Constantine, along with his coemperor Licinius, issued an edict granting freedom to all cults in the Roman Empire. This edict, the Edict of Milan, embraced the religion of the Christians. Christianity, beaten down by waves of persecution, became a permitted religion, *religio licita.* Belief in Christianity now became fully equal, before the law, to belief in the old gods of Rome. Thus was lifted from the hearts of Christian citizens the fear of being arrested, tortured, or put to death. This edict changed profoundly the attitude of Christian citizens to Rome. Their feeling of immense gratitude was a factor in their putting aside of their nonviolence of the first three centuries and in their willingness to take part in the military defense of their protector, the Roman Empire.

The Edict of Milan, the Imperial Ordinance of Constantinius Augustus and Licinius Augustus, stated in part:

> We decided to establish rules by which respect and reverence for the Deity would be secured, i.e., to give the Christians and all others the liberty to follow whatever form of worship they chose, so that whatsoever divine and heavenly deities exist might be enabled to show favor to us and to all who live under our authority. This therefore is the decision that we reached by sound and careful reasoning: no one whatever was to be denied the right to follow and choose the Christian observance or form of worship;

and everyone was to have permission to give his mind to that form of worship which he feels to be adapted to his needs, so that the deity might be enabled to show us in all things His customary care and generosity.[10]

The imperial rescript was to be published by the governor of each province so that the "generosity" of the edict might be known to all citizens throughout the empire. Among the provisions was one ordering the return to Christians of confiscated properties.[11]

There is much in the edict of a quid pro quo, pointing to the favors to be expected from "whatsoever divine and heavenly deities exist" as a result of the edict.

The edict was certainly designed by Constantine. This was proved later when Licinius relapsed into the old practice of the Roman Empire in persecuting Christians. Constantine put his mark on the armies of Rome through his emblem. The emblem designed in accordance with his dream has been variously described. From Eusebius we might conclude that it was a cross such as the one on which Jesus was crucified, but without the corpus. Another description asserts that it consisted of the first two letters of the word "Christ" in Greece, the Chi Ro (XR). This could consist of the interlocked X and P. It might well have served as an emblem of the god whom Constantine referred to as "the unconquerable sun, my companion." The emblem was placed on the shields of the soldiers and on the large battle standards. It was, as Burckhardt remarks, "an emblem which every man could interpret as he pleased but which the Christians would refer to themselves."[12] According to Burckhardt, Constantine reported the vision to Bishop Eusebius long after 312, when it was supposed to have occurred. The emperor swore an oath that it had happened as he described it. Eusebius accepted the emperor's account and referred to the sign as "the victory-giving cross."

The emblem played an important part in battles, the standard being cared for by a special guard. It was enshrined in its own tent, to which Constantine retired before important engagements. Whether the emblem represented the cross of Christ, the sign of innocent suffering and spiritual victory, or the monogram of Christ, the religion of Christ was now linked to armies that shed blood. Constantine made no formal profession of faith, but he honored a sign that in his view brought him success. At the same time, he retained the old gods, especially the sun god, on the coin of his regime. He was represented as the "Pontifex Maximus," the high priest of the Roman religion and of its college of priests.

Emperor Constantine desired an empire of peace. To the Christians, the Constantinian peace was a matter of gratitude and rejoicing. No one expressed the gratitude with more joy than Eusebius. He himself knew what it was to be persecuted, having been exiled for the faith under Diocletian.

Eusebius wrote:

> Thanks be to God, the Almighty, the King of the universe, for all his mercies; and heartfelt thanks to the savior and redeemer of our souls, Jesus Christ, through whom we pray that peace from troubles outside and troubles in the heart may be kept for us as stable and unshaken for ever....

> Come hither and behold the works of the Lord. What wonders He has wrought in the world making wars cease to the ends of the world; the bow He will break and will shatter the weapon, and the shields He will burn up with fire. . . .
>
> The power of God's spirit penetrated all the members of the church; all were of one mind, one faith, one voice raised in praise of God. Our Bishops performed ceremonies with full splendor, our priests offered their sacrifices in the Church's majestic rites. People of every age and both sexes gave themselves to whole-hearted prayer and thanksgiving, and joyfully worshipped God, source of all blessings.[13]

Eusebius continually lavished praise on Constantine:

> Men had now lost all fear of their former oppressors; day after day they kept dazzling festival; light was everywhere, and men who once dared not look up greeted each other with shining faces and smiling eyes. They danced and sang in city and country alike, giving honour first of all to our Sovereign Lord, as they had been instructed, and then to the pious emperor with his sons, so dear to God. Old troubles were forgotten and all irreligion passed into oblivion: good things present were enjoyed, those yet to come eagerly awaited. In every city the victorious emperor published decrees full of humanity and laws that gave proof of munificence and true piety.[14]

Eusebius, always the flatterer, referred to Constantine as the "lawful son" of his father, though it was known that he was the son of Helena, a concubine.

In the new relationship between Christians and the empire, divisions between the Christians insofar as they threatened the unity of the empire became a concern of Constantine. Hosius, bishop of Cordoba in Spain, is said to have suggested that settling the controversy could be achieved by bringing the bishops together. The Council of Arles, in France, was convoked by Constantine in 314 with the result that Donatism was duly condemned by the bishops. In brief, the Donatist heresy arose in 312 in the North African church when Bishop Donatus objected to the consecration of a bishop who, during the persecution, had handed over the scriptures. He led a movement against the acceptance of those who had handed over the holy books to be presiders at the Eucharist. In another heresy, Arius, a priest of Alexandria, began teaching in 319 that Jesus Christ was not the "eternal" son of God become man. Jesus therefore does not share the same substance with the Father. Arianism was growing among the faithful, as well as among priests and bishops. Arius went about singing his belief in a song of his composition. Hosius considered that the same antiheresy medicine might be applied to Arianism as to Donatism, namely, a council.

The new council was called "ecumenical" because it brought together the bishops of the Roman world (except the pope). Some three hundred bishops came from east and west to Nicaea, in Bithynia, Asia Minor, in 325. Nicaea affirmed that Jesus was "begotten not made" and was of "the same nature with the Father." Nicaea also decided that all churches should observe Easter on the same Sunday and that bishops and priests who were ordained while unmarried should remain unmarried.

Of special interest to the Catholic tradition of peace are the contradictory statements that were issued from Arles and Nicaea. The third canon of the Council of Arles asserted, "Those who throw down their arms in peace time are to be excommunicated."[15] The harsh penalty for soldiers who decamp from military life, exclusion from the Lord's Supper, is hard to believe. It could be that Christian soldiers, now that peace had been achieved, simply abandoned military life and returned to their homes. There were borders still threatened by barbarian attack, and some provinces had shown signs of defecting from the empire. There was need to have an army in readiness. At Arles in the year following the Edict of Milan, the bishops were, in all probability, moved to exhibit their gratitude to Constantine. Whatever the reasoning behind the canon, it established that in return for governmental protection, the church accepted the civil and military obligation which the state might demand of its citizens.

A seemingly contradictory canon was adopted at Nicaea in 328. The famous twelfth canon decreed that "they who first proclaimed their faith and abandoned the soldier's belt, but who afterwards, like dogs returning to their vomit, go so far as to give money and presents to be readmitted into the army, must remain three years among the audientes (2nd degree of penance) and three years among the substrati audientes (3rd degree of penance)." The canon goes on to explain that "those who prove by penance and good deeds their sincere desire to be reunited with the church may eventually be readmitted to assist at the whole of the liturgy."[16]

This canon would indicate that the three hundred or so bishops were against army service. Most of the bishops came from the eastern empire, in which Licinius, despite his concurrence with the Edict of Milan, had instituted a fierce persecution of Christians. All soldiers were obliged to offer sacrifices to the gods of Rome. Many soldiers, however, seemed to have been able to leave the ranks rather than make the act of sacrifice. The bishops knew that many Christian soldiers had suffered, including, in one case, forty soldiers who had been martyred together. The canon thus refers to a particular circumstance in which Christian soldiers were ordered to forswear their faith by paying homage to pagan gods.

The Arian Heresy

The larger concern of Nicaea, however, was the Arian heresy. During the discussions, which were opened by Bishop Hosius, Constantine presided from his throne, his royal purple shining with jewels and ornaments of gold. Despite the fact that he towered over the bishops of the church from an imperial throne, Constantine was no more than a "hearer," having taken the first step toward entering the Catholic Church. Constantine knew that the decision of an anathema against Arianism was almost unanimous at Nicaea. Yet some years later when Arius falsely swore to him that he was truly Catholic, Constantine was persuaded to reconsider Arianism, Nicaea or no Nicaea. Constantine ordered Alexander, the patriarch of Constantinople, to receive Arius in complete communion with the church. A clash was in the making since the patriarch adamantly refused the emperor's demand. The emperor threatened to remove him from his see. Arius and his followers, supported by Eusebius of Nicomedia, might have come close to prevailing over orthodox doctrine. Unexpectedly, however, Arius died. One year later, 337, Constantine himself died, asking for baptism when he knew his end was near. He was baptized by an Arian bishop, Eusebius of Nicomedia.

The ambiguity of Constantine's beliefs was indicated after his death when he was mourned by Catholics and Arians and by those who still followed the old gods of Rome. His body was conveyed to his capital, Constantinople, where prayers were offered in the Basilica of the Holy Apostles for the man who had ended years of fear and persecution. In Rome, part of the memorial to Emperor Constantine consisted of the customary tribute to a dead emperor by the senate; he was apotheosized and declared a god, thus joining the line of imperial gods which included his own father.

•

That a price was paid for the peace, the so-called Constantinian peace between the Roman Empire and the Christians, is made clear by Donald Senior. "Certainly," he writes, "it was good that the pain and division inflicted on the church by three centuries of intermittent persecution came to an end with the fourth century." Senior adds a significant qualifier, however: "But," he continues, "the Edict of Milan in A.D. 313 also began an unholy embrace between state and church that would often hobble the freedom of the gospel and compromise the church's integrity down to our own day."[17] This would seem to be the consensus of twentieth-century Christians who see in this period of history the beginnings of Caesaro-papism, with all its implications for the involvement of Christians in wars.

Ambrose: Roman of the Romans

Ambrose fully merited the description "Roman of the Romans." His birth was not in Rome but in Trier, in Germany, where his father had his residence as prefect of Gaul, administering a vast territory reaching through part of Germany to include France, the Iberian peninsula, and Britain. Ambrose himself, as a member of the ruling class of the Roman Empire and a lawyer, was made governor of Liguria and Emilia in northern Italy. This eminence gave him the patrician self-confidence, as bishop of Milan, to "speak truth to power" and to defy encroachment of imperial power over the church.

"The Emperor is within the church not above it"[18] was his famous statement of defiance. The empress-regent Justina, mother of the emperor, attempted to have a new basilica and other Catholic sanctuaries turned over to the Arians. Ambrose refused, and Justina, influenced by an Arian bishop, contrived to place Ambrose in a cruel dilemma; he was ordered to submit the claims of the church (against the Arians) to be judged by the emperor or, if he refused, to abandon Milan. The Arian bishop who had devised the trap was Auxentius, who had lost the see of Milan to Ambrose when the people had rejected him and acclaimed Ambrose their candidate.

Ambrose escaped the trap by choosing a third way. He rejected the proposition that the claims of the Catholic Church could be judged by a layman. Justina's plan was that the Arian claim be given a favorable judgment. At that point, the Catholic sanctuaries of Milan would be turned over to the Arians. Surrounded by the faithful, Ambrose occupied the new basilica. It was a response of nonviolent resistance.

Soldiers stood guard, not permitting anyone to leave, though some were allowed to enter, probably bringing food supplies. Day after day and night after night, the people defended their sanctuary, shouting their rage-fed hostility to the Arian bishop, Auxentius.

Ambrose calmed them, leading them in psalms and hymns which he composed himself. During the week of the occupation of the basilica, he had time for many sermons. The people came to the point of assuring their bishop that they were ready to die with him. They accepted his passionate conviction that no violence could be used to defend the church. At the end of the week the imperial court yielded; the people kept their sanctuaries. It was at this juncture that Ambrose uttered his famous dictum, reminding the imperial court of its position within, not above, the church.

There is no doubt that the Roman Empire left its mark on the church, a mark that was only too evident. When Emperor Gratian, ruler of the western empire from 375 to 383, refused the title of Pontifex Maximus, it was not discarded, but taken over by the pope in Rome. Similarly, the concept of the college of the cardinals was instituted, reminiscent of the college of pontifices, priests of the gods and goddesses of Rome. It was during the rule of Gratian that a feud between Rome and Constantinople moved the reigning pope, Damasus I, to describe the church, hitherto known as the Catholic Church, as the Holy Roman Catholic Church. During the same period, in the year 382, the traditional religion of Rome was disestablished, which meant that public funds would no longer be channeled into public ceremonies honoring the gods and goddesses of Rome.

Does it need to be emphasized that as the number of Christians grew, Christianity produced incredible changes in humanizing the culture of a violent and often barbarous empire from Armenia to Spain? Ambrose was at the center of this humanizing influence, being chosen as one of the advisers to Emperor Gratian. The fate of the empire was linked in the mind of Ambrose with the expansion of the kingdom of God on earth. The Roman Peace, with its network of roads and safer travel, had been achieved in time to allow the apostles to carry the message of Christ to peoples at the far end of the known world.

It was not surprising that Ambrose saw the wars of fourth-century Rome as just wars when they were fought against Arian Goths, sworn enemies of the Catholic Church. As an adviser to Emperor Gratian, Ambrose felt called upon to address him before the battle of Adrianople in 378: "I ought not to detain you with many words, O Emperor, intent as you are on war and preoccupied with victory over the barbarians. Go forth, then, protected by the shields of faith and holding the sword of the spirit. Go forth to the victory promised in days past by the Scriptures. For in those days Ezekiel prophesized the devastation we have suffered and the wars with the Goths."[19]

The Roman army was disastrously defeated by the Visagoths even before Emperor Gratian came on the scene with the troops. The address to the emperor is notable on two points: it cites the Old Testament as a precedent, and it employs the "shield and sword" used figuratively in the New Testament in their literal meaning. Ambrose had just begun his twenty-second year as bishop of Milan when this patriotic address was made. Such a patriotic gesture was never repeated, although Ambrose did serve the empire as a peace emissary to his birthplace, Trier, to forestall a military invasion into Italy.

In *On Duties,* he reminds Christians that "in the matter of war, care must be taken to see whether the wars are just or unjust."[20] He echoes Cicero in pointing out that there is no such thing as a lawful war without a just cause. Cicero made it clear that even in wars undertaken for conquest or for glory there needed to be a just cause.

In discussing dutifulness, Ambrose acknowledges the duty to one's country and sees it as superior to the duty owed to parents. "Dutifulness, binding in justice, is owed primarily to God, secondarily to one's country, thirdly to parents and then to all others."[21] Fortitude, seen in earlier times as, above all, the quality needed by martyrs, had become, for Ambrose, the quality needed in warfare and in civil life: "For fortitude, which in war preserves the country from the barbarians, or helps the sick at home, or defends one's neighbors from robbers, is full of justice."[22] Clearly, participation in war was taken for granted.

What of the nonviolence of the gospel? It is taught by Ambrose but is transposed into personal life; in fact, it is demanded of personal life. In *On Duties,* he gives the famous example of a Christian caught in a shipwreck. Should he surrender his life-preserver to a fellow passenger, a foolish one at that? Yes, the Christian should give up his life-preserver so as not to save his life at the expense of another's. In another example, the Christian, when attacked by an armed robber, should not respond with blow for blow, "lest in the act of protecting himself, he weaken the virtue of love."[23]

Ambrose then cites Matthew's gospel (26:52): "Put up your sword; everyone who kills by the sword will be killed by it." Christ, he points out, desired to cure everyone by his wounds and did not want to be protected by harm done to his persecutors. A nonviolent response to an attacker is demanded of the individual Christian, he points out in a discourse on the Psalms, because as a Christian he must prefer a divine, and therefore eternal, good to a human, and therefore temporal, one.

A nonviolent response is also required of priests. "Interest in matters of war seems to be foreign to our role," he tells priests in *On Duties,* "because we are concerned with matters of the soul rather than of the body, and our activity has to do not with weapons but with peaceful deeds."[24] It is clear that priests are to eschew violence. Yet because of his assertion, echoing Cicero, regarding the duty of coming to the aid of people unjustly attacked, Ambrose, more than any other hierarch, is considered the father of the Crusades.

Might he have responded as uncritically as Pope Urban II? Insofar as the first Crusade was a response to an appeal of the Byzantine emperor on behalf of persecuted Christians, Ambrose might have approved the military enterprise. Insofar as it was an appeal for military aid because of the despoliation of the Holy Sepulcher, he might well have demurred, if we judge by his own actions.

There was another attempt made by the empress-regent to take over a Catholic basilica for the use of the Arians, including herself and members of the court in Milan. When a group of Arians attempted to occupy the Basilica of the Apostles, the infuriated people captured an Arian priest. Ambrose, to avoid bloodshed, sent his priests and deacons to save the priest from the crowd. Ambrose, in relating the incident, said, "When I was told that the church was surrounded by soldiers, I said, 'I cannot give it up, but must not fight.'"[25]

In a work entitled *On Widows,* Ambrose states, "The church, however, does not conquer the forces opposed to it with temporal arms but with the arms of the spirit, which are capable in the sight of God of destroying fortresses and heights of spiritual wickedness.... The church's weaponry is faith; the church's weaponry is prayer, which overcomes the adversary."[26]

In one aspect of war, Ambrose differed from the view of Cicero, namely, that of meting out punishment to the wrongdoers at the end of hostilities. The old Roman law and the old covenant both countenanced vengeance. Ambrose, in a discourse on Luke's gospel, provides the Christian contrast: "The law calls for reciprocal vengeance; the gospel commands to return love for hostility, good will for hatred, prayers for curses. It enjoins us to give help to those who persecute us, to exercise patience toward those who are hungry, and to give thanks for a favor rendered."[27] These sentiments, need it be said, can be expressed only after the just war had run its course. The just war seemed firmly in place.

Augustine

To Carthage Then I Came

It would be hard to find two persons more different in origin and temperament than St. Ambrose and St. Augustine. Ambrose, the patrician, started out at the top of Roman society. He was a lawyer, measured in his words, governor of an important province, and destined for an honored, successful life. He was learned in Greek. St. Augustine started out as a provincial in a small town in North Africa, passionate in Latin, his second language, after Punic. He was too poor to pursue his education without a patron. He thirsted for glory, through eloquence and achievement. Like Ambrose, Augustine relied on Cicero, who served as the awakener of his youth to wisdom and the good. To Cicero, he owed the concept of the right to war.

"To Carthage then I came," wrote Augustine regarding his eighteenth year, "where a cauldron of shameful lusts seethed and sounded about me on every side."[28] He plunged into the sensate life about him, took a concubine, and at nineteen fathered a son. Carthage, then the second city of the Roman Empire, could not contain the soaring ambitions of the brilliant young man. On he went to Rome, center of the world, against the wishes of his devoutly Catholic mother, Monica. She prayed all night at the shrine of St. Cyprian while Augustine, having lied to her, slipped away and took ship to Italy.

After teaching rhetoric for a decade, he won, through his extraordinary gifts of intelligence and eloquence, the post of master of rhetoric of Milan, a city which boasted the presence of the court and of a pulsing intellectual life. Then he met Ambrose, Milan's bishop, fourteen years his senior.

Augustine, having set aside the faith taught him by his mother, lingered on the fringes of Manicheism. It was a work of Cicero's, *Hortensius,* which turned the heart of the young man to a yearning for "undying wisdom." His mind was turned to God by the good pagan Cicero. Out of curiosity he joined the crowds who attended the sermons of Ambrose.

It was in Milan that Augustine became convinced of the truths of the Catholic faith. The decision to break with the joys and attractions of his former life caused him torments, so that he tore his hair and beat his forehead. After his conversion to the Catholic faith, unforgettably related in his *Confessions,* he presented himself for baptism to Ambrose as bishop of Milan. He was thirty-two years of age.

The catechism of Milan had been an "awesome discipline" according to Peter Brown: "On the eve of the feast of the Resurrection," writes Brown, "Augustine and the throng of other 'competentes' of all ages and both sexes would troop to the Baptistery behind the Basilica of Ambrose. Passing between the curtains, Augustine would descend alone, stark naked, into a deep pool of water. Three times Ambrose would hold his shoulders beneath the gushing fountain. Later, dressed in a pure white robe, he would enter the main basilica ablaze with candles . . . for a first participation in the mysteries of the risen Christ."[29]

Adeodatus, Augustine's son, was also baptized. Monica, now a widow, had come to be with her son and grandson. Sending his concubine back to North Africa, Augustine joined an ascetic, monastic type of community near Milan. His brother, his son, and some friends gathered about him.

Return to Africa

As a baptized Christian, Augustine returned to his birthplace, Thagaste, and formed an ascetic community similar to that of Milan. Adeodatus died there at sixteen.

Visiting Hippo, a coastal town, to talk to a candidate for the monastic community, Augustine entered the cathedral. People who knew him shouted that they wanted him as a priest, pushing him forward to the bishop's throne.

"I was grabbed," he wrote later. "I became a priest and from there I became your Bishop."[30] From Hippo, forty miles from Thagaste, he never moved, except for church meetings.

His entry into holy orders had been unprevisioned, like that of Ambrose. For thirty-five years, as bishop of Hippo, he led his people, preaching sermons, some soaring, some didactic, every Saturday and Sunday and also on special weekdays. Around him he gathered a monastic community of priests who lived by a simple ascetic rule. Together they gave an example of what it was to be tireless servants of God. The Christian community was threatened by heresies that were aggressive and even violent, the Donatists, the Pelagians, the Arians, and the Circumcellians. Through it all, Augustine poured out a steady stream of spiritual works that nourished not only his own flock but the whole Catholic world.

Apostolic Letter to the Catholic World

The conversion of Augustine to the Catholic Church was of such moment that its sixteen hundredth anniversary was marked by an apostolic letter from the Holy See to the Catholic world. In 1986, Pope John Paul II issued the fifteen-thousand-word "Apostolic Letter on Augustine of Hippo" on the date of the annual feast of St. Augustine, August 28. The pope asserted that St. Augustine "has been present ever since [his conversion] in the life of the church and in the mind and culture of the whole Western World."[31]

"I am happy," wrote Pope John Paul II, "that the propitious circumstance of his conversion and baptism offers me the opportunity to evoke his brilliant figure once again." The pope makes mention of Augustine's great contributions to theology and indicates how his wrestlings with the relationship between faith and reason, and the great questions of life, are pertinent to the present age.

"It was above all in studying the presence of God in the human person that Augustine used his genius," the pope states. The pope points out that "it is no small merit [of Augustine] to have narrowed all Christian life down to the question of charity." The pope recalls the words of the saint regarding the gifts of the Holy Spirit, "Have charity and you will have them all; because without charity whatever you have will be of no benefit."

The subject of peace enters into the apostolic letter at several points, while war is only mentioned once. Struggle, however, is mentioned, and the citation chosen by the pope is from Augustine's *Confessions.* "God has willed that our struggle should be with prayers rather than with our own strength." The pope calls the *Confessions* "a work that is simultaneously autobiography, philosophy, theology, mysticism and poetry."

The pope states that Augustine recommends above all to those who have control over the destinies of the peoples that they love peace and that they promote it, not through conflict, but with the methods of peace. The pope quotes Augustine: "There is more glory in killing the wars themselves with a word than in killing men with the sword; and there is more glory in achieving or maintaining peace by means of peace than by means of war."

Pope John Paul II writes:

> It is good to mention here some of the definitions of peace which Augustine made according to the various contexts in which he was speaking. Starting with the idea that "the peace of mankind is ordered harmony," he defines other kinds of peace, such as "the peace of the home, the ordered harmony of those who live together, in giving orders and in obeying them," likewise the peace of the earthly city and the "peace of the heav-

> enly city, the wholly ordered and harmonious fellowship in enjoying God and enjoying one another as God"; then the universal peace which is the "tranquillity of good order," and finally the order itself that gives its place to each of the various equal and unequal things.

The citation contains the phrase of Augustine most commonly quoted on peace: "the tranquillity of order."

The apostolic letter does not focus on war or on the conditions which would justify Christian participation in it. Not mentioned is the contribution which reached into the lives of countless people who know nothing else about Augustine but one thing: his contribution to the tradition of "justified warfare." Through the centuries, Augustine, along with Ambrose, has been linked with the teaching that Christians in good faith can take human life in a "just war."

The Tranquillity of Order

The tranquillity of order is peace and is the way to peace according to Augustine. "Peace between man and God," he says, "is the well-ordered obedience of faith to eternal law. Peace between man and man is well-ordered concord." The "tranquillity of order" is an abstract and a high goal, a shining goal on a far horizon, calling for a congeries of external conditions and internal dispositions. It is a situation that, as Augustine unfailingly reminds us, would be difficult to achieve by fallen human nature in a fallen world.

Augustine's discussions of "justified warfare" is not contained in any one treatise, though he produced, by his "sleepless diligence," to use Pope John Paul II's words, an immense corpus of work. His references to the subject of war represent a very small proportion of his writings. His references to war and peace are scattered throughout his works, including his treatise on the Sermon on the Mount and his lengthy work *The City of God.*

He confronted the issue of war in response to situations in his life and above all in four letters called forth by the problems presented to him. These are the letters to Faustus the Manichean, to Boniface, to Marcellus, and to Darius. He wrote the first of these letters (to Faustus) when he was forty-three; the last, to Darius, was written when Augustine was seventy-four, a year before his death. At that time, the Vandals, having conquered North Africa, were almost at the gates of fortified Hippo.

Augustine relied on Cicero for the concepts of the right to war. "Just wars," states Augustine, "are usually defined as those which avenge injuries, when the nation against which warlike actions are to be directed has neglected either to punish wrongs committed by its own citizens or to restore what has been taken from it."[32]

Augustine considered that while humankind moved toward the ultimate peace of ordered tranquillity, the injustices and wrongs of the world would demand unremitting human struggle. War arises from sin. According to Augustine, war serves both as a consequence of sin and as a remedy for it. Running through Augustine's writing is the concept of the City of God and the City of Man. Cain, the firstborn, belonged to the City of Man, while Abel, his brother whom he killed, belonged to the City of God.

Augustine divides human beings as belonging to the City of God or to the City of Man, according to what captures their love. Those who love God, believers who obey the word of God, are part of the City of God; those who are lovers of self are part of the City of Man. This is the undergirding of his spiritual outlook, including his outlook on war.

"[That] kind of war is undoubtedly just which God himself ordains," asserts Augustine.[33] Augustine buttresses his support of justified warfare by referring to the God-ordained wars of Hebrew Scriptures. He gives subdued emphasis to the Sermon on the Mount, which contains the distinction between the attitude toward the enemy in the old and the new dispensation. As Pope John Paul II points out, there is in Augustine an overarching emphasis on charity, and in his own way, Augustine explains how one can go to war with an enemy without dispensing with charity.

First of all he stands with Ambrose on the duty of nonviolence toward a personal enemy. "I do not approve of killing another man to avoid being killed oneself," he asserts. A person who kills on his own volition and authority is guilty of the crime of homicide, in Augustine's view.[34] He explains the crucial difference between killing in self-defense and killing in defense of the community. A person can kill when "one happens to be a soldier or public official and thus acting not on his own behalf but for the sake of others, or for the city in which he lives. Provided, of course, that one possesses legitimate authority and acts in accordance with his position."[35]

In *The City of God,* Augustine elaborates on the exceptions to the commandment against killing:

> The divine law made certain exceptions to the principle that one may not kill a human being. Included in this category are individuals whom God by means of some law or an explicit command, limited to a particular time and person, has ordered put to death. Anyone who acts as a delegate in this regard is not himself the slayer since he is like a sword that is a tool in the hands of its user. Accordingly, men who have waged war at God's command or who have put criminals to death in their capacity as agents of the state in accordance with its laws, that is to say, on very justifiable grounds, have not violated the commandment, "Thou shalt not kill."[36]

Augustine confronted the issue of war in response to the reality of his time and place, an empire which provided sinews of order for a great expanse of the ancient world, including North Africa. In "The Just War Ethic," J. Bryan Hehir comments on the dual aspect of his response: "Augustine provided the basic rationale for other just-war theorists by utilizing a moral argument which legitimized the use of force as a means of implementing the gospel command of love in the political order. Augustine's argument was political in the sense that his moral judgement on warfare emerged from an assessment of the possibilities and requirements of order in the political community."[37]

Letter against Faustus the Manichean

Augustine's response to four letters gave him the opportunity to deal with various aspects of war. Many of Augustine's key concepts on war are contained in the letter "Against Faustus the Manichean." Faustus regarded the wars of Hebrew Scriptures as irreconcilable with the teachings of the New Testament. The elect among Manicheans were said to be pacifists.

Augustine explains that in the wars led by Moses, the prophet was carrying out the orders of God and was thus acting out of obedience rather than savagery. Those who received punishment were the ones who deserved it. "What is blameworthy about war?" asks Augustine, and answers that the blame in war consists in "the desire for harming, the cruelty of avenging, an unruly and implacable animosity, the rage of rebellion, the lust of domination and the like."[38] It is to punish these evils, Augustine makes clear, that wars are justly undertaken by " 'good men,' acting under the command of God or legitimate authority."[39]

In the same letter, Augustine refers to New Testament examples that honor the necessity for military service and war. When soldiers came to John the Baptizer for baptism, says Augustine, he did not order them to throw down their arms. John, knowing that in carrying out military orders the soldiers were "not murderers, but ministers of the law, not avengers of their own wrong, but defenders of the public safety,... said to them, 'Do violence to no man, neither calumniate any man, and be content with your pay.' "[40]

Augustine continues with an example from the conduct of Jesus: "Jesus Christ Himself, ordering to be rendered to Caesar the pay which John declares to be sufficient for the soldier: 'Render,' He says, 'to Caesar the things that are Caesar's; and to God the things that are God's' (Matt. 22:21). And taxes are levied for this purpose that pay may be found for the necessary soldiers."[41]

The commendation given by Jesus to the centurion (Matt. 8:9–10) was just, according to Augustine, who noted that Jesus did not command him to desert the military calling.[42]

Even the dilemma of the soldier, presumably a Christian soldier, serving under a sacrilegious commander is discussed in the letter. "A just man," in the opinion of Augustine, "can fight justly for maintaining public peace at the command even of a sacrilegious man, when what is commanded him is either not contrary to God's precept, or not certainly against it; thus, although the king may commit sin in giving the order, the soldier obeying the order to fight is innocent. How much more innocently therefore are wars waged when God commands, for He can command nothing evil as everyone who serves Him knows."[43]

Augustine offers one example where conscientious objection is called for:

> The Emperor Julian was an infidel, an apostate, a scoundrel, an idolater. Christian soldiers obeyed their emperor despite his lack of belief, but when it came to the issue of Christ, they acknowledged only Him who was in heaven. If Julian wanted them to honour idols or throw incense on the altar, they put God before him. But whenever he said "form a battle line," or "attack that nation," they obeyed instantly. They distinguished between

an eternal and a temporal matter, but at the same time, they were subject to their temporal master for the sake of their eternal one.[44]

The letter "Against Faustus" was written in 398, over eight decades after the Edict of Milan, a period of prodigious growth of the church throughout the Roman Empire, and of its assumption of a significant place in society. The days of Christians as a persecuted minority were hardly within living memory. The empire was seen as a beneficent force, despite its occasional intrusion into church affairs, such as that by Justina.

A military autocracy, for that is what Roman rule had largely come to mean, was the basis for stability and order. Christianity was particularly strong in North Africa, where such places as Carthage had been blessed by the blood of the martyrs. Christians were aware that proconsular Africa was a narrow populated band between the sea and the vast mysterious land to the south, a land from which people speaking in strange tongues occasionally launched raids on the carefully tended estates and small provincial towns. The "public peace" alluded to by Augustine had become an urgent concern of the Christian community. It was to this concern, and the duty implied in it, that he was addressing himself. His teachings on wars, when undertaken to make or restore peace, were accepted by Christian citizens and served as a bulwark of the moral order in society.

"Peace" is a good so great, said Augustine, "that even in this earthly and mortal life there is no word that we hear with such pleasure, nothing we desire with such zeal, or find to be more thoroughly gratifying."[45] Later in the same text he states: "Whoever gives moderate attention to human affairs and to our common nature, will recognize that if there is no man who does not wish to be joyful, neither is there anyone who does not wish to have peace. For even they who make war desire nothing but victory — desire, that is to say, to attain peace with glory. It is therefore with the desire for peace that wars are waged."[46]

Regarding the methods of warfare, Augustine does not give much guidance. He uses an example from Hebrew Scriptures in which Joshua is told by God to use ambushes. For those engaged in a just war, therefore, whether victory is won in open combat or by ruses is of little moment in Augustine's thinking.

Despite the linking of his name to the justification of war, Augustine states that the wise man, even while engaged in just wars, will lament their necessity and will want to be delivered from all wars. Only the wrongdoing of the enemy compels the resort to warfare.[47]

The great paradox in Augustine is his manner of translating into action Jesus' command to love the enemy and to do good to the persecutor. In a lengthy treatise on the Sermon on the Mount, Augustine discusses the Christian rule of conduct toward the persecutor: "When an indignity has been received, the mind is not to be excited into hatred, but taking compassion on the weakness shown, it is to show itself ready to suffer more; nor is it to neglect correction, to which it may employ either advice or authority or force." Always in Augustine is reserved the right to correct, through force if necessary. He offers the example of the correction, by beating if called for, given by the father to the erring son.[48]

Loving the enemy is only a seeming paradox to Augustine, since one can preserve the inward disposition to love while carrying out just punishment. "Right love could include war, for Augustine," writes William R. Stevenson Jr.[49] War, according to Augustine, should be waged by those moved not by rage or lust for domination but rather by the love that corrects and disciplines.

"You may chastise but it is love," states Augustine. "You may strike but you do it for disciplinary reasons because your love for love itself does not permit you to leave the other person undisciplined. At times what comes from love and hatred seems self-contradictory. Hatred sometimes comes out in sweet tones and love in hard tones."[50]

In discussing the command to "turn the other cheek," Augustine states, "Here there is no prohibition aimed at correction; for this, too, is of the nature of mercy and does not stop the way of that disposition by which one prepares to endure still more from him whom he wishes to amend. But only he is competent to inflict this sort of punishment who by the greatness of his love has overcome the hatred with which those are wont to be inflamed who desire to be avenged."[51]

Letter to Marcellinus

Fourteen years after his letter to Faustus, Augustine had occasion to write to Marcellinus — a pious man, friend, and Roman commissioner of proconsular Africa — regarding a dilemma he raised. Was Christian teaching, especially that contained in the antitheses of the Sermon on the Mount, inconsistent with the duties and rights of citizens. Augustine replied, "What is 'not rendering evil for evil,' but refraining from the passion of revenge — in other words, choosing, when one has suffered wrong, to pardon other than punish the offender, and to forget nothing but the wrongs done to us?"[52]

Augustine recommends patience as "the habitual discipline of the heart" in suffering. Yet when there is need for correction, that correction should be administered even though it seems evil. He maintains that if the precepts of the Christian religion were followed, wars would result in the enjoyment of peace marked by "the natural bond of piety and justice."[53] The person corrected may see it as being healed through pain:

> Let those who say that the doctrine of Christ is incompatible with the state's well-being, give us an army composed of soldiers such as the doctrine of Christ requires them to be; let them give us such subjects, such husbands and wives, such parents and children, such masters and servants, such kings, such judges — in fine, even such taxpayers and tax gatherers as the Christian religion has taught that men should be, and then let them dare to say that it is adverse to the state's well-being; yea, let them no longer hesitate to confess that this doctrine, if it were obeyed, would be the salvation of the commonwealth.[54]

A tragedy came to Augustine two years after he wrote the letter to Marcellinus. Brown writes that Marcellinus fell victim to the violence of the politicians. After a revolt, Marcellinus was one of those arrested, and after a general purge, he was put on trial. Augustine and church leaders asked for mercy for all those arrested. It seemed as though the feast of St. Cyprian would be a suitable time for an amnesty. Instead, on September 13, 413, Marcellinus was taken to the public park and beheaded. The influence of the church was unavailing against a corrupt and violent political elite. Brown remarks that Augustine "had lost his enthusiasm for the alliance between the Roman Empire and the Catholic Church, at just the time when it had become effectively cemented."[55]

Letter to Count Boniface

Boniface, commander of Roman forces in Africa, wrote to Augustine asking for advice. Boniface was at a crisis in his life; his wife had died; and he was considering separating himself from the military service and possibly entering a monastery.

In response to Boniface's dilemma, namely, that the monk's life is closer to God than that of a layman, Augustine assures Boniface that the soldier's life is also blessed: "Do not think that it is impossible for anyone to please God while engaged in military service." As an example, Augustine cites "holy David," an indication of his frequent use of examples from Hebrew Scriptures when wars, as willed by God, were surely just.[56]

The letter then becomes more general and offers a rich source concerning Augustine's view of war and victory. Even "Blessed are the peacemakers" is cited by Augustine in persuading Boniface to continue in military service:

> Peace should be the object of your desire; war should be waged only as a necessity and waged only that God may by it deliver men from the necessity and preserve them in peace. For peace is not sought in order to be the kindling of war, but war is waged in order that peace may be obtained. Therefore, even in waging war, cherish the spirit of the peacemaker, that, by conquering those whom you attack, you may lead them back to the advantage of peace; for Our Lord says: "Blessed are the peacemakers, for they shall be called children of God."

Augustine reiterates the absolute importance of the will, which should never stoop to hatred of the enemy: "Let necessity, therefore, and not will, slay the enemy who fights against you. As violence is used toward him who rebels and resists, so mercy is due to the vanquished or the captive."

Persuaded to return to military service, Boniface became a changed man. He married again, this time to an heiress who was an Arian, took a concubine, and surrounded himself with a large retinue. He treacherously changed sides, allying himself with the conquering Vandals.

Letter to Darius

The next letter gives evidence of how close Augustine was to the drama of his time, and to the leading actors in the drama. The letter was to Darius, the military commander sent from Rome to negotiate a truce with the Vandals. He had succeeded in his mission. It was 429. Augustine was to die the following year.

The Vandals, unstoppable, had swept across North Africa. By the tens of thousands they, with "the Gothic bible of Ulfilas at their head," had confronted the Roman legions.[57] Their savagery after each conquest knew no bounds; they tortured all who might have hidden treasures — landowners, clergy, and bishops. One bishop was buried alive, another laid on red-hot plates of iron. After thoroughly looting churches and monasteries, they put them to the torch. Into besieged and fortified Hippo streamed the wounded refugees denuded of all resources. The tragic cries of the war-afflicted and homeless at Augustine's door occasioned the last letter of Augustine dedicated to war and peace. It contained his strongest statement against war and the shedding of blood.

After praising Darius for his achievement of a truce, he adverts to the words "uttered by Him who is truth: 'Blessed are the peacemakers, for they shall be called the children of God.' "[58]

Augustine commends warriors who by "consummate bravery... and the blessing of the divine protection" have obtained peace for the state. "But," he adds, "it is a higher glory still to slay war itself with the word, than men with the sword, and to procure or maintain peace by peace, not by war. For those who fight, if they are good men, doubtless seek peace; nevertheless it is through blood. Your mission, however, is to prevent the shedding of blood."

Commenting on the fact that the "just war" had been fashioned within the mold of the Roman Empire, Paul Peachey raises the issue of the aftermath of the empire when competing nation-states "each arrogated to itself the attributes of the former Empire." "Thus," he says, "if in some measure a case for a 'just war' could be made in a barbarian-molested Empire, that case collapsed when the enemy was an equally 'Christian' nation, and yet this difference was hardly noted; the old habits remained."[59]

Followers of Jesus have, over the centuries, wrestled with Augustine's strictures on war and bloodshed. Does discipleship lead them to accept his teaching that, for Christians, violence is not morally acceptable in defense of oneself or one's goods, but acceptable and even a duty in the defense of the community (the state)? Their spiritual wrestling has often been agonizing, as they try to understand how one can maim or kill an enemy with a disposition that does not depart from the love that, in Jesus' teaching, belongs to the enemy.

Augustine, as Peachey points out, formulated his theory of the just war in a particular set of circumstances. Though a holy man, he could not know the future in which the conditions of war would be changed beyond all recognition, in which armed legions would no longer be fighting massed men at borders. Instead, weaponry would be utilized that had the power to destroy cities without making the crucial distinction between soldiers and noncombatants.

"The true emphasis of the Saint's writings, as a whole," says Eppstein, "is far less upon righteous battles than upon peace and mercy, forbearance and long suffering."[60]

There is a mystery in how from St. Augustine's enormous number of works, including some eight hundred sermons, the one teaching of the "just war" was divulgated through the entire church and became part of the teachings of other Christian churches. This mystery will be examined in chapter 4. There is no mystery in the fact that Augustine, and to a great extent Ambrose, opened the door to justified war and bloodshed for the followers of Jesus, a door that has never been closed.

Chapter 4

Is It God's Will?

They are to be reconciled with their neighbors, and to restore what belongs to others.

They are not to take up lethal weapons, or bear them about, against anybody.

All are to refrain from formal oaths unless where necessity compels in the cases excepted by the Sovereign Pontiff in his indult, that is for peace, for the faith, under calumny and in bearing witness.

—St. Francis of Assisi, *Rule for Lay People*

It is in some sense a mystery how the scattered references to "justified warfare" in the voluminous works of a theologian could exert lasting influence on the universal church and eventually on other church bodies.

As discussed in chapter 3, one key to the mystery is that the theologian in question was Augustine of Hippo, one of the four Fathers of the Western church. Because he was considered "the theologian," his teachings carried enormous weight in the church. Even so, two questions remain: How did the references to war and peace, widely distributed throughout an immense corpus,[1] come to be codified? And how were they conveyed to the entire teaching church? The answer lies with one man, Gratian of Bologna, who was a Camaldolese monk and master of canon law. His great work, a synthesis of church law prepared about 1148, came to be known as the *Decretum*, or the *Decretals*. The *Decretum* formed the first part of the *corpus juris canonum*, the body of canon law, of the church.[2]

The Influence of Augustine

In the section dealing with war, Causa 23, Gratian brought together almost all the key texts of Augustine on war and peace. "The influence of Augustine suffused the entire Causa," observed Frederick Russell. "It would be difficult to fault Gratian for the comprehensiveness of his selection of Augustine's texts."[3]

In point of fact, it is necessary, from my point of view, to fault Gratian. He saw fit to omit a key statement on war made by Augustine toward the end of his life. Gratian's lack of comprehensiveness in this matter may have had tragic consequences because the statement in question came as close as Augustine ever came to gospel nonviolence. The statement makes it clear that it is "a higher glory still to slay war itself with the word, than men with the sword, and to procure or maintain peace by peace, not by war." This strong statement has been cited earlier as part of Augustine's letter to Darius, a Roman army officer.

Was it possible that an assertion by the great Augustine opening the door to a nonviolent response was antithetical to the temper of the time? Europe was in the throes of the Crusades, the period of "redemptive violence," of the "holy violence," to redeem the Holy Places. The First Crusade began in 1095; the Third Crusade, 1189 to 1192, was in progress.

The *Decretum* was used as a textbook in all the faculties of law and schools of the church, so that Augustine's texts were transmitted far and wide, eventually reaching such theologians as St. Thomas Aquinas (1225–74). Russell observed further, "To Master Gratian we owe the introduction of the just war into modern international jurisprudence. For centuries, Gratian reigned as the foremost auctor in the jurisprudential speculation about war."[4]

To Augustine's texts as bedrock, a number of speculations were added and conditions formulated. These became, in time, the conditions under which a just war could be fought. Gratian also gave attention to the war/peace thinking of such theologians as St. Ambrose of Milan and St. Isadore of Seville. He cites the latter to the effect that the repelling of violence by violence was justified by natural law, as derived from Roman law.

Wars Can Be Pacific

Gratian himself considered that just wars could be "pacific." He wrote, "With the true servants of God, even wars are pacific [*sic*] as they are entered upon not through cruelty or greed and have as their object peace, the repression of the wicked and the deliverance of the good."[5] There are of course many Christians who would take issue with Gratian on this matter.

Gratian's *Decretum* asserts that "soldiers can please God with warlike arms," and that "it is not a sin to serve as a soldier." He also asserts that "whoever dies in battle against the infidels is worthy to enter the heavenly kingdom."[6] Echoing Cicero and Augustine, Gratian points out that a war is just when it is declared by a formal declaration and waged to regain what has been stolen or to repel enemy attack.

Gratian's codification of Augustine's teachings was arranged in a series of statements under the overall section, mentioned earlier, Causa 23. In it the assertion is made that anyone who kills with the sword without lawful authority commits murder, while the soldier who wields the sword in obedience to authority is not guilty of murder.

Physical punishment of sin, even punishment by war, was in accordance with the gospel precept of patience, according to the *Decretum*. Following Augustine, the *Decretum* taught that love of enemy could be preserved as an inward disposition when punishing enemies, even killing them in a just conflict. Russell comments that "Gratian's wholesale acceptance of Augustine's doctrine so convinced succeeding canonists and theologians that on this issue they were stimulated not to vigorous debate but to endless, unoriginal and tedious repetition of the assertion that love and patience did not prohibit warfare and killing."[7]

In a similar vein, Roland H. Bainton remarks that Augustine "sought to restrain war by the rules of the justum bellum and the dispositions of the Sermon on the Mount."[8]

Empire Becomes the Patria

The church of Augustine's day had given the Roman Empire its support to oppose the double threat of heresy and barbarism. The battle waged by Emperor Constantine against Licinius (in 324) not only solidified Constantine's power but served to uphold the rights of Christians against the persecution launched by Licinius. Constantine's triumphal arch carried the inscription "By the inspiration of the Divinity he avenged the state through the just use of arms." The word "Divinity" would be accepted by both pagan and Christian citizens. The empire had become the "patria," the kingdom, worthy of the loyalty of its Christian citizens and of their service as soldiers.

Theodosius the Great, who was baptized in 380, issued a momentous edict in the same year. Henceforth, Roman citizens should accept the Christian faith professed by Pope Damasus in Rome (366–86), and by Bishop Peter of Alexandria. Both church leaders, clearly separating themselves from Arianism, asserted that belief in the Trinity was the test of orthodoxy. At the death of Theodosius I in 395, the Empire of the West was severed from the Empire of the East. His grandson, Theodosius II (401–50), ruling over the Empire of the East from Constantinople, decided that those men defiled by idolatrous worship would not be admitted to the army. Thus the task of serving as warriors, guards, and policemen in protecting Roman society, and the threatened imperial borders, fell to Christians. The armed Christian, unthinkable two hundred years earlier, became commonplace.

The last emperor of Rome was deposed by Odoacer, a tribal chieftain, in 476, and the Roman Empire ceased to exist. The city of Rome, already sacked several times, was a place of sorrow and desolation. The papacy did its utmost to mediate conflicts and avoid bloodshed. The senate and the consuls disappeared. In 593, it was to Pope Gregory the Great, who had been prefect of Rome before he became a monk, that the benighted city turned in its chaos and distress. Agilulf, the king of the Lombards, had brought his armies to the walls of Rome. It was said that it was as much by his prestige and presence as by the promise of an annual tribute that Gregory won an agreement from Agilulf not to destroy what remained of the "mistress of the world." For nine years Gregory had made efforts to arrange a treaty between the emperor of the East and King Agilulf. In the end, he negotiated the treaty himself and achieved a special truce for Rome and its surrounding districts.[9]

The habit of being in readiness to settle differences by resort to the sword became a plague in the dark ages that descended after the fall of Rome. It was tribal chieftains who absorbed the power that leaked away from Rome as its garrisons no longer secured the frontiers. Each petty chieftain or lord in the developing feudal system could be the judge in his own case and decide that his cause in a conflict was the just one. The retainers and cultivators on his land had no choice but to fight at his behest.[10]

There were times when the papacy appealed to military campaigners for help in repelling invasions of Rome. No longer "patria," the city itself continued its preeminence, if only symbolic, as the seat of the Catholic Church. It was at the request of Pope Leo III, threatened by Lombard armies, that Charlemagne came to Rome.

After the success of his attack on the Lombards, Charlemagne was crowned by the pope as Roman emperor of the West. In the crowning, on Christmas Day, 800, the iron crown of the Lombards was placed on his head. Charlemagne asserted that it was his duty to defend the church with weapons against attacks by pagans and against devastation by infidels. The pope, for his part, would pray for the success of the Carolingian campaigns, which could and did result in mass conversions as well as massacres. The church of Christ now had as its defender the leader of a warlike Germanic tribe, the Franks, who claimed to be the successor to the Roman emperors.[11]

Could the promulgation of Gratian's *Decretum* be in part a response to the rampant and uncontrolled violence of the period? Certainly Causa 23 on war seemed an attempt to communicate Augustine's passion for order, as embodied in just war teachings, to a disordered time. The promulgation of the *Decretum* in the middle of the twelfth century reflected that period in pointing to a new class of wars already introduced as Spain and the South of France met the forces of Islam from North Africa. The new class of war was that called holy. It took its name from the cross of Christ, in the French form *croisade,* in the Spanish form *cruzada.* The church had reached the period of the Crusades when violence was blessed to free Jerusalem and the Holy Places.

Gratian included in the *Decretum* a crucially important dictum, "He who does not ward off an injury done to his fellow man is like him who does the injury." In this, Gratian is paraphrasing a statement of Ambrose. Gratian, however, omitted the phrase "when he can" after the words "injury done to his fellow man."[12]

Another key statement of the *Decretum* could apply to a war then in progress, the Second Crusade: "They are not immune from crime who do not liberate those whom in fact they have the power to free." A complete turnabout from Christianity's early centuries was contained in the statement that "whoever dies in battle against the infidels is worthy to enter the heavenly kingdom."

The Crusades

End Private Wars: Become Soldiers of Christ

The First Crusade was a response by the West to an appeal by the Byzantine emperor Alexius I for help against the Seljuk Turks. Christian refugees from Jerusalem and other parts of the Holy Land brought appalling stories of persecution. Though Jerusalem had been under Muslim rule since the eighth century, the pilgrimages to the Holy Places had rarely been impeded. After 1076 pilgrimages became unsafe, and the Holy Sepulcher itself was being despoiled. The new masters of Jerusalem referred to the Holy Sepulcher as the "dung hill." The followers of Jesus were being set upon by the followers of Muhammad.

A year after the envoys of Alexius had met with Pope Urban II, and had received from him a formal promise of help, a council was held in Clermont, central France. Urban, a monk of Cluny, who followed the pope-reformer Gregory VII, had continued the reforming thrust, using the council as a means of spiritual renewal. The great council in Piacenza, to which clerics and the whole population were invited, was followed by the even greater council in Clermont in 1095. Besides two hundred eminent hierarchs, there was a large presence of knights and barons and thousands of laypeople. So numerous were the laity that tents had to be erected for them in the open fields. After other spiritual concerns had been addressed, Pope Urban preached a sermon from a platform in the Clermont marketplace. In that fateful sermon, he exhorted the Christians to take up arms to liberate their fellow Christians of the Eastern church and to free the Holy Places occupied by the infidels. The opponents were termed Saracens no matter what their provenance. The expression covered all Muslims, whether Arabs, Moors, or Seljuk Turks. Urban saw the Crusade as a means of unifying the Christians to fight for a higher ideal than that of local conquests.

No more powerful insight into the rampant violence of the time—a violence that the teaching of Jesus could not expunge from the hearts of warlike tribes—could be found than that contained in Urban's call to the faithful of Europe.

The pope cried out:

> Let those who in the past have been accustomed to spread private war so widely among the faithful advance against the infidels in a battle which ought to have begun already and which ought to end in triumph. Let those who were formerly brigands now become soldiers of Christ; those who once waged war against their brothers and blood relatives fight lawfully against barbarians; those who until now have been mercenaries for a few coins achieve eternal rewards; . . . those who are about to depart must not delay, but when the winter is over and spring has come they must get eagerly under way with the Lord as their leader, after setting their affairs in order and collecting money for their expenses on the journey.

Calling the fomenters of local wars the oppressors of orphans and the ravagers of widows, the pope went on to warn his assembled hearers: "The end of the world is near. The days of the anti-Christ are at hand. If, therefore, when He comes, He shall find no Christians in the East, as at this moment there are scarce any such, then there will be no man to stand up against him."[13] The response was immediate, especially to the pope's recital of a grave report from Jerusalem. The Christians were being murdered, and some were enslaved and forced into the practice of other rites. Altars were polluted, said the report, and women violated. Men had their navels cut open and their entrails torn out before they were killed. Some were tied to posts to be shot at with arrows or attacked with drawn swords in attempts to behead them in one stroke.

It has been pointed out that many historians fail to envision the mind of the

Christians of the time regarding the Crusades. Their reverence for Jerusalem, where the blood of the Savior had been shed, nourished their willingness to sacrifice their possessions and suffer the penitential journey of over two thousand miles across land and sea.

The faithful took to heart Urban's words, "Whoever for devotion alone, not to gain honor or money, goes to Jerusalem to liberate the church of God can substitute this journey for all penance." Pope Urban assured them that the homes and lands of all who took part would be safe, protected by a truce. The pope urged them to the heroic endeavor under the battle cry of *Deus Vult.* The emotional tone of Urban was more than matched by that of Bernard of Clairvaux, who told the pope, "You ordered; I obeyed.... I opened my mouth; I spoke; and at once the Crusaders have multiplied to infinity." The crusading soldier was told to see himself as "not a man slayer but a slayer of evil, and an avenger of Christ against those who do wrong."[14]

Deus Vult

The faithful were inflamed with the desire to embark on the journey which would give them the same spiritual benefit as if they had made the pilgrimage to the Lord's empty tomb. This, however, would be an armed pilgrimage, in which they might have to lay down their lives. Pope Urban himself traveled over France and Italy calling upon the faithful not to the bearing of the cross but to the bearing of the sword for the cross — a spectacle that has fascinated and appalled people over the centuries.

By 1096 an enormous trained force of nobles, knights, retainers, and convinced faithful was assembled to "take the cross," a cross of heavy wool material attached to the outer garment. Raymond, count of Toulouse, led the First Crusade, whose participants were drawn chiefly from France. He was clad like other knights in a hauberk of mail over heavy woolen garments. Over the hauberk he might wear a wool or leather surcoat. The horses were the finest in the land. Lower orders mounted farm horses or donkeys, and those at the lowest rung of the ladder went on foot.

Masses, public prayers, and the display of banners emphasized the divine purpose of the journey as the Crusaders went forth from such places as Cologne, Clermont, Paris, Le Puy, Toulouse, and Genoa. The leave-takings were marked by the tears of mothers, wives, and children who might not be so gripped by crusading fever that they did not realize that they might never see the departing heroes after that day.

All over Europe, townspeople encouraged the First Crusade with their prayers. As they left Europe and made their way through inhospitable terrain like Anatolia (now Turkey), soldiers of the cross needed all the prayers that had been offered up, and were being offered up, for them in the homes and churches of Europe. Over five hundred of the crusading band died, chiefly of thirst, shortly after arriving in Anatolia. In the Anatolian mountain passes, baggage animals slipped and fell into ravines from paths muddied by rain. Knights had no choice but to throw away the heavy armor that encumbered their movement. Dysentery was a plague that weakened nearly all the Crusaders as they encountered the East. Having captured Nicaea, the Crusaders besieged Antioch, the ancient city where the followers of Jesus were first called by the name Christians.

Jerusalem Liberated

On June 7, 1099, the Crusaders, now greatly reduced in number, laid siege to Jerusalem. Without food, and with water brought from miles away in the sewed hides of oxen, the Crusaders battered the walls of the holy city. Finally, an emir, realizing that nothing could be gained by continued siege, threw open the gates and allowed the forces of Raymond of Toulouse to enter.[15]

Through streets that became mounds of corpses and pools of blood, the Crusaders reached the Temple site. On its flat roof, a large number of Jerusalem's people had taken refuge. They were chiefly old men and women and children. Tancred, a pious south-Italian Norman, held back his attacking force and promised the noncombatants his protection. Before he could return on the following day, another Crusader had put them all to the sword. Jews as well as Muslims were butchered.

The recovery of the Holy Sepulcher in mid-July 1099 through a long tide of blood was the occasion of prayers of gratitude at the empty tomb of the risen Lord. Godfrey of Bouillon, barefoot and garbed in a white linen garment, made his way to pray at the shrine. The Latin kingdom of Jerusalem was established with Godfrey being offered the kingship by the other Crusaders. He refused to become king, asserting that where Christ had worn a crown of thorns on his head, he did not wish to wear a crown of gold. He preferred the title of Defender of the Holy Sepulcher. His brother Baldwin I, who did not share such qualms, accepted the kingship. Two other mini-kingdoms were established at the same time, both headed by Crusaders belonging to European nobility.

It took time for word of the liberation of Jerusalem to reach the areas of Europe from which the Crusaders had set out two years earlier. The jubilation was great. Only in Rome was the joy muted. Pope Urban never received the word. He had died on July 29, 1099, before the news reached the city.

The enthusiasm for what was accepted as holy violence, fed by the success of the First Crusade, led to the mounting of the Second Crusade in 1147, when Jerusalem again seemed threatened. It ended in dismal failure in 1149, despite the impassioned preaching of the great Cistercian monk Bernard of Clairvaux and the leadership of two monarchs, Louis VII of France and Conrad III of Germany.

Less than four decades later, Western Christians were filled with horror at the news that Jerusalem had again been captured by the Saracens. The name of the Kurd, Saladin (Salah-ad-Din), who had led the conquest in 1187 inflamed Europe. He had united the Muslim world extending from Cairo to Baghdad. The capture of Jerusalem was only an incident in a jihad he had declared against all Christians. The war cry was, "For Islam, forward the monotheist army."

The Third Crusade (1189–92) evoked the same fervor, especially among the leaders, as had the first. Three kings responded to the call of Pope Gregory VIII, Frederick the Holy Roman Emperor, Philip II of France, and Richard I of England. The first, Frederick I ("Barbarossa"), died on the way in Asia Minor; the second, Philip, turned back to Europe; and only the third monarch, Richard ("Lionheart"), reached Jerusalem. Although he was an intrepid warrior, Richard could not retake Jerusalem from the Saracens. The Latin kingdom of Jerusalem had shrunk to a small section of the Palestine coastline. Saladin, a militant Muslim warrior, was a cultivated man concerned with literature and theology. He respected the bravery and ruthlessness of Richard, as Richard respected the same qualities in him. In an early battle Saladin beheaded all the captured prisoners of Richard's forces, and Richard treated the prisoners from Saladin's side in like manner. In the end the leaders concluded a treaty which permitted Christians free access to the Holy Sepulcher and other sacred places. Richard left the Holy Land and the Third Crusade ended.

Apex of the Third View of Warfare

The first three Crusades have been glimpsed here because the Third Crusade evoked a response of conscientious objection that was reasoned and detailed.

The Crusades marked the apex of the third view of warfare held by the followers of Jesus. The first view was that of gospel nonviolence, the refusal to inflict violence for any reason after the example of Jesus, who accepted suffering and death for the ransoming of humankind from sin. The second view, that of the "just war," sanctioned killing not for personal defense (for which gospel nonviolence had to be maintained), but for the defense of the community (for which certain conditions had to be met). The third view, that of holy war, went beyond the "just war" in uniting warfare with a holy cause.

The Holy Sepulcher marked the site on earth from which the Son of the Most High ascended into heaven. As the scene of the resurrection of the Crucified One, and the locus of the empty tomb, it was at the very heart of the belief of Christians. Faith in the Resurrected One was the foundation of their belief in their own resurrection.

As in all wars, the enemy was pictured as "other," as dehumanized. The cruelty of the Saracens, their depredations against churches, and their holy war against Christianity resulted in their being relegated to the outer reaches of "otherness." They were not simply dehumanized but desacralized, the image of God in them totally erased. In wars that were simply just, the cruelty necessary for conquest could be mitigated by mindfulness that the opponent was also a child of God, albeit a mistaken, wrongheaded, evil, or criminal child of God. The Saracens were simply God's enemies. The mandate of the Christians, a mandate mediated to them by a pope, by bishops, and by priests was to spare nothing to free the Holy Places and liberate the Eastern church. Thus the Christians could aim in good conscience at total mercilessness.

On the opposing side, the Muslims of the time had declared a jihad against the Christians, whom they considered as infidels. Their belief in the Trinity was viewed as an affront to the pure monotheism of Islam. Though jihad is most often understood in terms of war, it has a wider meaning that comes from its root, which refers to exertion or struggle. "The Jihad in the broadest sense of exertion," says Majid Khadduri, "does not necessarily mean war or fighting, since exertion in Allah's path may be achieved by peaceful as well as violent means."[16]

In theory, Islamic jurists distinguish four ways by which the jihad might be practiced — the jihad of the heart, of the tongue, of the hands, and of the sword. According to Khadduri, all four ways are means by which the believer may fulfill his obligation to Allah. The first or greater jihad (as ascribed to Muhammad) is that of combating the devil and his evil power over the heart. The jihads of the tongue and the hands call for the believer to support what is right and correct what is wrong. The jihad of the sword is concerned with the struggle, through war, against unbelievers and the enemies of the faith. In such a war believers are obliged to sacrifice their wealth and their lives.[17]

In the early days of Islam, its conquering armies moved over the world with the mission of making believers of all peoples, by persuasion or by the sword. A territory as large as that of the Roman Empire was won before the forces of Islam met opposition in the West at Tours and in the East at the borders of India. The world was divided by Muslims into the *dar-al-Islam,* the territory under Islam, and the *dar-al-harb,* the world of war, the territory of unbelievers.

Jihad, in its meaning of holy war, has been identified as the mark of Islam from its earliest sweep of conquest. Also at the heart of Islam is the conviction that a war is righteous if its purpose is to reconquer for Islam any part of the *dar-al-Islam* which had been wrested from it. Those who die in such a war are promised paradise. Jerusalem, the place of pilgrimage second only to Mecca for Muslims, wrested from Muslim rule by crusading forces, was therefore to be recaptured by a righteous war, a jihad. Along with Jerusalem, the other territories of the Holy Land were to be reconquered for Islam.

For the Christians, however, the Crusades constituted a reversal of the teachings given by Jesus to his followers. Far from the "love your enemies" of the Sermon on the Mount, even far from the original concept of the just war, were the practices of the Crusaders. Violence in the defense of places called holy became indiscriminate. Violence indeed began to capture the Christian view of life. It has been noted that in any war each side, whatever its reasons for engaging in war, becomes like the opposing side, even its mirror-image. The methods used by one side, however monstrous, begin to be used by its opponents, since otherwise defeat would be invited. Even the underlying concepts of one side influence the other side, as in the Crusades, where not only Muslims but Christians fought, killed, and died secure in the belief that by dying in battle against the opponents of God, they would be worthy to enter the heavenly kingdom.

The revulsion against violence and coercion in the early church stemmed in the spiritual order from the teaching of Jesus; such revulsion was heightened by the coercion and barbarous violence unleashed against the early Christians by the Roman state. Yet their response then had been that of nonretaliation, of the nonviolence found in the gospel of Jesus.

The Church and Islam: Similar Views

The Crusades occurred at a time when the church and Islam shared similar views on the relationship between religion and the state, and of the function of the state to spread religion. One of the popes during the crusading period, Pope Innocent III, held the extreme view that as matters of the spirit take precedence over matters of the body, and as the church rules the spirit, and as earthly kings rule the body, earthly kings must be subject to the papacy. The state was therefore to be the enforcer of divine law. In this, the church of the time shared the Islamic view of the state and its ruler. "The Caliph, as head of Islam," states Khadduri, "stood in some respects as both pope and Emperor, whose chief functions were the universalization of Islam and the enforcement of the divine law."[18]

The Crusades served as an epiphany of the problem of Caesaro-papism in the church. Just before the Crusades, Pope Leo IV in person led an army against the Norman invaders of Italian lands, though St. Peter Damian and others criticized the Vicar of Christ for this act. Innocent IV, as an instance of the reliance on violent methods, permitted the use of torture in cases of heresy tried by the tribunal of the Inquisition.

The brief account given here of the first three Crusades and their times points to an alliance of the church with violence. It points to a nadir in the history of the church, in its prophetic ministry, and in its transmission of the discipleship taught by Jesus to his followers. It is helpful to know that there was not unanimity in such positions and that there were voices calling upon the church to return to the teachings of its founder. One such voice has been preserved through the scholarship of George B. Flahiff, a priest of the Order of St. Basil associated with St. Michael's College of the University of Toronto.

Conscientious Objector to the Crusade

In the face of nearly universal acceptance by Western Christianity of the holy violence of the Crusades, a voice was raised in objection by Radulphus, or Ralph Niger. Flahiff relates that Niger was an Englishman and most probably a monk. Something is known of his life from his letters to patrons and various ecclesiastics. He was born about 1140, and his lifetime englobed the period of tension between King Henry II and St. Thomas à Becket, which culminated in 1170 in the murder of Thomas by the knights of Henry's household.

Ralph Niger was one of those expelled by the king, probably because he expressed himself as opposed to Henry's policies. In his letters, Niger often referred to Henry as "the king under whom blessed Thomas, the Martyr of the English, suffered." This pictures Ralph Niger as a person who was not awed by authority, and who could question the action of a king capable of ordering an assassination. There is evidence that he spent his time of exile in Paris as a teacher and writer on theological and biblical topics, including a commentary — based on the work of St. Jerome — on the first seven books of the Hebrew Scriptures. His contemporaries referred to him as master.

The work by Ralph Niger dealing with the Crusades seems to have been started about 1187, the date of the fall of Jerusalem to Saladin. Niger's return to England most likely occurred when King Richard Lionheart was on the throne. The king became the hero of the people of England when he left them for the far glory of the land whereon the Savior's feet had trod.

Niger is mentioned by Bainton[19] and others, all of whom base their references on the research of George B. Flahiff, CSB. He studied the Latin of Niger's texts, reproducing and translating extensive sections of them for *Medieval Studies,* the publication of the Pontifical Institute of Medieval Studies of Toronto, Canada.[20]

After reading his research, I met with Flahiff at the University of Toronto in 1960. He hoped to expand on the work of Ralph Niger and possibly produce a book on the man and his work. When I saw Flahiff in Rome in 1965 in the course of the last session of the Second Vatican Council, he had become the archbishop of Toronto. He encouraged me in my concern for the recognition of conscientious objection by church bodies and by the laws of various countries. He expressed regret that his ecclesiastical duties prevented him from continuing the research necessary for further publication on Ralph Niger. We last discussed Niger in 1976 during the Eucharistic Congress in Philadelphia. By this time conscientious objection as a human right had been placed on the agenda of the UN Human Rights Commission. Archbishop Flahiff died without being able to return to his research on conscientious objection to the Crusades.

Spiritual Pilgrimage to Mystical Jerusalem

What follows is a rendering of Ralph Niger's arguments through the translations of George B. Flahiff. Flahiff remarks that the tone and the method of Niger's writing in *Deus Non Vult: A Critic of the Third Crusade* combine to produce a unique insight into the crusading period.[21] Flahiff states that Niger not only shows himself unfavorable to the expedition at a moment when so many others were urging it, but also proceeds to marshal systematic arguments drawn from scripture, reason, and experience to support his position.

The occasion of his writing, Niger explains, is that Palestine, holy beyond all other places by the life and work of the Savior, had fallen to the Saracens. The king of Jerusalem was imprisoned, and other leaders were prisoners or already slain. The relic of the cross had been stolen by the infidels.

Niger recounts how the princes of the West had responded by taking up the cross to repair the evil done to the earthly Jerusalem, and in so doing, to obtain remission for their sins. A pilgrimage, or Crusade, will profit nothing, Niger remarks, unless it is joined to a spiritual pilgrimage, an inner journey to the mystical Jerusalem. This inner journey, according to Niger, "can be made without ever leaving one's home, and in many cases, it should be." Niger refers to three examples of the journey to Jerusalem: the journey of the people of Israel to Jerusalem after captivity in Egypt; the return to Jerusalem after exile in Babylon; and the return of St. Peter following imprisonment by King Herod.

If men decide to go on a Crusade to the earthly Jerusalem, they should first attend to the inner spiritual journey. He makes clear the necessity of capturing the Jerusalem of the heart for Christ before embarking on the struggle to capture the earthly Jerusalem.[22]

According to Flahiff, Niger is not a pacifist but rather a noninterventionist since he believes that under certain conditions war is justifiable. He puts in question, however, the advisability of any Crusade and, in particular, the expediency of the Crusade then being preached to the faithful. He is concerned about the dangers to the church in the West from heresy and about the prudence of interfering in the problems of the Eastern church. He questions whether the extreme hardship and danger involved in crusading could be justifiable. To Niger, leaving the church in Europe to help the church in the East is like "rushing off a great distance to extinguish a neighbor's fire when one's own house is in flames." The flame to which he refers is the heresy of the Cathari, spreading both in France and England. This threat to the Christian faith and the church, states Niger, has already emptied churches, while the fanatic Cathari practice their own obscene rites. He asks whether any good could come of ridding the Holy Land of Saracens if unbelievers were multiplying at home. "Why do our princes rush off to Palestine to recover the wood of the cross while the Crucified One Himself is being abandoned here at home?"

Do the Holy Land Christians Deserve Help?

Niger wonders whether the situation in Palestine really deserves help from the West in the form of a Crusade. He inveighs against the excesses and luxury of the lives of the eastern Christians as described by people who returned from the Crusades. He is especially repelled by the spectacle of Heraclius, the patriarch of Jerusalem, who with his followers had made a tour of western Europe in an appeal for aid. "Adorned with gold and silver," Niger points out, "their garments actually reeked with exotic perfumes. Such a retinue as he had, no Patriarch of the West could ever afford such wealth and luxury, no prince could equal. Yet these men were seeking aid, begging for pecuniary assistance against Saladin."

Niger emphasizes that among Palestine's population were a large number of criminals. It might well have been so since, in preaching the Second Crusade, St. Bernard had urged adulterers, robbers, and perjurers to join the Crusade in expiation of their sins. Niger repeats an accusation, frequently made, that Palestinian Christians had been guilty of acts of treachery against the Crusaders who had come to help them.

The Hebrew Scriptures are cited by Niger as examples of God's punishment on evildoers. The calamities God allowed to befall the Jewish people to bring them to repentance might be likened to the calamities brought on the people of Palestine. To hasten to relieve their plight might interfere with God's mysterious designs for their punishment and correction. The argument of God's providence, of viewing the calamity that befell the Palestinian Christians as God's judgment on them, is a daring one, since it cuts across the argument of ecclesiastics that God's providence was with those who went to the defense of the eastern Christians.

Arguments against being swept up in the wave of enthusiasm for the Crusade, according to Niger, include the dangers to which the participants exposed themselves. The dangers arose not only from physical obstacles, including exposure to the burning sun and to the elements, but also from spiritually corrupting influences. Experiences on the journey might vitiate the highest motives with which the pilgrimage was begun. The Crusade might be for the vulnerable an occasion of sin rather than of grace. Niger makes a special point that anyone who needlessly exposes himself to death is guilty of sin. Embarking on a journey full of perils with the possibility of death at the hands of a barbarous enemy should only be undertaken after due deliberation, Niger insists.

Clerics Already Bear the Cross of Christ

In his analysis of those who should not participate in the Crusade, Niger includes poor people, women, and above all clerics. He treads lightly here, since the pope urged the Crusade on all men alike, with the promise of remission of sins. While asserting that he does not have the intention of putting in question the papal decision, he makes the point that the decision must be correctly interpreted. The pope, he explains, would never decree anything contrary to justice or reason. A thief, for example, could only take advantage of the indulgence granted by the pope if he had made restitution. Satisfaction as well as penance are necessary for the sinner to be received into the grace of God.

The question is whether the shedding of blood can be viewed as atonement for sin, or whether the Crusade can be considered as satisfaction for all sins without discrimination or without prior restitution. Clerics who by their profession are obliged to "turn the other cheek" would therefore have no business engaging in a Crusade and polluting their consecrated hands by killing Saracens. Since religious under vows already bear the cross of Christ on their bodies, why, asks Niger, should they add the Crusader's cross? If they wish to preserve their vows, religious should refrain from going on the Crusade. On the practical level, the monk, having no resources of his own, could not finance his participation in the Crusade. Either, as a mendicant, he will consume much-needed resources, or, if he goes at the expense of another, what merit can he expect from the pilgrimage?

Also excluded, according to Niger, are the poor in general, simply because of the scarcity of resources. The Saracens were skilled in cutting off food supplies at critical places. Among the poor, a number of younger men might accompany the Crusaders to carry out the daily tasks of meeting their needs. Women should also be excluded, since the presence of women was reported to have led to immorality in the camps.

Those who are justified in joining the Crusades are the bishops, archbishops, abbots, chaplains, and, above all, knights. Hierarchs may be counted among the Crusaders, Niger states, as long as their flocks are spiritually served in their absence. It would be an ill-ordered act of charity if by caring for the few, the salvation of the ninety-nine sheep at home would be put in danger. While clerics in general should not be part of the Crusade, a certain number would be needed for confessions and for the celebration of the Mass.

God Allowed the Holy Land to Be Occupied

Niger's strongest argument against a too hasty mounting of a Crusade is the fact that God himself in his inscrutable designs had given the Holy Land into the power of the Saracens, or at least had allowed them to occupy it. He asks if Christians should seek to destroy the Saracens for what God himself has allowed to them. Given that the Saracens might not be pleasing in God's sight, still he does not will the death of the sinner. As they are human beings like ourselves, he points out, they are to be treated accordingly.

Niger, in contrast to many Christians of his time, does not dehumanize or demonize the Saracens, but includes them in a common humanity. He does make it clear that the law allows one to resist force with force, but as with all remedies, this should be utilized with moderation. In his view, it would be preferable to strike with the sword of the word and bring the Saracens to the faith by their own volition. God, he says, is not pleased with forced service. Anyone who wishes to spread the faith by violence transgresses by that very fact the discipline proper to the faith.

It is not known how many others shared Ralph Niger's conviction, but it is clear that he was in good standing with the church despite the fact that he put in question the church's acceptance of the Crusades.

Flahiff concludes: "What makes Niger's position truly extraordinary is that he, a cleric, voices and defends it in the very midst of a Crusade that hardly needed the call of the popes to launch it, so spontaneous and enthusiastic was the response of Western Christendom to the plight of the Holy Land. . . . The traditional Deus Vult must have been on many Christian lips as Niger dared for the first time to cut across it and proclaim Deus Non Vult."[23]

The Lord Give You Peace: Brothers Francis and Illuminato

Deus non Vult (it is not God's will) was expressed in dramatic action when two men clad in dun-colored, undyed garments crossed the no-man's-land between the camp of the Crusaders and the Muslim camp. It was in 1219, in the wake of the battle of Damietta in the Nile Delta. The battle, in the Fifth Crusade, was one in which the crusading forces had been defeated with immense losses. The two men, Brother Francis of Assisi and Brother Illuminato, were considered half-demented for their insistence on going unarmed to meet the leader of the Muslim forces, Sultan Melek-el Kamela. Probably the papal legate gave his permission because Francis had urged against the disastrous battle in the first place. The Muslim soldiers who allowed free passage to the emissaries and conducted them to the sultan's tent might have considered them harmless fools "touched by God."

Francis and Illuminato were courteously received by the sultan, who listened as they made witness to the faith of Jesus. He refused the offer of the two visitors to undergo the ordeal by fire as proof of their faith, but had them escorted with honor back to the Christian side. The two poor men refused the lavish gifts offered by the sultan.

Francis of Assisi and twelve companions had journeyed to Damietta on a mission which, they fully believed, might have ended in their martyrdom. They reached the "enemy" side of battle by way of Cyprus and Acre, avoiding Constantinople. That city, becoming a pawn between the Venetians, the Franks, and the eastern empire, had been mercilessly sacked by the crusading forces in the Fourth Crusade. This tragedy had preceded the disaster at Damietta by a decade and a half.

Francis and his companions returned to Acre on the coast of the Holy Land. It is known that they made a peaceful pilgrimage to the Holy Places, but there is no record of their experiences.

Francis, by leading a weaponless band of men to Damietta, was offering the witness of nonviolence, the witness of their willingness to suffer after the example of Jesus. Every action of Francis and his followers bore out their distinction from the Crusaders. The very fact that they went by foot, rather than on horses, to meet the sultan separated them from the knight-combatants.

Some fifty years earlier, Ralph Niger had questioned the use of violence in the period of the Third Crusade. While Niger was a thoughtful objector and critic, Francis was a prophetic peacemaker in the spirit of Jesus, a Christian who saw the person called "enemy" as a child of God to be respected and treated as such. The licentiousness of the soldiers in the Christian camp at Damietta and the maiming and killing of human beings in both the Christian and Muslim camps could only strengthen his revulsion against bloodletting.

Counterimage to Chivalry

The high adventure of the Crusades did not capture Francis or his followers. Their life provided a counterimage to the crusading, chivalric knight. There could have been no more striking contrast than that between the lowly men, who went out as lambs among wolves, and the knights, armed and riding steeds covered with rich trappings.

In his youth, Francis, the son of a prosperous cloth merchant, shared the chivalric dreams of the young men of Assisi. As a mounted knight, the twenty-year-old joined the Assisians in armed conflict against the Perugians. He also fought in the battle of Collestrada. Taken prisoner, he was laid low by illness, an illness which persisted until his release and return to Assisi. On his recovery, "being still ignorant of the Divine Will," according to his biographer and fellow Franciscan Bonaventure, "he determined to go into Apulia to enter the household of a certain Count of great magnificence and liberality who dwelt in the country, hoping in his service to acquire military honor and renown."[24] The pope was at war with Apulia, in the south of Italy. Count Gentile had put out a call for volunteers to join him in the papal forces. Though equipped for the battle, Francis was troubled by a dream, and returned to Assisi.

An encounter with a leper was pivotal to his future life. Francis not only gave him a kiss but overcame his revulsion and embraced him. From that time forward, he served the lepers of the countryside. In his *Testament,* he recounted, "The Lord gave to me, Brother Francis, the grace to do penance in this way. The Lord himself led me in among them and I practiced mercy with them. And when I came away from them, what seemed bitter to me, was changed to sweetness of spirit and body for me. And after that, I did not wait long and left the world."[25]

Francis found his place in the church by incorporating into his life "the commands and counsels of Jesus Christ." As young men joined him in following this life, he formed a rule. To a church marked by what has been termed "the virtue of magnificence," Francis and his followers presented a church that was stripped of wealth and luxury, but also of the legacy of violence.

The influence of Francis does not lie simply in the fact he attracted thousands of followers who committed themselves to living according to the gospel, owning nothing, and supporting themselves by manual labor, or failing that, by begging. Of greatest importance to those concerned with the church's marriage to violence was his opening to laypeople of the option of nonviolence.

For Brothers and Sisters in Their Homes

For "brothers and sisters living in their own homes," Francis gave a rule that called for simple living, including garb of humble, undyed cloth, and a program of prayer, fasting, penance, and the works of mercy. This took form in 1221 in the first rule of the Third Order.

Above all, it provided specific rules that made for nonviolent living by lay Christians:

> They are to be reconciled with their neighbors, and to restore what belongs to others.
>
> They are not to take up lethal weapons, or bear them about, against anybody.
>
> All are to refrain from formal oaths unless where necessity compels in the cases excepted by the Sovereign Pontiff in his indult, that is for peace, for the faith, under calumny and in bearing witness.[26]

These rules, innocent-sounding though they may be, undercut the whole basis of the feudal system, a pyramid of mutual obligations. At the peak of the pyramid was the nobility, essentially a military class, each nobleman with his complement of knights. The upkeep of the warrior-knights called for support from vassals and underlings. Oaths recognized by the noble, generally the landowner, bound the vassal to fulfill his obligations, which included service in military campaigns. Many towns and communes engaged in military campaigns against each other, as had Assisi and Perugia in 1202. Bishops and the pope himself called out their knights to take the field to settle disputes, as the pope had done in Apulia.

The people living in their own homes were now united by a new brotherhood, a religious relationship which superseded their relationship to a lord or to a city.

The refusal to bear arms was a condition of membership in the Franciscan Third Order. That the faithful flocked to Francis as vowed monks or nuns, and later as laypersons, was a spiritual phenomenon. It spoke to the thirteenth century and to every succeeding century. The Franciscan revolution presented a needed correction to the opulence and luxury of too many church leaders of the day. Heroism took on a new face, the opposite of chivalric heroism. Franciscan heroism consisted of resisting violence and of performing the works of mercy, of visiting the sick, including lepers, of committing resources to a common fund for the poor and for burying the dead. It involved mirroring in Christians' lives something of gospel poverty and simplicity.

"My Brothers [referring to the Friars Minor] are called lesser so that they may not presume to become greater," Francis explained. Francis applied to daily life the words of the Beatitudes:

> Let us Brothers all take note that Our Lord says, "Love your enemies, and do good to those who hate you" (Matt. 5:44). For also our Lord Jesus Christ, whose footsteps we should follow (1 Pet. 2:21), called his betrayer friend and offered himself voluntarily to his crucifiers. All therefore who unjustly put us on trial and... [put upon us] shame and insult, grief and pain of various kinds, together with martyrdom and death, are our friends whom we ought to love much, because through what they put on us we obtain life everlasting.[27]

He reminds the brothers that they should not resist evil by violence: "Rather, if anybody hits them on one cheek, let them offer him the other, and if anyone takes away their outer dress, let them not refuse their tunic either; let them give to everybody who asks, and if anyone takes away what belongs to them, let them not demand it back (Luke 6:29–30)."[28]

There were those among laypeople whose lives were transformed by Francis's evocation of the gospel of Jesus. Among the thousands of laymen who followed the "little poor man" were those who took to heart the rule about not taking up deadly weapons or "bear[ing] them about, against anybody." There were those who, despite the vassalage of the feudal system, knew that being a member of the Third Order freed them from the feudal oath to bear and use weapons at the behest of the overlord.

Francis died in 1226 with the marks of the Crucified One on his hands, feet, and side. The two-hundred-year period of the Crusades, from Pope Urban's appeal in 1095 to the final fall of Acre in the Holy Land in 1291, had over six decades to run. In this era in which nine Crusades inflicted a deep wound on Christian history, Francis led his followers in a peace church. It was a fellowship within the church which revived the gospel peacemaking of the first centuries of the church. These followers of Jesus took him at his word, that his kingdom was not of this world. Since this was the case, how could his followers maim or kill any human creature for whom the Son of God had laid down his life? During the life of Francis, the reasons for which Christians were willing to engage in bloodshed ranged from the concept of Christendom, in regaining the places sacred to the life of Jesus, to the lesser loyalties involved in securing the borders of the tiny kingdoms ruled by many types of overlords. Without announcing in so many words *Deus non Vult,* Francis, on the Fifth Crusade, gave the example of it in deed. He treated the one called "enemy" as a child of God and met him weaponless, whether Muslim or Christian, sultan or citizen of a nearby city-state.

"The Lord revealed to me," said Francis, "that we should speak this greeting: 'The Lord give you peace.'" He added, "While you are proclaiming peace with your lips, be careful to have it even more fully in your hearts. For we have been called for the purpose of healing the wounded, binding up those who are bruised, and reclaiming the wrong."[29]

Universal Kinship

The Third Order of St. Francis has existed since its foundation by the saint. Called in the twentieth century the Secular Order of St. Francis, it forms part of the Franciscan family of laity, religious, and priests.

The members of the Secular Order of St. Francis have in their rule:

> Mindful that they are bearers of peace which must be built up increasingly, they should seek out ways of unity and fraternal harmony through dialogue, trusting in the presence of the divine seed in everyone and in the transforming power of love and pardon....
>
> Moreover, they should respect all creation, animate and inanimate, which bears the imprint of the Most High, and they should strive to move from the temptation of exploiting creation to the Franciscan concept of universal kinship.[30]

Chapter 5

St. Thomas and the Angels

May eternal concord, already established by nature, and more wondrously re-established by Christ, unite all things. And finally, let every man zealously aspire to that which pertains to the concord of all men. —DESIDERIUS ERASMUS

Aquinas

Spiritual and Physical Weapons

"Free will exists in a nobler manner in the higher angels than it does in the lower," wrote St. Thomas Aquinas, "as does also the judgment of the intellect."[1] This occurs in the treatise on angels by the preeminent Scholastic theologian. He dealt with such subjects as "the substance of the angels, their mode of knowledge and their perfection in the order of grace and glory." Thomas had respect for the compatibility of human reason, the intellect, and the gift of faith. In the matter of angels, he allowed his soaring intellect free rein in the fourteen questions and seventy articles devoted to them in the *Summa Theologica.* It has not gone without notice that in the question on war, the great thirteenth-century saint was constrained, dealing with it in one question and four articles.[2] On this issue, he was limited by one significant premise. For Thomas, St. Augustine had already covered the subject of war in the fifth century and had said just about the final word on the matter.

Thomas Aquinas (1225–77), as the younger son of a family of lesser nobility, could not inherit the family landed estates. He was destined by the family to become a Benedictine monk at the nearby abbey of Monte Casino and was placed there as a child to be cared for by this monastic community. When Monte Casino was in danger of suppression by Emperor Frederick II, the monks sent him to Naples for study. There he encountered the newly founded Dominican Order. It is said that he was attracted to the poverty of the Dominicans, which began, like the Franciscans, as a mendicant order. Thomas defied his family and decided to receive the Dominican habit. Despite the fact that his father, Count D'Aquino, went to the length of imprisoning his son for a year in an attempt to alter his decision, Thomas persisted in his determination to remain in the Dominican Order. The count wanted his son to be part of a long-established congregation more in keeping with the status of the family.

The master of the Dominican Order sent the young Thomas to Paris for study. It was there, the very center of the church's intellectual life, that he came under the tutelage of Albert the Great, the influential scientist-theologian. Becoming a master of theology at the University of Paris, Thomas taught from biblical texts, engaged in public debates on theological topics, and preached in the university church.

For seven years he devoted himself to a synthesis of Christian theology grounded in scripture and the Church Fathers. The works of Aristotle, the Greek philosopher of the fourth century B.C.E., were reintroduced into western Europe in the Middle Ages through Arab and Jewish scholars. Much of Scholastic philosophy, including that of Aquinas, was influenced by Aristotle. Often in Aquinas's synthesis we find references to "the Philosopher," meaning Aristotle. The result of Aquinas's seven years of labor was the *Summa Theologica,* a monumental work described as "for the instruction of beginners." It was left unfinished.

Despite the fact that during the lifetime of Aquinas war was a constant, it clearly was not one of his priorities. While he lived, four Crusades took place, the Sixth Crusade originating three years after his birth and the ninth cut short in 1272, three years before his death. Aquinas made only one reference to the Crusades, in question 40 ("On War") in the *Summa.* It arises in the second article dealing with whether bishops should fight. He refers to the order given by Pope Leo IV (d. 855) that all the people should gather to oppose the Saracens who were rumored to have come in secret to the port of Rome. While Pope Leo gave the order to fight, Aquinas pointed out, he excluded bishops and clerics from the role of fighting. Question 40 occurs in the second division of the second part of the *Summa,* where war is considered under the sins contrary to charity.

Aquinas asserts that the command to the apostle Peter to put his sword in the scabbard applied to all bishops and clerics:

> Warlike pursuits are altogether incompatible with the duties of the bishops and clerics for two reasons. The first reason is a general one, because, to wit, warlike pursuits are full of unrest, so that they hinder the mind very much from the contemplation of divine things, the praise of God, and prayers for the people, which belong to the duties of the cleric. Wherefore just as commercial enterprises are forbidden to clerics, because they unsettle the mind too much, so too are warlike pursuits.... The second reason is a special one, because, to wit, all the Clerical Orders are directed to the ministry of the altar, on which the Passion of Christ is represented sacramentally. Wherefore it is unbecoming for them to slay or shed blood, and it is more fitting that they should shed their own blood for Christ, so as to imitate in deed what they portray in their ministry. For this reason, it has been decreed that those who shed blood, even without sin, become irregular. Now no man who has a certain duty to perform, can lawfully do that which renders him unfit for that duty. Wherefore it is altogether unlawful for clerics to fight, because war is directed to the shedding of blood.[3]

Prelates are limited to the spiritual weapons against those who bring spiritual death or bodily harm to their flocks. "The weapons of our warfare are not carnal but mighty through God." Such weapons, according to Aquinas, are salutary warnings, devout prayers, and, for those who are obstinate, the sentence of excommunication:

> Prelates and clerics may, by the authority of their superiors, take part in wars, not indeed by taking up arms themselves, but by affording spiritual help to those who fight justly, by exhorting them and absolving them, and by other like spiritual helps. Thus in the Old Testament (Jos. VI.4) the priests were commanded to sound the sacred trumpets in the battle. It was for this purpose that bishops or clerics were first allowed to go to the front: and it is an abuse of this permission, if any of them take up arms themselves.[4]

Many laypeople would want to point out that they receive from the ministry of the altar the living bread and the saving cup. Since in these also the passion of Christ is represented sacramentally, how is it becoming to them to shed blood of fellow human creatures made in the image of God? The willingness of laypeople to shed their own blood for Christ, as witnessed in the early church, was not mentioned in the question "On War." While gospel nonviolence was enjoined on the clergy, laypeople were expected to live by the just war criteria.

The first article in the question "On War" lists the objection, "It would seem that it is always sinful to wage war." Aquinas points to the teaching of Jesus in the gospel of Matthew (26:52) that all those who take up the sword shall perish by it, to buttress the objection that "all wars are unlawful." The objection is disposed of in the words of Augustine, who pointed out that if the Christian religion forbade war altogether, those who asked for advice in the gospel would have been told to put aside their arms and to quit soldiering altogether. "On the contrary, they were told: 'Do violence to no man . . . and be content with your pay.' If he commanded them to be content with their pay, he did not forbid soldiering."

It is of great significance that the "he" who gave the command was not Jesus but John the Baptizer, and the words were spoken (Luke 3:14) before Jesus began his public ministry. To extrapolate from this one sentence that soldiering was acceptable to the followers of Jesus is questioned by many. The fact that lay Christians were not called to gospel nonviolence, as treated in the first article of question 40, became almost inextricably connected with church teaching regarding participation in war.

Three Conditions of Just War

Aquinas proceeds to explain the alternatives that the layperson has to gospel nonviolence and nonretaliation. He lists the three conditions that make a war just.

He states:

> In order for a war to be just three things are necessary. First, the authority of the sovereign by whose command the war is to be waged. For it is not the business of a private individual to declare war, because he can seek for

> redress of his rights from the tribunal of his superior. Moreover it is not the business of a private individual to summon together the people, which has to be done in wartime. And as the care of the common weal is committed to those in authority, it is their business to watch over the common weal of the city, kingdom or province subject to them. And just as it is lawful for them to have recourse to the sword in defending that common weal against internal disturbances, when they punish evildoers, according to the words of the apostle (Rom. 13:4): He beareth not the sword in vain: for he is God's minister, an avenger to execute wrath upon him that doth evil; so too, it is their business to have recourse to the sword of war against external enemies.

He supports his argument with the famous citation from Augustine's letter to Faustus (22.75): "The natural order conducive to peace among mortals demands that the power to declare and counsel war should be in the hands of those who hold the supreme authority."[5]

Aquinas goes on to the second condition: "Secondly, a just cause is required, namely that those who are attacked, should be attacked because they deserve it on account of some fault." This is supported by Augustine's definition of a just war (quoted earlier). Aquinas continues: "Thirdly, it is necessary that the belligerents should have a rightful intention, so that they intend the advancement of the good or the avoidance of evil."[6]

Aquinas adverts to the possibility that a war declared by legitimate authority and for a just cause may be rendered unlawful through a wicked intention. Again, Augustine comes to the aid of Aquinas, who employs the famous citation from the letter to Faustus: "The passion for inflicting harm, the cruel thirst for vengeance, an unpacific and relentless spirit, the fever of revolt, the lust for power, and such like things, all these are rightly condemned in war."[7]

With regard to the precept of refraining from personal resistance or self-defense, Aquinas departs from the absolute nonviolence of Augustine in personal response to evil. In other words, self-defense by violence or retaliation is, according to Aquinas, not denied to the Christian. Aquinas quotes Augustine to the effect that such restraint involved in nonretaliation should "be borne in readiness of mind."[8] Aquinas turns it around. "Nevertheless," he says, "it is necessary sometimes for a man to act otherwise for the common good or for the good of those with whom he is fighting." This section is open to other interpretations. It is generally accepted that Aquinas means to justify self-defense in the context not only of self-preservation but of preserving others whose rights are in jeopardy. Aquinas recalls Augustine's letter to Marcellinus: "Those whom we have to punish with a kindly severity, it is necessary to handle in many ways against their will. For when we are stripping a man of the lawlessness of sin, it is good for him to be vanquished."[9]

Relying on Augustine's letter to Boniface, Aquinas concludes: "We do not seek peace in order to be at war, but we go to war that we may have peace. Be peaceful, therefore, in warring so that you may vanquish those whom you war against, and bring them to the prosperity of peace."[10]

"It would seem that it is unlawful to lay ambushes in war," asserts the Objection in the third article of the question "On War." Ambushes, since they are a kind of deception, seem to pertain to injustice. A further objection arises from the gospel imperative of not doing to our neighbor what we would not want to be done to ourselves. This would certainly include ambushes and deception. Aquinas quotes Augustine to the contrary, to the effect that a just war may be carried on openly or by ambushes. As often in his arguments, Augustine leaps over the new covenant to the old covenant. Joshua, he maintains, was commanded by God to lay ambushes for the city of Ai (Josh. 8:2). In the end, the article decides that it is concealment that is meant by ambush, and this may be employed in a just war.[11]

The only other issue raised in the question "On War" is "whether it is lawful to fight on holy days." "It would seem unlawful," states the objection, "to fight on holy days. For holy days are instituted that we may give our time to the things of God." After some searching reasoning, the decision is reached that, "for the purpose of safeguarding the common weal of the faithful, it is lawful to carry on a war on holy days, provided there be need for doing so: because it would be to tempt God, if notwithstanding such a need, one were to refuse to refrain from fighting."[12] It is on this note that the question "On War" is concluded. In a work devoted to the just war, Joan D. Tooke states:

> Aquinas' direct teaching on war then is slight and unoriginal. Derived more or less wholesale from Augustine and Gratian, it is abstract and theoretical, and inspired by no personal emotion or thought. It is related neither to contemporary political and ecclesiastical conditions nor to the rest of his thought. It is dealt with in a treatise on charity, but is considered as a problem of justice. Nevertheless, it is considered there, in both individual and social aspects, as a specifically religious problem, and Aquinas professed to give a clear-cut Christian ruling.[13]

Other questions regarding the taking of human life command the attention of Aquinas, including that of putting a sinner to death. Beginning with the objection that it does not seem lawful to kill the sinner, the article ends with the conclusion that it may be good to kill a man who has sinned since he is more harmful than a beast. Here the common good of the community is subsumed. A private individual may not take it on himself to kill an evildoer since only the public authority can condemn him to death for the common good. Under no circumstances, says Aquinas, may a cleric or ecclesiastical prelate engage in corporal slayings.[14]

An issue foreign to a modern reader is discussed under the title of "Whether a Religious Order Can Be Directed to Soldiering." It is an indication of how time-bound Aquinas's concerns may be, since he lived in the era of the Knights Templar. Starting with the objection that no religious order can be directed to soldiering, the argumentation takes into account not only Augustine but also Judas Maccabeus of the Old Testament. The conclusion is reached that a religious order may be "fittingly established for soldiering, not indeed for any worldly purpose, but for the defense of divine worship and public safety, or also of the poor and oppressed."[15]

There has been a tendency to treat Thomas Aquinas in the same manner as Aquinas treated Augustine, as the *terminus ad quem,* the end of all questioning. Cardinal Emmanuel Suhard made it clear that rather than the *terminus ad quem,* the *Summa Theologica* is a beacon of light whose beams supplied clarification for theologians and for all followers of Jesus. Thomas came to be called "the Angelic Doctor," yet in his day his work was considered controversial. Even his openness to the philosophical thought of a pagan, Aristotle, raised some questions.

Accompanying the Christian Classics edition of the *Summa Theologica* is a sage comment by a fellow Dominican, A. G. Sertillanges: "The church believes that Thomism is an ark of salvation capable of keeping minds afloat in the deluge of doctrine." He points out that the church "knows that it is fallible, and that in respect of passing theories, it has shared the errors of different times." The church, according to Sertillanges, notes that "both knowledge and faith converge on it [Thomism], because it has taken up its position between them like a fortress at a meeting of the roads."[16]

Bernard of Clairvaux: Malicide, Not Homicide

The "deluge of doctrine" mentioned by Sertillanges was certainly evident in the discussions swirling about the issue of violence and war in the Middle Ages. Most bishops, abbots, theologians, decretists, and decretalists took Gratian's *Decretum* as their springboard; others, however, introduced fresh insights into the discussion.

Such a new approach came from St. Bernard of Clairvaux (mentioned above). He used the term "malicide," instead of "homicide," to describe the taking of life in defense of God and church. The destruction of evil, malicide, appeared to cleanse the act of killing in a church-blessed war from any sense of guilt or reason for remorse. The opponent was robbed of humanity and seen solely as a perpetrator of evil, so that the taking of his life was a praiseworthy and meritorious act.

Bernard was referring in particular to the war service of the Knights Templar, a religious order which he helped found in 1128. The Knights took their name from their house in Jerusalem on the site of the Temple of Solomon. Originally formed to protect pilgrims to the Holy Places, the order soon participated in the actual fighting of the Crusades. The new type of knight was one who took vows of poverty, chastity, and obedience and who belonged to a new type of order in the church, one that combined religion and the military. The ban on killing by clerics and ecclesiastics (maintained by Augustine, by Gratian, and later by Aquinas) was lifted in the case of the Knights Templar. In a sense, they had the best of the medieval worlds. On the one hand, they embodied the spiritual ideals of monasticism, and, on the other, they were chivalry incarnate, riding on their horses to heroism and glory. Many men of noble houses were attracted to their ranks. They exemplified the union of the church with what was considered holy violence. Over their armor, these knights wore the distinctive mark of the red cross on a white ground.

The red cross distinguished the Knights Templar from the Knights Hospitallers of St. John of Jerusalem, whose symbol was a white cross over a black ground. The Hospitallers, also founded in the early twelfth century, began their work in caring for the ill and infirm pilgrims at the hospital attached to a Catholic church in Jerusalem. They were divided into three classes, true knights (termed knights of justice, who engaged in fighting), chaplains, and Brothers (who served the needs of the community).

Bernard urged them to battle in lengthy sermons. Albert Marrin cites one, a few sentences of which give the tenor of his approach:

> Fearless certainly is a Knight, and safe from all surrounding danger, who, as he clotheth his body with a breastplate of iron, also hath his soul clad with the breastplate of faith.... But when the actual conflict hath commenced, they at last put aside their former deliberateness, as if they should say: "Do not I hate them, O Lord, who hate thee, and am I not at war with those who are hostile to thee?"... They know at least not to count too much on their own strength, but to hope for victory from the Lord of Sabaoath, with whom, according to the words of Maccabeus, they know that it is very easy for many to be shut up in the hands of few.... Such hath God chosen for himself and gathered from the ends of the earth as servants from amongst the bravest of Israel, so that they may faithfully guard the resting place of the true Solomon — forsooth the Holy Sepulcher — all with swords in their hands, all prepared for battle.[17]

There is more than an echo here of Israel, the chosen of God, going forth to a God-inspired battle.

Death Seldom Mentioned

The writings of the medievalists on warfare are remarkable in their portrayal of death as a glory — from the point of view of the Crusaders or Christians engaged in "just war." The actual experience of war is a distant reality. The maimed and bleeding bodies of the battlefield, the severed limbs, the agonized cries of the injured enter very little into the "deluge of doctrine" about war. The rotting corpses and the burials after battles are unworthy of description.

Frederick H. Russell in *The Just War in the Middle Ages* comments that "it may seem puzzling that the Decretalists seldom mentioned the most obvious consequence of war, death. It was as if they felt that wars could be fought without killing, and while that might have been the case in petty feudal wars it was becoming less so in large scale wars between princes."[18] Russell's work is invaluable in its treatment of medieval thought on war.

After Gratian came a series of commentators called decretists, who expanded on his work and on the "dicta Gratiani," Gratian's own interpolations. Following the *Decretals* of Raymond of Penafort, minister general of the Dominican

Order, came a number of interpreters called decretalists. The concerns of the decretists and the decretalists contributed to the growth of the canon law of the church.

What did these legally minded clerics find to discuss over the years? One of the practical issues concerned the duty of a vassal to his lord, or prince, in a real war, not a small skirmish. That a vassal had no choice but to aid his lord militarily was agreed upon by most decretists. The vassal was bound even if the lord went to war against the brother or other member of the vassal's family. One decretist opinion held that the vassal was not obliged to fight if a war declared by his lord was clearly an unjust one, or if his participation involved the vassal in outright sin or perjury. If, however, the vassal merely had doubts about his lord's cause, he was bound to take part in the war. If he was called upon to carry out unjust orders, the vassal and all military participants were not to be blamed. The guilt rested with the lord. Vassals were forbidden to obey the orders of a lord who had been excommunicated, or who was in schism against the church, or who was a heretic.

A decretist, Stephen of Tournai, raised the dilemma that a war might be just on both sides, or just on only one side, or unjust on both sides. It might be that a just war would be waged unjustly. There might have been examples of each type of war in the experience of the decretists themselves, but in their discourses they took their examples from the wars of Israel. King David and Judas Maccabeus were often cited.

The vassal or subject could, in effect, be a conscientious objector to a war if he was certain that it was unjust, or that it obliged him to commit sin. Another reason for conscientious objection entered the discussion through Rolando Bandinelli, later Pope Alexander III (1159–81). He offered the opinion that lay Christians who aspired to perfection should not be forced to take part in armed violence. Such laymen wished to join clerics and all ecclesiastics in the refusal to shed blood.

Most of the decretists contended that all wars of the Saracens against Christians were unjust. One decretist disagreed, asserting that Saracens could wage just wars. He put in question the accepted conviction that wars of conquest for the conversion of infidels could be just. Another pointed out that as infidels had the right to defend their lives and kingdoms, they could repel an attack with a just war.

One of the decretalists, Hostiensis, commenting on a decretal of Pope Innocent IV (1243–54), posited that the bond of solidarity uniting all Christians should exclude wars among them. The wars between Christian princes should be condemned, while the only licit wars were those against non-Christians. If the Christians were united, only the Holy Roman Emperor, who considered himself God's temporal vicar, and the pope himself could declare a just war. If such a law prevailed, the wars of feudal lords would be unjust. If, however, a feudal lord insisted on going to war, Hostiensis proposed that his cause be judged by the emperor, the pope, or an appointed judge.

Innocent IV, however, asserted the right of lords to use the violence of war in their own defense. The pope himself was the inheritor of a war between the

Holy Roman Emperor, Frederick II (1194–1250), and Pope Gregory IX (1227–44). Pope Gregory had used his mighty weapon, that of excommunication, while Frederick used his, the armed might of his soldiers. Most of the papal states were seized by Frederick.

In the continuing popes versus emperor struggle, Innocent deposed Frederick and strengthened Frederick's enemies by conferring on them the privilege of Crusaders. This resulted in plunging Germany into civil war. The military power of the allies of the papacy brought the struggle to a close with the result that Frederick's royal house of Hohenstaufen was ruined. Little credit redounded to the pope, who had relied on the secular arm. The taxes levied to supply the forces supporting the papacy made Innocent IV an unpopular figure. So much for the just war when it was a matter of conflict between emperor and pope. Is Russell too harsh in his comment, "The theocratic megalomania of Hostiensis and thirteenth-century popes provided secular officials with the example of high theory in the service of low cunning"?[19]

A thorny issue in which the temporal power invaded spiritual autonomy needs to be mentioned, that of lay investiture. Under feudalism, the prince or landed proprietor claimed the privilege of investing the spiritual leader in his domain with the ring and staff, the symbols of his spiritual duties. From this arose a multitude of dilemmas, including the duty of a bishop or abbot so invested to supply soldiers in the event of a declaration of war by the lord. The decision as to whether the conflict was just was out of the hands of the spiritual leader, who was a vassal to the lord.

The struggle between church and temporal rulers involved a series of Holy Roman Emperors, in particular Henry IV, who was ruthless in insisting on his powers over spiritual rulers, whom he wished to treat as vassals. The issue was complicated by the fact that some bishops and abbots possessed temporal power in their localities. The church under various popes defied the emperors and insisted on the ending of lay investiture. An agreement was reached in 1122 between Pope Calixtus II and the Holy Roman Emperor Henry V, who renounced the right to invest prelates with the staff and the ring of their spiritual ministry.

A Third Party: The Human Community

While in general the conditions of declaring and waging a just war, and the concept that war is an instrument of justice, were preserved in the work of such Spanish theologians as Franciscus de Vittoria (ca. 1480–1546) and Franciscus Suárez (1548–1617), their views go beyond the matter of war between two parties. They introduce the well-being of a third party — the wider human community. Humanity is represented in the following statement of Vittoria in *On Civil Power:* "And so, since any one state is part of the world as a whole, and since especially a Christian province is part of the state, if war is made with advantage to one province or republic, but with loss to the world or to Christendom, I think that war would be unjust."[20]

A native of Burgos, Spain, where he became a priest of the Order of St. Dominic, Vittoria studied at the University of Paris. In time, he became professor of theology at the University of Salamanca, to which he attracted some of the preeminent theologians of Europe. Salamanca at that time vied with Paris as the intellectual heart of Europe. At the peak of his influence, Vittoria was invited by the emperor to be his theologian at the Council of Trent, but his death intervened.

Society had reached a different stage of development, so that Vittoria found it necessary in *On the Law of War* (*De Jure Belli*) to define the word "state," or "sovereign prince." In his definition, the state is a perfect community, meaning it is complete in itself, not part of any other community. It must have its own laws and legal system. The examples he offers are the Kingdom of Castile and Aragon and the Republic of Venice. He points out that many perfect states could be under one prince. Only a perfect state, or the prince at its head, has the authority to declare war. Though Vittoria accepts the fact that a wrong received is sufficient to resort to war, he adds a warning that since the evils of war are of such dreadful character, it is hardly lawful to respond with warfare when the wrong is not of the most serious kind.[21]

Vittoria urges that before resorting to war, the prince should consult with wise and good advisers who are free from bias, anger, or greed. He also points to the availability of the papacy as mediator to forestall violence. Where Christian princes are engaged in war which would injure religion, the pope could forbid them to fight and could use his authority to make a judgment for or against one of the contenders. Among his canons for a just war are first that a prince should not go looking for occasions of war, but rather he should, if possible, live at peace with all men. It is inhuman, according to Vittoria, to rejoice in reasons for killing men whom God created and for whom Christ died.

A second canon states that even a just war should not be waged so as to ruin a people, and a third canon deals with clemency for the vanquished. In general, the fault lies with the prince, but it is his subjects, who fight under him in good faith, who suffer in a harsh peace.

In *On Civil Power*, Vittoria stresses the importance of proportionality in war. Even when all the conditions for a just war are present, the war is not just if the harm it brings is greater than the benefit or advantage to be achieved. This introduces an emphasis on the *jus in bello*, the laws in actual warfare, as well as the *jus ad bellum*, the right to war. He is clear on the point that if a subject is convinced of the injustice of a war, he must refuse to fight in it. This would involve killing innocent people, and the subject must refuse this, even in defiance of the prince's command. Vittoria indicates that the expedient of allowing the prince to rush his people into war should be hedged about with limitations and restrictions that tend to discourage the choice of war.

Because during Vittoria's lifetime his co-nationals were colonizing Latin America, he addressed himself to the question of violence in colonial enterprise in *On the Indies* (*De Indis*). The extensive work, marked by convoluted arguments and time-bound reasoning, does establish some clear principles. Pointing out that the Indians were in peaceable possession of their lands and goods before the Europeans arrived, he argues that the Indians as the true owners should not be despoiled of the lands, and he reminds the people of his time that the emperor is not the lord of the entire world. Even if he were, he would not be entitled to take possession of the lands of the Indians, unbaptized or uncivilized though they might be. The native peoples have rights against the colonizing power. One of their rights was to refuse conversion to Christianity, since forced conversion is against the laws of God. Vittoria decries the fact that the native peoples were induced without full knowledge to accept the Spanish colonizers as their lords and to cede their lands to them.

The breakthrough of Vittoria was his enunciation in *On Civil Power* of a law above princes and kingdoms, a law of nations. He saw the whole world as a single commonwealth, capable of making laws to which the whole world would adhere and which any one kingdom would not have the right to disobey.

Like Vittoria, Franciscus Suárez attempted to put hindrances in the power of the princes to resort to war. A native of Granada, he was a member of the Society of Jesus. He studied at Salamanca and went on to achieve eminence as a professor of theology in Rome, Salamanca, and Coimbra. He is considered the last of the Scholastic philosophers. His teachings are contained in a work described as monumental, *On Laws and God the Law-Giver* (*De Legibus ac de Deo Legislatore*).

He treats war under the general subject of charity (*de caritate*). As one hindrance to the precipitous choice of war by a prince, he urges contending parties to put the issue before the pope. Though the pope does not have temporal power, he does have the power to examine the causes of the war and to pass judgment. The parties would then be bound to adhere to the judgment, though the pope had no power other than moral.

The human race, according to Suárez, though divided into so many different peoples and nations, still has a certain unity. The various communities of the world are never completely sufficient unto themselves and need some law to guide them in relationship with each other. This comes in part from natural law, but in addition law has arisen from the customs of the peoples. This can be called the law of nations. Such accepted customs include the receiving of legates under a law of immunity. Suárez makes the supremely simple and rational statement that "it is contrary to prudence and the common good of the human race that all contentions between supreme rulers ought only to end in war."[22]

Another humanist from Spain, Luiz Vives (1492–1540), a friend of Erasmus and critic of the conventions of Scholasticism, contributed to Christian renewal in such works as *On Pacification* and *On Concord and Discord in Mankind.* Following Spanish humanists, the Dutch Protestant scholar Hugo Grotius (1583–1645) is considered the founder of international law. He stressed the operation of natural law and shared with Aquinas the acceptance of the just war. Teaching that the individual transcends the state by virtue of God's ownership of life, he upheld the right of conscientious objection to warfare on the basis of man's duty to obey God rather than man. His seminal work is entitled *Concerning the Law of War and Peace.*

Erasmus: A Catholic Humanist Addresses War's Reality

The Catholic humanist who inveighed against war in words that are ringingly relevant, and still cited in the twentieth century, was the philosopher known to the world as Desiderius Erasmus (ca. 1467–1536). His references to war displace the bloodless abstractions of theologians with descriptions that bring to life the bloody, debasing reality of organized killing. The steel blade piercing the body of the human person often intrudes in his narratives. Above all, Desiderius Erasmus helps provide the grounding for a theology of peace.

Desiderius Erasmus of Rotterdam was born illegitimate to a future priest. His father, living with Margaret, a widow whom he planned to marry, had two sons by her. His father's family, wanting their son to take holy orders, had him go to Italy and informed him falsely that Margaret had died. In his grief, he was ordained. After he returned to Holland, he remained faithful to his vows as a priest.

Desiderius Erasmus was orphaned at the age of thirteen when his parents died in the plague. During his schooling, he was influenced by what was called the Modern Devotion of the Brothers of the Common Life. The movement was close to the Third Order of St. Francis and emphasized simplicity in life and reform of church practices. Entering the Augustinian Canons at Steyn, in Holland, he took holy orders in 1492. With the approval of his prior, he left monastic life and became secretary to the bishop of Cambrai. Realizing the unusual intellectual gifts of the young priest, his superior gave him much latitude in entering the theological, ecclesiastical, and cultural world of his time.

Erasmus studied at the University of Paris, where he was so poverty-stricken that he had to take private pupils to support his own studies. The regimen imposed on the poorer theology students in Paris was so harsh that his health broke down. He had a lifetime concern for the reform of education, in particular education for Christian living. He maintained himself and his wide travels through tutoring royalty and through grants from the eminent persons to whom he dedicated his work.

Erasmus saw the church at its best and at its worst. The church at its best he met in the persons of his friends, who included Thomas More; John Fisher, the archbishop of Rochester; and John Colet, dean of St. Paul's Cathedral. They were humanists all, men deeply concerned with the potentialities of the human person in the world while taking into account his eternal destiny. John Fisher arranged for Erasmus to lecture in Greek at Cambridge University, and Thomas More opened his home to him. There he began to write *In Praise of Folly,* a work he dedicated to More with the words, "I swear nothing has brought me more pleasure in life than companionship like yours."[23]

What peace-loving Erasmus considered to be the church at its worst was embodied in the pontificate of Pope Julius II (1503–13). This was the pope who joined in warfare, riding into battle encased in full armor and brandishing his sword. His cause was the return of the papal estates. Erasmus turned his own rapier against Julius in *Julius Excluded,* published anonymously. Erasmus describes the pope presenting his credentials to St. Peter along with a band of his soldiers. St. Peter explains that certain works, like those of mercy and peace, will gain him entry. These works are contrasted with the boastings of Julius concerning the victories that he and his soldiers have achieved, including the return of papal territories. St. Peter rejects these boastings and explains that it is deeds like feeding the hungry and clothing the naked that will gain him entry. He does not open the gates.[24]

Thoroughly dedicated to the reform of the church and the eradication of corruption, Erasmus did not break with the church as did other priests, among them Martin Luther, Menno Simons, and Ulrich Zwingli. Erasmus, during the thirty years of his writing career, a time of intense religious turbulence, maintained a position of "loving disagreement" with his church. An irenicist in a polarized time, he aroused invective from both sides, the reformers finding him too understanding of the church, the Catholics seeing him as too close to the followers of Luther and Calvin.[25] It was a time when stakes were high and the headsman's ax was sharp for those considered heretics by either side.

In the religious turbulence, even fury, of the time, Erasmus was a voice of moderation and reconciliation. He corresponded with Luther, addressing him as a "younger brother in Christ" (Luther was sixteen years younger than Erasmus). Erasmus insisted that Luther be given a fair hearing for his theses even after Luther had reached what to Erasmus was the breaking point. This was the assertion by Luther that there were only two rather than seven sacraments, these being baptism and the Lord's Supper. When the diet held in the city of Worms in 1521 became, in effect, a church council, Luther appeared before it with his writings on the table. Three parties took part in the dispute over Luther's teachings, the defender of the papal view that Luther should be put under ban by the emperor, the defenders of Luther, and a third party formed of those who had hoped for some compromise. The third party was so influenced by the spirit of Erasmus that it was referred to as the Erasmians. Emperor Charles V, on behalf of the diet, ordered Luther to leave the city of Worms. Some weeks later, in the absence of the emperor, the diet declared Luther an outlaw. The Christian body, the universal church, to the immeasurable grief of Erasmus, was to be riven by schism. Even after the break from Rome, Erasmus kept up correspondence with Melanchthon, described as Luther's theologian.

Sweet Is War to Those Who Do Not Know It

The appalling evil of war, and in particular war between Christians, comes to the surface throughout the writings of Erasmus. He wrote thousands of letters, including the famous letter to Abbot Anthony a Bergis, the abbot of St. Bertin Abbey. Erasmus asks the abbot to use his influence with the emperor to bring to an end the conflict then raging. He asks a trenchant question as to why war is not forbidden as a means of defending even the faith. "Why," he asks, "on these occasions, do a few maxims handed down from one to the other by mere men suggest themselves to our minds, rather than many positive precepts uttered by Christ himself, by the Apostles, by orthodox and approved fathers, concerning peace, and patience under evil?"[26] At a time when the depredations of the Turks struck dread into the hearts of Europeans and fanned the fires of war, Erasmus wanted the Christians to give the Turks an example of Christian living. "What do you suppose the Turks think, when they hear of Christian kings raging at each other, with all the madness of so many evils let loose? And raging for what? merely on account of a claim set up for power, for empires and dominion."[27]

The reference to the Turks in this letter reflects many statements of Erasmus about the need to give an example of gospel nonviolence to non-Christians and thus give them the opportunity to come to Christ — the opposite of the mentality of the Crusades.

Through his letters, Erasmus attempted to bring reason and irenicism into situations that were inflamed or to assuage the fury of ongoing war. Among the thousands of adages of Erasmus is the one still frequently quoted: "Sweet is war to those who do not know it."

Three works of Erasmus present poignantly and often with painful realism the horror of war, *In Praise of Folly, The Education of a Christian Prince,* and *The Complaint of Peace.*

In Praise of Folly

Erasmus, his heart set on the renewal of the church, became caustic in dealing with the failures of the church of his day. *In Praise of Folly* began as a somewhat lighthearted essay on human foibles but later focused seriously on the enfeeblement of discipleship by the church.

Folly, in the person of a woman, ranges over a wide area of human and religious life. War and peace are prime focuses of her passion.

Horrified by the complicity of the church in justifying war, Folly refers to the learned churchmen who "put the name of zeal, piety and valor to this manifest insanity, and think up a means whereby it is possible for a man to draw a murderous sword and plunge it into his brothers' vitals without loss of the supreme charity which is in accordance with Christ's teaching."[28]

Once and for all Erasmus disposes of the opinion taught since Augustine that the act of killing a human being can be accomplished while maintaining a loving disposition. He refuses to accept the split between the inner disposition and the external act. Erasmus expresses his impatience with the Scholastics through his mouthpiece, Folly. She refers to the "worthy Doctors, the Subtle Doctors or Most Suitable Doctors, Seraphic Doctors, Cherubic Doctors, Holy Doctors and Incontrovertible Doctors," and then, in a serious vein, points out what they do: "They let fly at the ignorant crowd their syllogisms, major and minor, conclusions, corollaries, idiotic hypotheses and further scholastic rubbish."[29]

In Praise of Folly touched the critical and pacifist chord in Christian teaching. It was published in thirty-six Latin editions. Translations were made into Czech, German, and French during the lifetime of Erasmus.

A Kingdom Untainted by Blood

Invited to serve as tutor to Prince Charles, later to be Emperor Charles V, Erasmus put his counsels into a work entitled *The Education of a Christian Prince.* Much of his advice concerns building peace and avoiding war.

"War is sown from war," says Erasmus in "On Beginning War." "From the smallest comes the greatest; from one comes two, from a jesting one comes a fierce and bloody one, and the plague arising in one place, spreads to the nearest peoples and is even carried into the most distant places."[30]

Erasmus wants the prince to picture the evils in the wake of war, "even if it is the most justifiable war — if there really is any war which can be called 'just.'" Commenting that "the pontifical laws do not disapprove all war," Erasmus recalls that "Augustine approves of it in some instances, and St. Bernard praises some soldiers. But Christ himself and Peter and Paul everywhere teach the opposite. Why is their authority less with us than that of Augustine or Bernard?"

What he proposes to the prince and to all is the ending of the insanity of war and the achievement of peace as the highest art. Princes should be concerned about the common aim of peace, and put aside ambition and anger, while popes and bishops should put aside the concerns of wealth and power.

Learning how dreadful war is by experience is too costly an object for princes, Erasmus asserts. The prince is warned against nobles who provoke wars for their own gain. The good prince will look with suspicion on every war. "Let the good prince lean always toward that glory which is not steeped in blood nor linked with the misfortune of another," is his counsel.

The work is concluded with a prayer: "I pray that Christ, who is all good and supreme, may continue to bless your worthy efforts. He gave you a kingdom untainted by blood; He would have it always so. He rejoices to be called the Prince of Peace; may you do the same."[31]

The young prince, as emperor Charles V, went on to carry out the antithesis of the program described by Erasmus, embarking on wars from the time of his coronation by the pope.

The Complaint of Peace

In *The Complaint of Peace* (*Querela Pacis*), Erasmus chose to give his views as a humanist and believer in Jesus' teaching of love. He speaks through the mouth of Peace, and many of the comments are as valid and apt for later centuries as they were for the sixteenth century.

The Complaint of Peace begins with Peace surveying the animal kingdom with its examples of cooperation among animals. Animals, says Peace, have many kinds of armor and means of defense. Man can only live through cooperation with his kind. Man commands a wide variety of gifts which he must share in mutual love and friendship, as must the various nations of humankind. Born defenseless and helpless, man would perish without the care of parents and of other helpers in the community. Why does man reject peace and concord for unending strife and bloodshed? Then, Peace has a reason for hope in hearing about Christian man. She sees a city and feels that she can find there people living in peace and cooperation under law. Tragically, discord fills the city, and Peace is cast out.

From the turbulence of ordinary men, Peace goes to princes, to whom was given a special obligation by the Prince of Peace. Disappointed, Peace finds that the courts of princes are the true source of war. She next tries the scholars and theologians; again, Peace is disillusioned, seeing wrangling and the use of weapons that are "the deadly darts of their tongues."[32]

Peace turns in hope to religion. She is pleased by the fact that many in the religious orders wear white, her own color. Despite the cross they carry on their garment, despite their salutations of peace and the fact that they address each other as "Brother," they cannot agree among themselves, nor even with their bishops. From the monastery, Peace looks at marriage and finds that even a common home, common bed, and children do not keep out strife and disagreement. Trusting that she would find a haven in one individual, Peace is again deceived. There is not one without the war between passion and reason.

For Peace, the actions of people who are not ashamed to call themselves Christian belie their claim to follow the Prince of Peace. His coming was foretold by Isaiah and by angels who proclaimed peace at his birth, not the sounding of military trumpets. Christ greeted his disciples with the greeting of peace, and left his peace as a legacy to his followers. To preserve this peace, he enjoined upon them the commandment of love, not any love, but love as he loved, even love of enemies. His disciples should be known not by this or that uniform or any other distinguishing mark, but by that of love. He called himself a shepherd of love. If the sheep tear at each other, what will the wolves do? He calls himself a vine, and his followers branches, a picture of unanimity. If it is unthinkable for one branch of a vine to fight with another, it should be equally unthinkable for a Christian to fight another Christian.

"What a terrible thing it is," laments Peace, "that men receive the sacraments now administered in the army camps and then resort to combat."[33] The unanswerable argument of Peace for concord among Christians is contained in the question: "What does the communion of holy bread and cup decree but a new and indissoluble concord?" She returns to the argument in almost the same words, "Shall not the heavenly bread and that mystical cup unite Christians in charity when Christ has ordained it and they renew that sacrifice daily?"[34]

Yet Peace says that in the decade in which she speaks, men redeemed by the same sacrifice and fed by the same sacraments soak whole regions and dye rivers with Christian blood. How can the soldier dare to recite the Our Father as he goes out to kill his brother? How can he say "Hallowed be thy name" when he is dishonoring the name of God by fighting?

Peace warns against alliances, confederations, and dynastic marriages that often lead to war, and against nationalist feelings, such as those of Englishmen against Frenchmen. The Rhine dividing the French from the Germans cannot divide Christian from Christian; neither can the Pyrenees, which divide Frenchmen from Spaniards.

For Peace, the misfortunes occasioned by war should be paid for by those who caused it, not by farmers and civilians. They suffer from the destruction, while the rulers who brought the war suffer nothing.

Peace ends with a cry of the heart:

> I beseech you, princes upon whose nod mortal affairs depend, and who, among men, bear the image of Christ, willingly acknowledge the voice of your Creator, endlessly summoning you to peace.... I beseech you, consecrated priests of God, unflinchingly promulgate that which you most certainly know to be most pleasing to Christ and exorcise that which is most hateful to him. I beseech you, theologians, fervently preach the gospel of peace, and increasingly exhort the people to live by it. I beseech you, bishops, who excel all others in ecclesiastical dignity, see that your authority is properly exercised to eternally bind peace on all. I beseech you, magistrates, to be uncompromisingly faithful to the wisdom of kings and the mercy of bishops....
>
> May eternal concord, already established by nature, and more wondrously reestablished by Christ, unite all things. And finally, let every man zealously aspire to that which pertains to the concord of all men.[35]

Opponent of Nationalism and Justified War

Before his death in 1536, Erasmus suffered the loss of two friends to the murderousness of a king. Thomas More and John Fisher were beheaded by a monarch with whom Erasmus was acquainted. Regarding More, he lamented, "By his death I feel myself to be dead." Burnings and executions marked the years when the human person was dispatched from the earth for his choice of a path to God.

In his lifetime the tireless and inspired work of Erasmus seemed to have achieved nothing. The least nationalist of men, Erasmus saw virulent nationalism reaching into the religious realm with the formation of national churches.

Desiderius Erasmus, in the view of James McConica, saw as his purpose the transmitting of the moral experience of the Christian past. "In one respect, however," notes McConica, "it departed from the conventional opinion in denying the theory of the just war."[36] Erasmus insisted that the positive precepts uttered by Christ himself, by the apostles, and by orthodox and approved Fathers concerning peace should take precedence over the theories of Bernard of Clairvaux or Thomas Aquinas on the justness of wars.

If the soaring evocation of Peace and the searing presentation of war by Erasmus had found entry into canon law or at least into the textbooks of seminaries, the descent of the Catholic Christian community into the vortex of total war might have been hindered. Instead, the high speculations of "Cherubic Doctors, Holy Doctors and Incontrovertible Doctors," along with their syllogisms, prevailed.

Erasmus attempted to infuse into the arid and bloodless treatment of war by the Schoolmen the teaching Jesus brought to the human family, the way of resolving conflict without bloodshed.

Though the work of Desiderius Erasmus failed to make its imprint on the teaching church of his time, it has nourished generation after generation of seekers after peace. "It has left a legacy of common loyalties and ideals perhaps more durable than he dreamed," asserts McConica, "and wherever the cause of international peace is admired, the name of Erasmus is still invoked."[37]

Chapter 6

Just War Challenged

Peacemaking is not an optional commitment. It is a requirement of faith. We are called to be peacemakers, not by some movement of the moment, but by Our Lord Jesus Christ. —Catholic bishops of the United States, *The Challenge of Peace: God's Promise and Our Response*

"Either the just war was a moral and religious doctrine, in which it was deprived of coercive but not normative force, or it was a legal concept that served as a cloak for statism," asserts Frederick H. Russell in *The Just War in the Middle Ages.*[1]

History indicates how the just war served as a cloak for statism. From its beginnings as a concept, tied to a religious body, it could not achieve coercive force. Its function was to establish a set of norms or conditions for judgment to be made as to whether wars were just. This raised the question as to when the judgment would be made — before, during, or after the war — and who would deliver it and oversee that it was put into action. When Europe was predominantly Christian, the pope could deliver such a judgment, as he did in the Crusades in declaring a war that transcended the condition of "just" into "holy."

Even the pope, however, could only bless the cause. He did not control the means. The warfare that was just needed to be just in means as well as in cause. Because the enemy was the infidel, the question of means did not come up for discussion in the Crusades (except for proponents of gospel nonviolence like St. Francis of Assisi).

John Howard Yoder comments on a historical turning point. It occurred when the Protestant princes, having opted for Lutheranism, were ranged in battle against the Catholic powers. "Just when Catholic tradition," he states, "began to disentangle itself from the no-holds-barred crusade, . . . internecine battles restored to the heart of Christendom the reality of the crusade without the name. During the Thirty Years War (1618–48) half of the middle of Europe was destroyed in the name of God."[2]

The Thirty Years War became a vivid reality for many twentieth-century audiences through the powerful, ironic drama *Mother Courage* by Bertolt Brecht. With every spring the carnage of the war years was renewed. In the play, the chorus announces:

Christians awake! Winter is gone.
The snows depart! Deadmen sleep on!
Now all of you who still survive
Get out of bed and look alive!

Plying her trade, Mother Courage then reminds the men:

> For they can get from Mother Courage
> Boots they will march in till they die.[3]

At the peace of Westphalia, ending the Thirty Years War, nationalism was conjoined with religion when it was maintained that the religion of the prince was to be the religion of the subjects. Roland H. Bainton points out that "the churches of the reformation, with the exception of the Anabaptists, all endorsed the theory of the just war as basic."[4] It was after the Reformation that many state churches came into being, he explains. In a state church the work of the minister is so linked to the government that the exercise of prophetic ministry is diminished or blocked.

The just war tradition achieved creedal status in Lutheranism at the Augsburg Confession of 1530. In 1561, Anglicanism affirmed the just war in the Thirty-Five Articles, and in 1648, the Westminster Confession followed in affirming the just war.

Yoder stresses that there never has been a statement by a pope or church council that would make acceptance of the just war doctrine obligatory for Catholics.[5] Despite this, many Catholics believe that the doctrine of the just war has the force of dogma, if it is not actual dogma. The just war teaching has served the nations well in their war pursuits.

The conditions of the just war have been passed on through the textbooks of the Catholic theologians to each succeeding generation and have survived each new development of warfare into the nuclear era. In their fidelity to such abstractions as the just war conditions, the theologians have been described as "merchants of abstraction." The conditions remained, protected and inviolate, through the nineteenth and twentieth centuries. Catholic seminarians were taught the time-honored conditions of justified warfare rather than examining with their professors the glaringly unmet needs of new thinking about peacemaking and war-making.

A challenge to just war thinking raised its head when so-called civilized countries of western Europe entered the period of colonizing Asia, Africa, and Central and South America. Aristotle, having coined the term "just war," applied it to the wars conducted by Hellenes against non-Hellenes, whom he considered barbarians. Wars which enabled the cultivated, worthy Hellenes to wield power over less worthy men were, of course, just.[6]

Were the armies which went out from Spain, England, France, the Netherlands, and Portugal over the known world any more enlightened than the armies of the Hellenes who went out against the non-Hellenes? Not everything done by the colonizing armies was evil, and certainly in the wake of some armies there was an expansion of culture and religion. It was the sword, however, which provided the point of entry, and this was somehow justified by those who introduced it.

When the armies of Britain and France met on the soil of India, or the soil of what is now Canada, to establish hegemony during a colonizing enterprise, the issue of the justice of war did not enter the situation — what was at stake was simply which side would win and draw the disputed territory into its imperial sphere. The just war was a concept that could be used or ignored in any war action according to the wish of the warring state.

War as an Instrument of Justice

War was always presented as an instrument of justice from the time of its appearance in Christian theory in the works of Augustine and Ambrose. So embedded in Christian thinking did the just war theory become that many clerics referred to it as just war "theology."

The just war theory develops an analogy with the civil society of a nation-state in which an injured party has the right to justice through a judicial system. Provision is thus made for the redress of wrong and punishment for the wrongdoer. In international society, where states are considered supreme beings, there is no overall juridical system to provide for redress or punishment in the event of wrongdoing by one of them. The problem arises as to how one state, a being supreme in itself, can come to the point of punishing another state, which is equally supreme. Stanley Windass comments, "According to the usual explanation, a state which commits an offense becomes, by that very act, subject to the state which it injures, in respect of that particular act; automatically, therefore, the injured state becomes both judge and executioner of the offending one."[7]

The Conditions of Justified Warfare

The norms or conditions that should apply to the injured state when it becomes the instrument of justice are referred to as the conditions of the just war. They can be summarized as follows: the cause must be just; the war must be declared by legitimate authority (a sovereign state); and it must be fought with the right intention. In addition, some other conditions have been added. War must be the only means of securing justice, that is, a last resort. There must be a reasonable hope of victory, and the good achieved must outweigh even the probable evil effects of the war. A final condition refers to the waging of the war: only right means may be employed in the conduct of the war. A time-honored condition protected innocent civilians from attack. Thus the conditions relate to the right to declare war (the *jus ad bellum*) and the right conduct in war (the *jus in bello*).

To complete the rationale of just war thinking: all the conditions must be met for the war to be a just one. Another condition was proposed by a theologian, the right conduct after the end of hostilities. He called it "the *jus post bellum.*" It is treated elsewhere.

For the leaders of the sovereign state going to war, it would be necessary to take into account all of the preconditions, to give consideration to the probable evil effects, and to assess the means (including the type of weaponry) to be utilized.

Is it possible to imagine the leaders of a sovereign state, or a coalition, as in World War I or World War II, taking time to ponder the above conditions before unleashing artillery fire or dropping the first bomb? Could anyone suppose that if they were Christians, they would go to their spiritual advisers, to study with them the feasibility and morality of mounting a campaign that would involve punishing fellow human beings on the opposing side by mutilation and death? To the generality of people such an application of the just war conditions would constitute a bizarre utilization of the time of political leaders. Furthermore, such application rests on a dangerous presupposition, that war and its preparation are undertaken under moral norms and in a moral purview. Originally intended to limit wars, and to allow for the introduction of moral criteria into the conduct of war, the concept of the just war became the means of eliminating moral criteria.

One must not be unmindful of the fact that the just war conditions could only have been formulated in the first instance through a massive displacement, the displacement of the nonviolence of the Sermon on the Mount. Augustine, as has been pointed out, sometimes justified the participation of Christians in warfare by using the example of the wars of the Hebrew Scriptures. These wars, described as *deo auctore,* authorized by God, were righteous wars.

Since the time of Augustine, warring Christians and their leaders have paid lip service to the just war conditions in the wars in which they have had a leadership role or in which they participated. If the war is labeled just, then the minds and consciences of those concerned with prosecuting it are lulled into acquiescence. The flaunting of the word "just" permitted the nation-state to confer a brutal numbness on those called upon to perform acts of cruelty alien to ordinary human nature. The great opiate of the people in the twentieth century, a century punctuated by wars, is the concept of the just war.

In this way, the concept of the just war has served as a cloak for statism. The nation arrogates to itself the right to resort to the sword. The legal concept of the just war rests on no authority, moral or otherwise, above the nation-state. What is left is the self-interest of the state, which may be of the basest sort, such as seizing of territory, or of a somewhat higher nature, as defending a smaller power under attack by a more powerful neighbor. Even in the latter case, however, the state's motive may not be an unmixed and blameless one.

A President Talks Down to a Hierarch

A recent example of the operation of the relationship of the church to the state in a war situation is revelatory. When, in 1990, the president of the United States, George Bush, seemed close to preempting the role of the U.S. Congress, and was seemingly able to control the UN Security Council in involving the United States in the Persian Gulf War, he received a letter. It came from Roger Mahony, the Catholic cardinal of Los Angeles, and was endorsed by the body of American bishops at their 1990 meeting. The cardinal, in his role as president of the National Council of Catholic Bishops, listed the conditions of the just war and outlined briefly how they arose in Christian tradition. It might have been seen as a cautionary note to a political leader facing the dread decision to unleash death in a modern war.

President Bush took to a public podium what might have been construed as a reply to Cardinal Mahony and all those who shared in just war teaching. Citing Ambrose, Augustine, and Aquinas, President Bush declared that the war in the Persian Gulf was more than just — it was noble. He told a convention of religious broadcasters on January 28, 1991, when the war had been declared and the bombing of Baghdad was at its height: "The first principle of the just war is that it support a just cause. Our cause could not be more noble." The president said that the war's aim was "keeping casualties to the minimum." While the war was not concerned with religion "per se," the president asserted, "it has everything to do with what religion embodies, good vs. evil, right vs. wrong, human dignity and freedom vs. tyranny and oppression. . . . We will prevail because of the support of the American people, armed with a trust in God and in the principles that make men free."[8] An ovation greeted the president's speech, and the cloak of "just war" had been cast around the sixty-billion-dollar enterprise of the Persian Gulf War.

The stance of the American president was a forceful illustration of the accepted relationship of the leader of a nation-state over a church in the crucial matter of who had the right to declare a war as just. By citing the just war conditions (and adding his own blessing by calling the war noble), the national leader had the power to talk down to a hierarch representing the church which had given the just war conditions to the world. Nowhere was it clearer than in the case of the Persian Gulf War that the just war doctrine was a gift to the warring state.

The just war was a legal concept that could be utilized by those in power, in a word, those responsible for declaring and prosecuting a war.

Every War Declared a Just War

A former archbishop of Bombay, T. D. Roberts, who played a peacemaking role at Vatican II, asserted repeatedly, "Every war declared by every nation at every time is declared as a just war." He was echoing Erasmus, who simply dismissed all just war claims as meaningless. " 'Just,' indeed — this means any war declared in any way against anybody by a prince."[9]

Those who are called upon to fight a war have access, even in modern times, only to the arguments of their own nation-state. Thus they are fed on the depredations and dreadful wrongs committed by their opponents. When the United States embarked on an armed incursion into Panama, the whole enterprise was formally labeled "Just Cause." This official designation was questioned only by a few dissidents. Very few individual citizens have the courage or effrontery to question the justice of a war declared by their own state. In this way the just war, adopted by churchmen and developed by theologians, has turned out to be a lethal gift of the church and its people to the state, in particular to its war-making power.

The just war conditions, as a means of preventing war or of lessening its brutality, have shown themselves to be irrelevant to the war-makers. Christians, their role in war always presented to them by the state as the enterprise of justice through violence, have little choice but to join with their nation-state in judging and punishing. Does not this replace the teaching brought by their Savior, that of loving and forgiving?

I Seem to Hear a Child Weeping

The adage "War is sown from war" comes from Erasmus, but it does not need the keenness of his intellect to realize that each war has one sure outcome, namely, that it leads ineluctably to another war.

After World War I, up to then history's most destructive war, a cartoon appeared in the *Daily Herald* of London in May 1919. The Treaty of Versailles was being fashioned to complete the Treaty of Paris. Will Dyson, the cartoonist, pictures the four architects of the treaty — Georges Clemenceau, Woodrow Wilson, David Lloyd George, and Vittorio Orlando — leaving the Hall of Mirrors. Clemenceau, the "Tiger," remarks, "I seem to hear a child weeping." Behind a pillar stands a weeping child, hardly more than a baby. Above his head one reads "1940 CLASS." The cartoonist had the prescience, in which he was not alone, to realize that Versailles, harsh and vengeful, would give rise to a new war. He was mistaken on the date by a few months. The call-up date of those who were babies in 1919 was 1939. The protests to the Versailles Treaty by a defeated Germany, which had no part in the consultations leading to the treaty, were futile. The treaty went into force, bringing humiliation and deprivation to the German people. Some of its cruelest clauses, including the ruinous reparations payments, which were not suspended until 1931, prepared the way for the rise of National Socialism and Adolf Hitler.

The question which is supposed to silence and discredit proponents of nonviolence is "What would you do about Hitler?" One response is to utilize the peace treaties after hostilities for justice, compassion, and reconciliation, and even forgiveness. Peace treaties should provide the occasion to break the chain of violence, rather than, as more often occurs, forge a new chain of violence. Breaking the chain of violence, however, calls for qualities rarely if ever exhibited by governments, in particular, governments which are victors in war. Justice, compassion, and reconciliation are rarely invoked against those who have just been excoriated and dehumanized as enemies. Is it realistic to expect such qualities from nation-states?

World War II was undoubtedly just in cause. Drafted German and Austrian soldiers were dragooned into carrying out crimes of unspeakable horror against helpless human beings. Soldiers of the Western Allies in World War II were given the highest moral reasons on behalf of which they were to give their lives and take the lives of others. These moral aims had no relation to the conduct of the war nor to its conclusion in the peace agreements of the Crimea and Potsdam Conferences. The peace soon showed its failure in the friction that served to undo the Grand Alliance of East and West. Accustomed to resolving differences by resorting to arms, both sides entered into an arms race such as the world had never seen. In a few years they shared the possession of fifty thousand nuclear warheads. The sequel to World War II, the Cold War, gave the world a near half-century of nuclear terror, a world pulsating with nuclear death. The continent of

Europe was slashed by a barrier based on ideology, a barrier which cut through national and cultural boundaries and even through towns and villages. In Berlin, ideology was made visible in a barricade topped by barbed wire and shards of glass to draw blood from those who attempted to scale it, and also to give time to those on "watch towers" to shoot to kill.

Energy Enslaved

The visionary poet William Blake lived at a time when English life was disquieted by its wars on the continent of Europe. Blake's own life was profoundly disturbed as a result of his outspoken opposition to the Napoleonic Wars. He saw the Industrial Revolution feeding into an ever-growing war economy in which "all the arts of life they changed into the arts of death." In anguish and passion, he declared, "War is energy enslaved."[10]

All the energies of a people and their resources are corralled by the warring state, which, in so doing, aggrandizes itself to new heights of power and greater control over its citizens. In the more developed countries, the scientists whose energy could have gone into improved technological adjuncts for daily living are sucked into production for more efficient death-dealing. Over the centuries skilled workers produced the long bow and the more lethal crossbow. These gave way to the sword, the finest blades coming from Damascus in the Middle East, and the Toledo blade from Spain. Siege machines, catapults, and battering rams gave way to mounted cannons and heavy artillery when gunpowder entered the battlefield. Air war, developed during World War I, allowed the destruction of soldiery, but more importantly, the destruction of the cities and factories that supplied the men in the field. The energies of a multinational team of scientists brought the fire of the atom in bomb form to World War II. They kept all knowledge of the bomb's fiery power secret until it devastated two cities.

Another aspect of "energy enslaved" was exhibited when the sufferings of the Japanese under the bomb were accepted by large numbers of Americans as simply another price that had to be paid for victory. Emotional energy, harnessed during wartime, is focused on an enemy who must be expunged as soon as possible to save the lives of men in combat. Many, though not all, Americans were not too uncomfortable with the fate meted out to the citizens of Hiroshima and Nagasaki, or the fact that they were mainly civilians. They recited the argument of war then given by government agencies — that the nuclear bomb had saved as many as a million lives of American troops. Wartime secrecy had kept from them the fact that Japan, gutted and deprived of civilian provisions, had already started suing for peace. The emotional tone of a people in wartime, in particular that of a nation moving from success to success, is based on slogans, on the drumbeats of propaganda picturing the enemy as less than human. This empties the human heart of compassion and sympathy.

Stranger, perhaps, than the harnessing or enslaving of emotional energy is the harnessing of spiritual energy. If seminary teaching in successive generations had not been so securely anchored in the just war, how much creative energy might have been released? Nothing so closes the mind as the conviction that the last word has been said on a subject. That final word, church leaders agreed, had been said in the vastly different circumstances of fifteen hundred years earlier by two great saints. The forays into gospel nonviolence by Francis of Assisi and Erasmus seemed too radical for ordinary Christians. However noble the teaching and witness of Francis and Erasmus might be, they did not seem to be meant for emulation by ordinary believers. One could hope that if the followers of Jesus had been thoroughly acquainted with a developed theology of peace and peacemaking, one based on the nonviolence of the cross and the lives of the early community of disciples, there might have been many more examples of their "speaking truth to power" in wartime and even of their "witnessing unto blood."

In the Mass and stories of the saints are the examples of soldier-martyrs, Maximilian, Marcellus, and Martin of Tours. They might have been seen as relevant to the lives of Christians in each age. The nonviolent witness of men and women in the so-called historic peace churches has not been seen as relevant by the generality of Catholics.

The Pacifist Perspective Legitimized

In the mid–twentieth century the just war began to give way to the earlier belief and practice of the community of disciples that Jesus founded. "In the twenty years since John XXIII took up the papal ministry Catholic teaching on war has been in a state of movement," said J. Bryan Hehir, former secretary of the Office of the U.S. Catholic Conference for Peace and Justice. "The principle development," he pointed out, "has been the legitimization of a pacifist perspective as a method for evaluating modern warfare."[11]

That the legitimization of the pacifist position has occurred does not alter the fact that it is far from a majority position. A recent publication of the Catholic University of America presents the more commonly accepted position. "Individual pacifists," the publication points out, "are often called upon to give heroic witness to the value of human life. Nonpacifists should not only admire such witness but also be drawn by it to assess critically the putative justice of particular wars." At that point, the argument of the necessity for the just use of violence enters. "But what is admirable or virtuous on the part of individuals may not be admirable or virtuous on the part of statesmen. Supposing that human beings may justly resort to force, including killing by force, to resist wrongdoers, then those in charge of a community may have a duty to do so. The practice of pacifism by individuals directly affects only those individuals; the practice of pacifism by statesmen on the other hand would directly affect the entire community."[12]

The latter assertion tallies with the commonly accepted view of pacifism/just war carried in the *Catholic Encyclopedia.* It states that

> absolute pacifism is irreconcilable with traditional Catholic doctrine. Catholic exegetes likewise commonly reject the pacifist interpretation of Christ's teaching. His pronouncements on nonresistance to evil are taken as a counsel rather than as a precept, and for private individuals rather

> than public authorities since these would fail in an essential duty were they to offer no forceful resistance to violent aggressions from within or without. His warning to those who "take the sword" is commonly understood... to refer to those who usurp the function of rulers. Nor is there any intrinsic contradiction between a just war and Christ's command to love our enemies. A just war expresses hatred of the evil deed rather than of the evil doer.[13]

The Challenge of Peace

There came a time when the doctrine of the just war began to be challenged in the teaching church by the teaching of nonviolence. The signs of an emerging "peace church" could be discerned in the witness and statements by individual bishops and laypeople and in the peace pastoral of the body of American bishops. The updating of peace teaching by Vatican II is dealt with elsewhere.

In the peace pastoral, *The Challenge of Peace: God's Promise and Our Response,* issued in November 1983, the American Catholic bishops state, "Moved by the example of Jesus' life and by his teaching, some Christians from the earliest day of the church committed themselves to a nonviolent lifestyle."[14] The need to place peacemaking at the very heart of Christian living is emphasized. "Peacemaking," state the bishops, "is not an optional commitment. It is a requirement of our faith. We are called upon to be peacemakers, not by some movement of the moment, but by Our Lord Jesus."[15]

The bishops assert that what is needed is a more fully developed theology of peace. The pastoral itself serves as a step toward such a theology. In a section of the pastoral entitled "The Value of Nonviolence," the bishops cite two of the leaders of the early church, St. Justin and St. Cyprian, both of whom died for witnessing to the Lord's peace. Passing quickly over Martin of Tours in the fourth century and Francis of Assisi in the thirteenth century, they come to the twentieth century and point to "the nonviolent witness of such figures as Dorothy Day and Martin Luther King, which has had a profound impact upon the life of the church in the United States."[16]

While the bishops point out that "the just-war teaching has clearly been in possession for the past 1500 years,"[17] they offer a definition of a peace theology that would make room for those who reject the just war teaching, seeing it as a graft from the outside on Christian teaching. They state, "A theology of peace should ground the task of peacemaking solidly in the biblical vision of the kingdom of God, then place it centrally in the ministry of the church."[18]

In the section entitled "The Just War Criteria," the peace pastoral deals with Augustine, recalling that the love taught by Jesus suffered a sea-change when it was presented as a love which would use force and violence to protect the innocent from aggression and harm. "Faced with an attack on the innocent," the pastoral states, "the presumption that we do no harm, even to our enemy, yielded to the command to love understood as the need to restrain an enemy who would injure the innocent."[19]

Herein precisely is the vast crevasse between the love which justifies force, even lethal force, and the love which never veers from the reconciling, forgiving, suffering love which led Jesus to Calvary.

Dissuasion by Terror

Since they were addressing American Catholics in the nation-state which first used nuclear weaponry, the bishops presented a section on the moral assessment of deterrence, which has been referred to as "dissuasion by terror." They made room for the opponents of violence who would refuse to be defended by nuclear or other weapons, but in deference to those who "affirm a nation's right to defend itself," they discuss at length the problems of finding security through treaties and negotiations.

In the beginning of the pastoral the bishops allude to the crisis of the present age: "The crisis of which we speak arises from the fact that nuclear war threatens the existence of our planet; this is a more menacing threat than any the world has known."[20]

The examination of the nuclear deterrent is extended and detailed. The discussions over those texts are reported to have been bitter, necessitating the preparation of three drafts. Experts, scholars, and theologians were called in for consultation. Present were "a small and growing number of bishops committed to nonviolence and all-out opposition to nuclear weapons."[21] Well known for their stand on nonviolence were Bishop Thomas Gumbleton, Bishop Leroy Matthiesen and Bishop Raymond Hunthausen.

A committee was invited to Rome to consult with the Vatican and with the bishops from other NATO countries. The gist of their discussion was shared with concerned laypeople. When the relationship between the just war tradition and pacifism became part of the deliberations, a Vatican hierarch took the occasion to explain that there are not two traditions in the church, but rather one tradition, that of the just war. Within that option, pacifism is a legitimate option for individuals, not governments.

Echoing the assessment of the deterrent given by Pope John Paul II in his 1982 message to the United Nations Special Session on Disarmament, the bishops stated: "In current conditions, deterrence, based on balance certainly not as an end in itself but as a step toward progressive disarmament, may still be judged morally acceptable. Nevertheless, in order to ensure peace, it is indispensable not to be satisfied with this minimum, which is always susceptible to the danger of explosion."[22]

Even this strictly limited acceptance of a nuclear deterrence, however, constituted a grievous disappointment to pacifists like Bishop Gumbleton and most members of Pax Christi–USA.

In response to the possibility of a limited nuclear war the bishops pose six questions. They are directed to those who might consider such nuclear war in terms of just war teaching. The questions, if answered, would rule out even a limited nuclear exchange. One question is revealing: "How limited would be the long-term effects of radiation, famine, social fragmentation, and economic dislocation?"[23]

The pastoral emphasizes many pragmatic efforts to forestall nuclear warfare, and lethal nuclear accident. It gives strong support for a comprehensive nuclear test ban treaty. Much of the discussion on war, peace, and deterrence was destined to be part of the public debate in the United States. It exhibits the concern of moral leaders who recognize that deterrence had become the center piece of both the U.S. and Soviet policy in a period which they describe as that of "the superpowers in a disordered world."[24]

Disarmament of the Heart

There is, however, much that would support and nourish those who have chosen gospel nonviolence. "All of the values we are promoting in this letter," say the bishops, "rest utmostly in the disarmament of the human heart and the conversion of the human spirit to God alone who can give authentic peace."[25] Elsewhere the bishops remind us that "Christ Jesus, then, is our peace and in his death-resurrection he gives God's peace to the world."[26] "The words of Jesus," say the bishops, "would remain an abstraction were it not for two things: The action of Jesus and his gift of the Spirit."[27]

Though the just war undoubtedly has served as a gift to the nations, there are signs that it is being challenged by disciples of peace. It is being dethroned after fifteen hundred years. The "Challenge to Peace" of the American bishops is one of these signs. Difficulties, even suffering, await those who choose to live disarmed in war as well as in peace. The bishops address themselves to what followers of Jesus may expect if they weave nonviolence into every aspect of their lives:

> To set out on the road to discipleship is to dispose oneself for a share of the cross (cf. John 16:20). To be a Christian, according to the New Testament, is not simply to believe with one's mind, but also to become a doer of the word, a wayfarer with and a witness to Jesus. This means, of course, that one can never expect complete success within history and that one must regard as normal . . . even the path of persecution and the possibility of martyrdom.[28]

In the conclusion of the pastoral the bishops issue a statement that rings like a warning bell to humanity, and in particular to followers of Jesus to whom every person is a repository of the divine and the earth itself on which the Messiah stood is sacred. The bishops declare: "We are the first generation since Genesis with the power to virtually destroy God's creation."[29]

"In brief," the bishops assert, "the danger of the situation is clear; but how to prevent the use of nuclear weapons, how to assess deterrence, and how to delineate moral responsibility in the nuclear age are less clearly seen or stated. Reflecting the complexity of the nuclear problem, our arguments in this pastoral must be detailed and nuanced; but our 'no' to nuclear war must, in the end, be definite and decisive."[30]

People in general might consider the near destruction or destruction of our own small planet and of the human family as the ultimate tragedy. They may think that earth's passing away and the passing away of humankind could be part of the Creator's plan. But God's human creatures are imperishable. The tragedy would be if God's creation met destruction by the power and destructive will of humankind. The expression of this destructive will might well be a war undoubtedly declared just by its perpetrators.

Chapter 7

The Works of War

To remember Hiroshima is to abhor nuclear war. To remember Hiroshima is to commit oneself to peace....

Let us make a solemn decision, now, that war will never be tolerated or sought as a means of resolving differences.

Let us promise our fellow human beings that we will work untiringly for disarmament and the banishing of nuclear weapons....

Hear my voice, for it is the voice of the victims of all wars and violence among individuals and nations; hear my voice, for I speak for the multitude in every country and in every period of history who do not want war and are ready to walk the road of peace.

—Pope John Paul II at Hiroshima, February 25, 1981

The Diabolical Contrivance

The day after the nuclear bomb had destroyed Hiroshima, marking the entry into the nuclear age with the incineration of human beings and the obliteration of a city, a letter was written to J. Robert Oppenheimer, the man most responsible for the bomb.

"Dear Oppie," wrote his friend and colleague, Haakon Chevalier, on August 7, 1945, "you are the most famous man in the world today....There is a weight in such a venture which few men in history have had to bear. I know that with your love of man, it is no light thing to have had a great part in a diabolical contrivance for destroying them."[1]

Members of the Catholic Church were aware that the explosion of the "diabolical contrivance," whose violent flash lit the morning sky over Hiroshima on August 6, 1945, coincided with the feast of the Transfiguration of Jesus. This feast marked by Christians around the world is described in the New Testament. Jesus took Peter, James, and John to a high mountain, "and he was transfigured before them. His face shone like the sun, and his garments became as white as light." Moses and Elijah appeared with him, and a voice came out of the cloud saying, "This is my beloved son" (Matt. 17:1–5). Luke relates that Jesus' clothing became "dazzling white," and Moses and Elijah talked with him, appearing "in glory" (Luke 4:30–31).

The Transfiguration of Jesus anticipates his resurrection and prefigures his coming in glory, his parousia, at the end of time.

The flash of the nuclear bomb was a man-made light, brighter than the sun in the heavens. It became a huge column of fire, incinerating human beings and buildings alike. After the flash, the dust arising from the destruction plunged the city into darkness.

"The strangest thing was the silence," said a survivor. "You'd think that people would be panic-stricken, running, and yelling. Not at Hiroshima. They moved in slow motion, like figures in a silent movie, shuffling through the dust and smoke.... Many simply dropped to the ground and died."[2]

The dark cloud hung over the city until the afternoon. It was then that another kind of rain fell on the city, a black rain of heavy drops hard enough to cause pain to the victims. Besides bringing immediate pain, the rain brought a long-term threat of radiation exposure.

When it comes to picturing what transpired under that cloud, one becomes filled with the terror of what human beings can inflict on other human beings in wartime. Victor Frankl's aphoristic conclusion brings one's thinking on Auschwitz and Hiroshima to a full stop:

> Since Auschwitz we know what man is capable of.
> And since Hiroshima we know what is at stake.[3]

Many accounts have been written by eyewitnesses and by writers of all types regarding these two "signs" of the twentieth century and of its carnage. Perhaps such accounts will penetrate the hearts and minds of those who contemplate and plan wars, to convey what any modern war may promise.

The mayor of Hiroshima, Takashi Hiraoka, recalled, "The bomb reduced Hiroshima to an inhuman state utterly beyond human ability to express or imagine."[4]

Under the atomic cloud of August 6, 1945, children were cremated at their school desks as their schools collapsed in flames around them. Men and women, many with eyeballs liquefied, or hanging loose from their eye sockets, groped for places of water, pools, cisterns, the Ota River. Everywhere, the living found their skin peeling away from their flesh, hanging loose like bloody streamers from their arms and legs. Fingers flamed into fire. Burnt mothers suckled burnt babies at their breasts.

Radiation sickness brought the total casualties of Hiroshima to 190,000 by 1995, from about 130,000 on August 6, 1945. Can anyone plumb the abyss of misery of those who carried disfiguring scars on their faces and bodies for the rest of their lives or of those who feared that they carried within their bodies the seeds of lingering death? This is part of the promise of nuclear war.

Hiroshima and Auschwitz

Hiroshima and Auschwitz became irrevocably entwined in my mind on a visit to Poland in 1979. With a delegation of Pax Christi, I attended a commemoration of the fortieth anniversary of the outbreak of World War II, when Germany attacked Poland. We drove through remote and often desolate countryside to reach Auschwitz (in Polish, Oswieczem). There we were brought to the still-standing wooden barracks from which human beings were led to the fate prepared for them in the crematoria. We were then brought to another execution place, Block 11.

Abutting this building was a firing range with the thick burlap sacking still in place for deadening the sounds of shots. This was the execution place for inmates who committed the "crime" of trying to escape, or who breached camp discipline in some way. Fluttering before the execution range were small objects that looked at first like bright flowers, pink, blue, yellow, and red. They were in fact leis of folded paper cranes. Japanese visitors had placed them there, in that way establishing a bond between the victims of Hiroshima and the victims of Auschwitz, both consumed by fire.

In Japan, the paper crane is a symbol of long life and good fortune. According to legend, folding a thousand paper cranes will lead to the achievement of both. A Hiroshima school girl, Sadako, suffering from the radiation sickness caused by the bomb, began folding paper cranes but died before reaching a thousand. Her story became widely known when her classmates completed the thousand paper cranes in her memory. A statue of Sadako has been placed in the Peace Park of Hiroshima. It is regularly hung with leis of paper cranes, memorializing all, particularly the young, whose lives were taken by the bombing of Hiroshima or by its aftermath.

In Block 11 there was also a link with Nagasaki. A priest follower of St. Francis, Maximilian Kolbe, had spent some years, and even founded a magazine, in Nagasaki. He returned to his monastery, Niepokalanow, in Poland in 1936. The monastery was a virtual town of some five hundred professed monks. Shortly after World War II began, the monastery was disbanded and evacuated by the Wehrmacht. Among the many monks arrested by the Nazi occupiers was Maximilian Kolbe. Incarcerated in Auschwitz, he was given the number 16,670. An inmate of Block 14 had escaped, and his block mates were lined up before the commanding officer. Ten were to be selected for collective punishment, execution. One unfortunate man who had been so chosen cried out in agony that he had a wife and children who needed him. Kolbe stepped forward and asked to take his place, explaining he was a priest and had no family to be concerned about.

In addition to a place of execution by shooting, Block 11 was the starvation block, where condemned prisoners were simply left to die of hunger. Kolbe was deposited in that block, but lingered so long that he was put to death by an injection of carbolic acid before his wasted body was consigned to the crematorium. The man whose life he saved survived to tell the story. Kolbe was declared blessed by the Catholic Church in a ceremony in Rome. The declaration signified that his life and example were of the nature to be emulated by Christians.

Can Noncombatants Be Put to Death?

Hiroshima was, in the hope of millions, the last example of mass death from the air in World War II. The actual number of dead was not too surprising, given the number who had perished in the fire-bombings of Hamburg, Dresden, and Tokyo. People's minds became inured, even numbed, by word of over fifty-five thousand killed in Hamburg in one night, or close to one hundred thousand in Dresden, or between eighty-five thousand and one hundred thousand asphyxiated or incinerated in Tokyo. These figures, however, were achieved by fleets of bombers: 700 aircraft over Hamburg; 769 Royal Air Force bombers and 311 American B-17 bombers over Dresden; and 325 B-29 bombers over Tokyo. Obliteration bombing became to the general public an accepted fact of modern warfare. Moral awareness was engulfed in the necessity of destroying the enemy.

Those under the bombs and those moralists dissenting from obliteration bombing were not widely heard. One particular voice was heard chiefly in theological circles, that of John C. Ford, SJ. He addressed himself in a lengthy article to "the morality of obliteration bombing." He asserted, "I do not believe that any Catholic theologian in the face of conciliar and papal pronouncements, and the universal consensus of moralists for such a long time, would have the hardihood to state that innocent noncombatants can be put to death without violating natural law. I believe that there is unanimity in Catholic teaching on this point, and that even in the circumstances of a modern war every Catholic would condemn as intrinsically immoral the direct killing of innocent noncombatants." Ford described obliteration bombing as not the strategic destruction of a factory or a bridge, but the destruction of industrial centers with their populations. He raised the question of the bombing of civilians as typical of total war, pointing out that "if all modern war must be total, then a condemnation of obliteration bombing would logically lead to a condemnation of all modern war."[5]

Patricia McNeal points out that many members of the Catholic community joined Ford in his condemnation of obliteration bombing.[6] They included the editors of *Commonweal* magazine and James M. Gillis, editor of the *Catholic World*. The *Catholic Worker* published the words of Paul Hanly Furfey, who condemned the tactic as immoral. A Catholic bishop, Bishop Gerald Shaugnessy of Seattle, declared indiscriminate bombing and the killing of civilians as against all morality.

What was new with the bombing of Hiroshima was that a lone bombing plane with a single bomb could virtually obliterate a city and those within it.

On August 7, 1945, a press release from the White House announced to the American people and to the world: "Sixteen hours ago, an American airplane dropped one bomb on Hiroshima, an important Japanese army base. It is an atomic bomb. It is harnessing the basic power of the universe. The force from which the sun draws its power has been loosed against those who brought war to the Far East."

The press release was prepared by General Leslie Groves, director of the Los Alamos scientific laboratory, where the bomb was produced. The general could not resist adding, "We have spent two billion dollars on the greatest gamble in history — we won." The two billion dollars had been secretly siphoned from the American treasury without the knowledge of the American people. In a secret city, constructed in a remote mountain in New Mexico, scientists had converged from two continents to confect a device referred to as "the gadget," or "the thing." It was to be a device with lethal power so great that it was not estimable in advance of its actual use.

The Day after Trinity

The justification for exploding the Hiroshima bomb was saving the American lives that might have been lost in an attack on the home island of Japan. It was known, through access to Japan's secret code, that Japanese leaders were considering how an approach to peace could be made through Russia. Japan had already lost the war in the Pacific, and the plight of the home island was desperate. Could these peace feelers have been followed up so as to avoid the use of atomic fire? One answer is that on July 16, 1945, a test was made of what the secret city, Los Alamos, New Mexico, had produced—a bomb. The test was conducted on a stretch of open desert two hundred miles south of Los Alamos. The area was known as Jornada del Muerte (Journey of Death or Day of Death). Robert Oppenheimer named the site Trinity from a sonnet by the metaphysical poet John Donne:

> Batter my heart, three-personed God; for, you
> As yet but knock, breathe, shine, and seek to mend;
> That I may rise and stand, o'erthrow me, and bend
> Your force, to break, blow, burn, and make me new.

Donne was writing at a profoundly personal level, but the words "break, blow, burn" have a wider connotation in relation to the testing of the new power.

Before the test blast, General Groves had secretly arranged with the governor of New Mexico to have the area around Jornada del Muerte evacuated in case of catastrophe. One of the scientists recounted: "When it went off, we all were looking through a glass. And then we saw what was just a tremendous overpowering vision of this thing happening, seeing a mountain small beside it. . . . It made the whole desert light up as if noon . . . and appear small. A large desert rimmed by mountains appeared to be a small place. . . . I was a different person from then on."[7] Another participant remembered the cloud that hovered over all, "brilliant purple with all the radioactive glowing. . . . And the thunder from the blast, it bounced on the rocks. . . . It just kept echoing back and forth in that Jornada del Muerte."[8] A New Mexico resident, wakened by the thundering noise and the shivering of her adobe house, called to her husband, "I want you to come look. The sun's rising in the wrong direction."[9]

The scientists had won; they had a weapon. The military and political leaders had only to set the date to test it over a human habitation in the prosecution of the war.

If the first use of atomic power in bomb form could be justified, what could be the justification for a second bomb? One compelling reason from a scientific point of view was that the first bomb was fueled by uranium, while the second bomb was fueled by plutonium. The plutonium weapon was tested on Nagasaki, just three days after Hiroshima. The destruction was less, and the casualties were fewer, thirty-five thousand. The Nagasaki bomb was detonated over a suburb rather than over the center of the city, as in Hiroshima. It happened to be a site where some fifteen thousand Catholic Japanese made their home. Two out of every three of the Catholic Japanese perished, along with other Japanese and non-Japanese working in the city's factories.

Urakami Cathedral, the largest Catholic church in Japan, fell victim to the blast. In a hallway of the UN headquarters in New York City is a permanent display of artifacts from Hiroshima and Nagasaki — metal coins and vases fused by heat into shapeless lumps, shredded garments and pieces of military uniforms, and at the end of the hallway, a statue. Over seven feet high, the statue is from the ruins of Urakami Cathedral. It represents St. Agnes, venerated from about 300, when she was martyred in Rome for her faith. She is holding a lamb, a traditional representation; her face and the lamb are unmarked, but the back of the statue is pock mocked with atomic fire. It is clear that the statue fell forward. Followers of Jesus would be reminded of the Lamb of God, who takes away the sin of the world (John 1:29).

Two years after Hiroshima and Nagasaki, in a talk at the Massachusetts Institute of Technology, Oppenheimer made the moving statement, "The physicists have known sin."[10] The knowledge of nuclear power for destruction (as the press release announced triumphantly, using "the force from which the sun draws its power") had been let loose on the world.

When Robert Oppenheimer was asked about his reaction to the proposal of Senator Robert Kennedy that President Johnson initiate talks with a view to halting the spread of nuclear weapons, his answer pointed to one of the tragedies of the twentieth century: "It's twenty years too late," Oppenheimer replied. "It should have been done the day after Trinity."[11]

Humans as Matter

Would it have been possible for the scientists, the military, and the political leaders to decide on the day after Trinity, when the expenditure of two billion dollars was "justified," that exploding the bomb over human habitation was too risky a venture for the future of humankind? The physicist Robert Wilson put forward the idea of a demonstration rather than a "first use" in the war. "I suggested to him [Oppenheimer] that it would be a good idea if it would be used in a demonstration. For example, like the demonstration of Trinity, where the Japanese might be invited to attend."[12]

Nothing came of his suggestion. When news of the extent of the devastation of Hiroshima reached Los Alamos, Wilson felt "sick to the point that I would vomit." At the end of the war, Wilson separated himself from the world of science. "I gave up my clearance and have not worked on nuclear energy or any of its aspects or development or upon bombs."[13] The physicist Freeman Dyson, a member of the Los Alamos team, believed that the use of the bomb was inevitable, "simply because all the bureaucratic apparatus existed by that time to do it. The Air Force was ready and waiting. There had been prepared big airfields on the Island Tinian in the Pacific from which you could operate. The whole machinery was ready. President Truman would have had to be a man of iron will in order to put a stop to it."[14] Dyson faulted the Los Alamos team, since "nobody had the courage or the foresight to say no."[15] Isadore I. Rabi, Nobel laureate physicist commenting on Hiroshima, said, "It would bring home to one how powerful this is. It treated humans as matter."[16]

"I have been asked," Robert Oppenheimer related, "whether in the years to come, it will be possible to kill forty million people in the twenty largest American towns in a single night. I am afraid that the answer to that question is yes."[17]

Any Invention Portending Death Utilized

Was there ever a time in history when a ruler or a government decided against the use of a newly invented weapon against an enemy?

After the followers of Jesus learned from spiritual leaders that it was licit for them to take part in war, did they examine the weapons or contrivances for killing? Were these too cruel? Did they kill others than combatants? One example has been mentioned. Church leaders at the Lateran Council in 1139 passed an edict to condemn the crossbow as too cruel and barbarous to be utilized against human beings. Released from a bolt, the arrow tore at the inner organs of the body. "As perhaps man's first overt attempt at arms control, this edict deserves more attention than it usually receives," Robert O'Connell remarks.[18] In general, however, Christians utilized whatever weapons the technology of the time made available.

When Ambrose and Augustine opened the door to Christians to take part in the wars of Rome, they did not give guidance as to which weapons might or might not be used by Christian soldiers. The use of spears, swords, javelins, or other means was left to the officers of the Roman army. Augustine made it clear, however, that no spear, sword, or javelin was to be used against a personal enemy. The response to a personal enemy was to be that of nonretaliation, gospel nonviolence.

Throughout history, as death-dealing instruments were invented, they were put to use in war action against so-called enemies. Books dealing with the history of weaponry and war-making give evidence that any invention — gunpowder, for example, and its uses in muskets, cannons, and pistols — portending death to the enemy will be developed and utilized. No concept of shared humanity would stand in the way of fashioning weapons that meant swifter death, even more brutal death, to "the others," those considered enemies.[19]

As weapons became more efficient, and less controllable, "the others" came to include those who had no part in the conflict, women, children, and the aged. In European history, the treasuries of war-making governments, like that of the Hapsburgs, or of Sweden, France, Spain, or England, gave an eminent, if not first, place to the budget for armaments.

Thus even some of the most gifted of the twentieth-century scientists, laureates of the finest universities of Europe and the United States, did not interfere with the immediate use of new weapons, even of using the "diabolical contrivance" that they had fashioned. A demonstration of its destructive power might well have swayed the Japanese leaders into furthering their nascent moves toward peace.

Shared Humanity?

It was known by the end of World War II that the German effort to build a nuclear bomb had failed. If the suggestion of a physicist to hold a demonstration of the atomic bomb had been followed, the history of the age might not be the fearful one we know. A united proposal to President Truman and to his advisers by the Los Alamos team for a repetition of Trinity, rather than the immediate utilization over a human habitation, might have influenced the president's decision. There certainly were strong arguments for such a proposal, including the practical concern of the unknown long-term effects of unloosing radiation into the earth's atmosphere. There was also the argument of a shared humanity with an opponent already near the end of resistance. No concept of a shared humanity moved the intellectual giants. War showed itself, as always, the implacable enemy of any sense of shared humanity.

Throughout the ages, the resources of peoples have been poured into the weapons and accouterments of war. Never before, however, in the history of the world, had so much wealth, so many resources of human gifts and skills, so much of the finest and most exact in technology, been gathered in one place for a single aim as in Los Alamos. Never before had six thousand persons been brought together to a secret mountaintop to concentrate on one purpose — an instrument of death.

We Won

For some it was jubilation. For General Groves of Los Alamos it was, "We won." "Mr. Truman was jubilant," wrote Dorothy Day in *The Catholic Worker.* "He went from table to table on the cruiser that was bringing him home from the Big Three Conference, telling the great news; 'jubilant,' newspapers said. 'Jubilate Deo.' Regarding the number of Japanese dead, the *Herald Tribune* stated that the figure was not known for certain. The effect is hoped for, not known." Dorothy Day continued, "It is hoped they are vaporized, our Japanese brothers, scattered, men, women, and babies, to the four winds, over the seven seas. Perhaps we will breathe their dust into our nostrils, feel them in the fog of New York, feel them in the rain on the hills of Easton [site of the Catholic Worker farm]."[20] Many shared the same moral outrage as Dorothy Day.

The bomb fueled by atomic power was indiscriminate in a new and ominous way. It reached into the future with its killing power, bringing death to men, women, and children year after year through the new threat of radiation sickness.

The jubilation in the United States over its sole mastery of atomic power did not last long. Though the first resolution of the UN General Assembly in 1946 called for an atomic commission to make proposals for the elimination of weapons of mass destruction and the peaceful uses of atomic energy, other forces were at work. Britain and France followed as possessors of nuclear weapons, with China joining the nuclear nations in 1964.

A crucial date was August 1949, when the Soviet Union exploded its first atomic bomb. This launched the Cold War in earnest. The Soviets started with an advantage, using the American plans for the bomb exploded at Trinity. Despite a series of disarmament programs within and outside the United Nations, the most tragic race in the history of humankind had begun, the arms race. World War II, like the wars of history, had spawned its successor, the Cold War.

A race is presumed to have a goal, a finish point where the winner is honored and the losers are ranged in order of their achievement. The arms race between the two superpowers was a spiraling race of death, with each side pouring its resources into ever-more powerful engines of destruction. Genocidal planning proceeded apace by both superpowers.

The Dread Irony

The dread irony was that the same death that threatened the people of the enemy also threatened the population of the perpetrator of an atomic explosion. The radioactive cloud after a nuclear explosion could not be contained and packaged in a plastic balloon so that its contents could be decontaminated. It was free to roam the skies over the entire planet. Nevertheless, the scientists obeyed their nations in fashioning ever-more powerful instruments of destruction. The next in man's threat to his kind was the hydrogen bomb. This was claimed to be a cleaner bomb since there was less fallout. A neutron bomb was then envisioned by which the vast damage would be relatively confined, compared to a "conventional" atomic bomb, while the area of total radiation exposure was widened.

The superpowers attempted to keep their weapons technology secret, although it was the ghoulish silent partner at every meeting of the United Nations and every verbal collision between the superpowers. However, there were times when the bombs, at least one of them, became visible as well, as the missiles that would deliver them to their targets.

The World's Most Sought-after H-Bomb

In the spring of 1966, in an accident over Spain, an H-bomb of the order of twenty megatons was lost in the Mediterranean Sea near Palomares. Three other H-bombs of the same magnitude were its companions in the same carrier plane. A megaton represents the explosive force of one million tons of dynamite. Jerome Weisner and Herbert York, scientific advisers to the American government, reported that a one megaton bomb was about fifty times more powerful than the Hiroshima bomb. Ten megatons would be of the same order of magnitude as the grand total of all the high explosives used in all wars up to that date.

"A coming-out portrait of the world's most sought-after H-bomb" was a magazine caption under a color photograph of a silvery, narrow, ten-foot-long cylinder. Glimmering in the sun it was surrounded by the smiling men who had recovered it from the depths of the sea. The U.S. navy as well as aircraft executed a massive convergence on Palomares. Someone remarked that if the bomb had exploded, the coast of Spain would never have been the same.

If dropped over Columbus Circle, New York City, the missing bomb would have been capable of killing some six million New Yorkers, according to Dr. Tom Stonier. Stonier's book *Nuclear Disaster* describes the effects of such a bomb on New York City. The American nuclear arsenal in the mid-1960s bristled with more than thirty-five hundred bombs of over ten megatons.[21] Humankind shuddered at the revelation of crematoria in Europe during World War II. The crematoria of the United States could be said to be in the skies and come in neat packages.

This was the first time Americans had seen the product of their taxes. Weisner and York remarked at that time, "Both sides in the arms race are thus confronted with the dilemma of steadily increasing military power and steadily decreasing national security. It is our considered professional judgment that this dilemma has no technical solution."[22] Where would the solution come from to resolve the problem that technology had posed? Only, it might be hoped, from mediating and reconciling actions arising from moral decisions. None of these entered the strategy of the superpowers, whose goal seemed ever-more power, ever-more threat to humankind.

An earlier occasion of the unveiling of nuclear power had, in fact, occurred in October 1962. It concerned not bombs but missiles. Then, for thirteen days of October, the Cuban Missile Crisis froze the hearts of millions. Americans began to experience what the Soviets in particular were experiencing, namely, the horror of being helpless targets of nuclear projectiles. The order by the Soviet leader to send ballistic missiles to Cuba was verified when American reconnaissance planes photographed the construction of missile launching sites on Cuban soil. From ninety miles offshore, any American city could be targeted. The United States and the world were paralyzed by fear while the American president threatened the Soviet Union with massive retaliation if any missiles were launched against the United States from Cuba. On Sunday, October 28, the Soviet leader ordered the Soviet ships facing the blockade to turn back with their lethal cargo.

In four days, the Soviet leader agreed to dismantle the Cuban missile sites, and Americans could breathe easier.

Dissuasion by Terror

What Palomares and the Cuban Missile Crisis showed humankind was that its planetary home had become fragile. Americans realized how close they had come to an apocalypse of destruction when they read a statement of the secretary of defense, Robert McNamara, describing the events before the climactic Sunday of the Cuban Missile Crisis. In the *New York Times* he recalled, "On a beautiful fall evening, as I left the President's office to return to the Pentagon, I thought I might never live to see another Saturday night."

Still cruising through the skies were planes carrying weapons of megaton magnitude that could bring instant megadeath. Even the seas had been invaded by death-dealing instruments in submarines that moved silently in the ocean depths. "Massive assured destruction" was still threatened by one superpower on the other. People became accustomed to living under deterrence. Nuclear deterrence simply meant that the way to avoid war was to threaten the opponent with "unacceptable damage," meaning total, or near total, destruction. "Dissuasion by terror" was the order of the day. Year after year, people tried to persuade themselves that living under the "nuclear umbrella" provided security.

For some followers of Jesus, the threat to commit the crime of indiscriminate warfare, involving the killing of noncombatants, was evil in itself. Killing the innocent has always been murder, and intending to commit murder is sinful itself. Dissuasion by terror, for many followers of Jesus, contravened his teaching of love.

In the nuclear deterrent, followers of Jesus were confronted with an evil as insidious as the laws which persecuted their ancestors in the faith under the Roman Empire, as insidious but more monstrously lethal. According to the book *Preparing for Armageddon,* "the practice of nuclear deterrence sets up the state as a god since it demands that we be prepared to commit any degree of violence in its service. This is idolatry of the most extreme kind."[23] But like their ancestors in the faith, twentieth-century Christians had the suffering love of the Savior to sustain them in their dissent from the powers that ruled.

Pope Pius XII in a discourse to the International Office of Documentation for Military Medicine stated: "It is not enough therefore to have to defend oneself against simply any injustice in order to justify resorting to the violent means of war. When the damages caused by war are not comparable to those of 'tolerated injustice,' one may have the duty to 'suffer the injustice.' "[24] This statement could be construed as applying to nations as well as individuals, but what nation is willing to suffer injustice? It would also reinforce the principle that the right to defensive war is not a general right applicable to any and all wars of defense and to any and all degrees of the use of violence.

Absolute Moral Imperatives at Stake

Francis Stratmann, OP, a consistent exponent of gospel nonviolence, pointed out that even if an opponent armed with nuclear weapons started a war, a response in kind would not be morally licit, even in defense. He uses the simple argument of the end not justifying the means. As a theologian, his strictures apply to Christians, whether the nation-state observes them or not. It cannot be maintained, he points out, that the good achieved by the use of atomic weapons would outweigh the damage perpetrated by them. Stratmann stated:

> We know that we are faced with an unsatisfactory solution by renouncing a defense with atomic weapons. But there are situations also where we find ourselves at our wit's end, faced with the problem that is insolvable for the time being, with an impenetrable mystery of divine providence. But it is better to accept the darkness, to surrender ourselves to the all-holy justice

and mercy of God, than to take part in mass murder only because the other side commits it.[25]

In addressing "the hydrogen bombing of cities," the theologian John C. Ford, SJ, already cited on obliteration bombing, confronted the dilemma of Christians seemingly entrapped by the war ethos of their time. It was generally accepted that any method, even that of obliteration or vaporization, could be used against the forces of the "beast" of Nazism, or against the "evil empire" of communism. Ford stated that "if I assert that it's wrong to kill a million schoolchildren, I do not have to prove my assertion.... Catholic teaching has been unanimous for long centuries in declaring that it is never permitted to kill directly noncombatants in wartime."[26]

Ford concluded, "I think that problems are entirely oversimplified when reduced to the stark dilemma: either wipe them out or be wiped out yourselves. ... But if that were the dilemma, I would consider that we had arrived at the point where absolute moral imperatives were at stake, and the followers of Christ should abandon themselves to Divine Providence rather than forsake these imperatives."[27]

All Morality Unilateral

In ordinary living, the followers of Christ accept the principle that morality is unilateral; we cannot say to a wrongdoer, I will cease my part in wrongdoing if you will. One of the first moral principles to fall in war, along with the truth (called the first casualty), is that all morality is unilateral. The follower of Christ is held to this immutable imperative no matter what method of destruction the opponent may devise. The surrender to the "all-holy justice and mercy of God" and "divine providence" brings the followers of Christ back to their beginnings in the first three centuries under the persecution and death-dealing of an all-powerful empire.

Weapons So Fateful and Dishonorable

Large numbers of citizens whose taxes made possible the accumulation of weaponry of mass destruction accepted during the years of the Cold War that there was no alternative but to threaten terror for terror. Other citizens whose lives were guided by the imperatives of a love that was to be extended even to the enemy became conscious that they were living under occupation. They were citizens of occupied countries whose leaders were as far removed from them in the conviction of the sacredness of human life as were the leaders of imperial Rome from that of the first Christians.

The great divide — even abyss — which had grown up over the centuries since churchmen had accepted just war teaching allowed political communities to flout the moral law and impose on Christian citizens laws and actions absolutely contradictory to the message they had received from Christ. One of the revolutionary statements by Pope John XXIII confronted the abyss and overleaped it in the statement, "The same moral law which governs relations between individual human beings serves also to regulate the relations of political communities."[28] This statement, applying to the ongoing relationship of political communities in times of tranquillity, applies also in periods of war. The believer who condemns the no-holds-barred freedom from moral restraints that accompanies war, and refuses to comply, may have to pay a high price. Our century has given countless examples in every one of its conflicts of individuals being commanded by political communities to disregard and trample on the moral law in order to qualify as loyal citizens. Discipleship in time of war may result in imprisonment or the forfeiting of life itself.

At the height of the Cold War, when preparations for genocide were justified by both superpowers, the voice of morality and reason came from the Holy See and from the bishops of the Catholic Church. In the "letter to all humanity," *Peace on Earth,* Pope John XXIII asserted in April 1963 that nuclear weapons should be banned. In a discourse on August 8, 1965, Pope Paul VI referred to the twentieth anniversary of the destruction of Hiroshima and Nagasaki and reiterated John XXIII's assertion that atomic weapons should be banned. Paul VI stated:

> We pray that never again may the world see as disgraceful a day as that of Hiroshima.
>
> We pray that all may together and loyally outlaw the terrible art which knows how to produce, multiply and preserve [nuclear weapons] for the terror of peoples.
>
> We pray that men may never again place their trust, their calculations and their prestige in weapons so fateful and so dishonorable.
>
> We pray that the deadly device may not have killed peace, although it was to have sought it; may it not always impair the honor of science and may it not have extinguished serenity of life upon the earth.[29]

When the synod of bishops met in general assembly in November 1971, they stated that "the arms race is a threat to man's highest good, which is life.... It creates a continuous danger for conflagration, and in the case of nuclear arms, it threatens to destroy all life from the face of the earth."[30]

"The final goal of completely eliminating the atomic arsenal" was expressed by Pope Paul VI in his message to the UN General Assembly for its 1978 session on disarmament. "Nuclear weapons," he pointed out, "are the most fearsome menace with which mankind is burdened."[31]

A Moral About-Face

Pope John Paul II on his visit to Hiroshima on February 25, 1981, asserted:

> Our future on this planet, exposed as it is to nuclear annihilation, depends upon one single factor: humanity must make a moral about-face.
>
> The task is enormous: some call it utopian. But how can we fail to sustain the trust of modern man, against all temptations to fatalism, to

paralyzing and to moral dejection? We must say to the people of today: do not doubt, your future is in your own hands.

The moment is approaching when priorities will have to be redefined. For example, it has been estimated that about half the world's research workers are at present employed for military purposes. Can the human family morally go on much longer in this direction?

In a word, I believe that our generation is faced by a great moral challenge, one which consists in harmonizing the values of science with the values of conscience.[32]

Stewardship over Nuclear Weapons

Since Vatican II, Catholic bishops and lay leaders have been reexamining the question of warfare and, in particular, nuclear warfare. As the bishops of the United States, members of the Catholic community whose nation possesses the world's greatest nuclear capability, they have acquainted themselves with how this capability has been maintained. Bishop Walter Sullivan, president of Pax Christi–USA, along with Bishop Thomas Gumbleton, the organization's former president, paid a visit on October 7, 1997, to the Lawrence Livermore National Laboratories in Oakland, California. It is at Livermore that American taxpayers have been assessed the amount of forty billion dollars over ten years for the Stockpile Stewardship Program.

Led by David Robinson, editor of the *Catholic Peace Voice* and watchman for Pax Christi–USA on matters of nuclear weaponry, the two bishops, with twenty-four Pax Christi and other peace activists, had a five-hour meeting with the officials of the national laboratories.

In a Pax Christi press release of October 8, 1997, Bishop Sullivan stated:

For me, the visit was an eye-opener, a first hand look at nuclear technology. In *The Challenge of Peace,* our bishops gave only a conditional acceptance of the possession of nuclear weapons as long as meaningful disarmament takes place. Treaties have been signed, such as the non-proliferation treaty and the comprehensive Test Ban Treaty forbidding underground nuclear explosions and explosions in the atmosphere. What they are planning to do at Livermore National Laboratories, through the Nuclear Ignition Facility, is to conduct nuclear tests without leaving the Laboratories. This could skirt the spirit if not the letter of test ban treaties.[33]

Bishop Gumbleton, after the meeting at Livermore, which employs eighty-five hundred people, including scientists, physicists, biologists, engineers, and environmentalists, came to a less-than-hopeful conclusion: "The Livermore Laboratories will in fact enhance US efforts to continue development and testing of nuclear weapons. The Stockpile Stewardship allows us to explore nuclear technology using new methods, retooling and redesigning weaponry that would be useful in a conflict situation."[34]

The lesson for Bishop Sullivan was clear:

> The visit to the Livermore Laboratories made me more committed than ever to the Catholic Peace Movement that stands as a beacon light to raise our consciousness to the reality that we are living on a time-bomb ready to explode one way or another. We must find a way to bring peace and nonviolence into our societies and daily lives. Possessing nuclear weapons will not make us any safer or more secure. In fact, I view these weapons as a curse. For believers, the most urgent of tasks is to build up the peace of Christ.[35]

Undoing the Works of War

One aspect of the undoing of the works of war consists of ridding the planet of the dread legacy of old wars. The time arrived in the profound designs of God where the dismantling of the nuclear deterrent began to be accomplished. Nuclear wastes, the leftover of fifty years of Cold War weapons production, had to be dealt with. The degradation that such production had visited upon the earth had to be faced.

The scientists whose knowledge had made possible the fashioning of the weaponry now had to find ways to store and deactivate its waste products. Plutonium waste, for example, would be encased in molten glass, placed in steel cylinders, and buried. The study *Avoiding Nuclear Anarchy* related that one nation, the Ukraine, decided for denuclearization. For a brief period after the breakup of the Soviet Union, the Ukraine became the third nuclear power on the face of the earth through its inheritance of the Soviet nuclear arsenal. The Ukrainian parliament voted to eliminate the arsenal through the dismantling of intercontinental ballistic missiles comprising over eighteen hundred nuclear warheads. From Kazakhstan, another of the nations which comprised the Soviet Union, a stock of highly enriched uranium was flown to the United States to be buried for safekeeping in Oak Ridge, Tennessee. "The Russian Nuclear Archipelago," a section of *Avoiding Nuclear Anarchy,* makes disquieting reading with its descriptions of nuclear weapons stored in improvised facilities under poor security.[36]

The longtime wounds inflicted on the planet can only be healed by scientific planning, constant attention, and vast expenditures of funds. These expenditures should not be other than generous for the sake of humanity.

The legacy of past wars lurks in unexpected places, including the hold of a nuclear-armed submarine of the former Soviet Union. It remains where it was sunk, in the North Sea three hundred miles off the coast of Norway. An expedition was charged with sealing the vessel to prevent the leakage of plutonium and radioactive material into the North Atlantic.

The New Abolitionist Movement

Those who formed a significant part of the peace and justice movement in the United States found a fresh way of addressing the nuclear position of their country and of the world. Mary Evelyn Jegen, SND, former national coordinator of Pax Christi–USA and vice-chairperson of Pax Christi International, recalls that she met at intervals during 1982 with leaders of the Fellowship of Reconciliation, Sojourners Fellowship, World Peacemakers, and a New Call to Peacemakers. At their gatherings in Washington, D.C., they prayed together, asking for light in the task of awakening their members to the reality of "the deterrent." It was the time when the president of the United States in his travels overseas was accompanied by a military escort who remained at his side. The key member of his retinue carried an attaché case containing a mechanism which would allow the chief executive to activate at any time the arsenal of the United States, including the nuclear arsenal.

One of the group in Washington suggested that they model their antinuclear movement after the movement which fought for the abolition of slavery in the United States. The "new abolitionist movement" was born from the meetings of the five organizations.[37]

Jim Wallis, founder and pastor of the Sojourners Fellowship in Washington, D.C., and an author of note, wrote in 1983, "The nuclear danger is becoming an occasion of fresh conversion for growing numbers of Christians as opposition to nuclear weapons is seen as a matter of obedience to Jesus Christ. A new abolitionist movement is emerging based on prayer, preaching, commitment and sacrifice."[38]

In the view of the covenanters, the denial of the realty of the deterrent and its threat to humankind was the outward sign of a deep spiritual crisis. Jegen explained it: "The purpose of the covenant is to place before the churches the abolition of nuclear weapons as an urgent matter of faith. The nuclear threat is a theological issue, a confessional matter, a spiritual question, and it is so important it must be brought into the heart of the church's life."[39]

In *Christian Voices from the New Abolitionist Movement,* Wallis relates how twenty-five American followers of Jesus worked, hoping against hope, for the eventual abolition of nuclear weapons of mass destruction.[40] Bill Kellerman, pastor to a Methodist congregation in Detroit, found that he was facing a choice between being a pastor or a resister. "Given the gospel," he said, "given the bomb, given the calling of my heart, those two things [pastor and resister] seem to be inextricably one." For his public act of resistance inside the Pentagon, praying and refusing to move, he and his co-resisters were given a month-long sentence in Richmond County jail. "I hear it said, directly and indirectly," said Kellerman, "that our freedom to celebrate the Lord's Supper is won and secured, granted and guaranteed, by our nuclear arsenal."[41] His resistance aimed at distancing the freedom offered by Christ from that offered by political power.

A nuclear engineer, Robert Aldridge, after helping to design three generations of Polaris missiles, resigned from the Lockheed Company and surrendered a large salary. Breaking the tie with affluence was a struggle for Aldridge and his family as he took up work with handicapped children. "The overall effort of trying to live a nonviolent life," he said, "wife, husband and children together — is difficult because affluence has become so deeply ingrained."[42]

Only a unique and completely focused person would build his entire life around peacemaking. James Douglass, mentioned earlier, is such a person. He decided to live close to the naval submarine base in Bangor, near Seattle, Washington. The resistance community founded there was called Ground Zero, and its purpose was to maintain a continuous, nonviolent presence in Bangor, the home port of the Trident submarine, a vessel capable of firing over four hundred independently targeted missiles. This was called "the ultimate first-strike weapon." Near Bangor was the nuclear weapons storage area of the Pacific region. At Ground Zero, Douglass and others provided nonviolent training to those who joined the community. There were frequent public prayers and demonstrations at the locations in Bangor connected with the nuclear deterrent. The community aimed to make the arsenal visible to the American people.[43]

Through his life and acts of witness, and his books, Douglass has influenced many persons, including Catholic bishops. Douglass visited Raymond Hunthausen soon after he was named bishop of Seattle. The bishop found he agreed with Douglass's ideas on peacemaking and the nuclear arsenal. "I realized," said the bishop, "that what he was saying coincided with how I felt; but I wasn't doing anything about it." Many people in the Seattle diocese raised with their bishop the issue of nuclear weapons. Hunthausen asked them to consider refusing to pay about half of their income tax in protest of the amount of taxes spent on nuclear arms. Some of the people responded favorably to the challenge, but others expressed their great fear that they would lose all they had worked hard to acquire. Hunthausen emphasized that "it is the responsibility of each of us to look at the gospel, examine its implications, and take a personal stance." Finding persons who wanted to change their lives by separating themselves from work on the nuclear arsenal, Hunthausen made a response: "I have a fund started to help such people, . . . a small band-aid business."[44] If significant numbers of people needed aid, he realized he would have to appeal to the larger community. What the bishop was doing was an act of witness.

The influence of the new abolitionist movement permeated large segments of the American religious community through the leadership of the persons mentioned here and through the ever-widening network of those who spoke, wrote, lived, and witnessed to the absolute incompatibility between faith in "the deterrent" and faith in the Lamb of God. A disarmament policy allowing for the retention of thousands of nuclear warheads in the American arsenal was no longer a credible goal; the complete abolition of nuclear weapons was a far more credible and moral goal — something worth meditating on, praying and sacrificing for.

Manhattan Project I and Manhattan Project II

Alongside worldwide movements such as War Resisters International and the programs of peace associations like Pax Christi, which pressed for abolition of nuclear weaponry, many important antinuclear initiatives came from a group of activists and spokespersons whose proposals were gathered in *Critical Mass: Voices for a Nuclear-Free Future.* In the preface, "Disarmament, Peace and Compassion," the Dalai Lama urges that universal altruism be nurtured toward the achievement of global demilitarization. Antiwar activists including Daniel Ellsberg provide specific recommendations and plans to put the proposals into action. Ellsberg's idealism became known to the American public when he placed himself in harm's way by leaking the Pentagon Papers. By this act of conscience, for which he could have been convicted of treason, Ellsberg changed the minds of many Americans regarding unqualified support of the Vietnam War. With regard to the nuclear age, Ellsberg points out that the nuclear powers themselves, with their huge arsenals, underline that nuclear weapons are legitimate instruments of political power. He is convinced that while one phase has ended, another, perhaps even more dangerous, phase has begun. Manhattan Project I, he states, produced Hiroshima, Nagasaki, and the thousands of weapons of the Cold War. Ellsberg is of one mind with the leaders of Abolition 2000 when he asserts that the abolition of nuclear weapons is inextricably linked to nuclear disarmament. Many join Ellsberg in his proposal: "The fiftieth anniversary of the Manhattan Project and its lethal culmination is the right time to launch Manhattan Project II, with the practical aim of abolishing 95 to 99 percent of current nuclear stockpiles within a decade."[45]

Abolitionist Upsurge

An "abolitionist upsurge," expressing the groundswell of people the world over for the complete abolition of nuclear weapons, made itself known during the meetings of the UN Nuclear Non-proliferation Treaty Preparatory Committee.

While UN member states reviewed the Nuclear Non-proliferation Treaty within the walls of UN headquarters in April 1997, close to a hundred representatives of voluntary agencies, many of them enjoying the consultative status as NGOs (nongovernmental organizations), gathered nearby. They included peace activists from many nations, affiliated with various religions or with none, pacifists or not strictly pacifists — but all were united in one goal under the flag of Abolition 2000. They ranged from War Resisters International, a humanist organization following Gandhian principles; to the Fellowship of Reconciliation, an organization of faith groups comprising (among others) Baptist, Orthodox, and Muslim peace fellowships; to the World Council of Churches.

Pax Christi International was one of the founders of the overarching agreement whose purpose was to lead the struggle against the maintenance of the nuclear umbrella. Too many of the world's peoples still believed even after the Cold War's end that nuclear deterrence was essential to a nation's security.

An earlier gathering of the voluntary groups had been held in Moorea, Polynesia, in January 1997. It gave the Pacific peoples the chance to vent their frustration and helplessness in confronting what France, the United States, and the United Kingdom had inflicted on them in their nuclear testing. "The anger and tears of the colonized people arose from the fact that there was no consent, no involvement in the decision when their lands, air, waters were taken for the nuclear build-up, from the very start of the nuclear era."[46] When the peace activists from nuclear nations shared such cries of pain with UN delegates, the activists also expressed their own pain at the nuclear testing, the practicing for

Armageddon, of their own countries in the deserts, atolls, seas, and bowels of the earth.

At the next meeting of the UN Nuclear Non-proliferation Treaty Preparatory Committee held in the Palais des Nations, Geneva, from April 27 to May 8, 1998, representatives of voluntary groups and NGOs were again on hand. Supporting them were close to a thousand voluntary groups and organizations in a coalition loosely coordinated by the Nuclear Age Foundation of Santa Barbara, California. Their function was to keep reminding UN member delegates that while the aims of the Nuclear Non-proliferation Treaty were sound, the only aim that would eventually safeguard the planet would be the banishing of nuclear weaponry altogether. It was true that 149 nations had signed the Comprehensive Test Ban Treaty, but that treaty allowed certain member states, including the members of the Security Council (the United States, the United Kingdom, France, Russia, and China), to retain their nuclear superiority, while denying nuclear status to other member nations.

Before the opening of the preparatory committee, two of the NGOs representing peoples from the ends of the earth presented a statement addressed to the Nuclear Non-proliferation Treaty Preparatory Committee: "Act Now for Nuclear Abolition." The presenters were Dr. Konrad Raiser, general secretary, World Council of Churches, and Cardinal Godfried Danneels, president, Pax Christi International. The statement asserted:

> The time has come to rid planet earth of nuclear weapons, all of them, everywhere. The Nuclear Non-proliferation Treaty (NPT) Preparatory Committee has a remarkable opportunity at its upcoming meeting to set the course resolutely for the achievement of this goal.
>
> Nuclear weapons, whether used or threatened, are grossly evil and morally wrong. As an instrument of mass destruction, nuclear weapons slaughter the innocent and ravage the environment.... When used as an instrument of deterrence, nuclear weapons hold innocent people hostage for political and military purposes.

Calling the doctrine of deterrence "morally corrupt," the statement pointed out that "just ends cannot be achieved through wrongful means." Among the recommendations made to the preparatory committee were: "Cease all research, development; cease production and deployment of new nuclear weapons. Refrain from modernizing the existing nuclear arsenal and increasing the number of deployed nuclear weapons."[47]

The statement cited the words of Archbishop Renato Martino speaking for the Holy See at the First Committee of the UN General Assembly on October 15, 1997:

> Nuclear weapons are incompatible with the peace we seek for the 21st Century. They cannot be justified. They deserve condemnation. The preservation of the Nuclear Non-Proliferation Treaty demands an unequivocal commitment to their abolition.... This is a moral challenge, a legal challenge and a political challenge. The multiple-based challenge must be met by the application of our humanity.[48]

The members of the Nuclear Non-proliferation Treaty Preparatory Committee completed their deliberations on May 8, 1998. Their agenda included neither a call to unite for Abolition 2000 nor a call to launch Manhattan II toward the abolition of nuclear weaponry. The committee was constrained by the 1966 agreement limiting possession of nuclear capability to five member states, all members of the Security Council.

Nuclear Power for Self-Confidence and Self-Respect

Three days after the end of deliberations by the preparatory committee, the land of Gandhi surprised the world by exploding nuclear devices under the desert of Rajasthan. They were detonated on May 11, 1998, five devices in all. On May 13, a sixth was detonated. The Indian government, led by a coalition whose base was the Hindu Nationalist Party, had won the approval of the other political parties, including the Congress Party, for tests. It was the Hindu Nationalist Party that had nurtured the man who assassinated Gandhi in 1948.

The approval of the Indian people was expressed on the streets of Delhi and other cities with marches, the setting off of flares, singing, and shouts of "Victory to Mother India." "All that India has done," said India's prime minister, "is conduct five nuclear tests." He contrasted that number with the hundreds of tests conducted by the nuclear powers. He claimed that India was forced to conduct the tests because of threats to its security. These threats, he explained, came from Pakistan and from China, a nuclear power possessing hundreds of nuclear weapons.

Always smoldering in the relationship between India and Pakistan was the dispute over Kashmir, a province whose people were overwhelmingly Muslim, but which had fallen under the dominion of India through the action of a Hindu ruler at the partition of the subcontinent.

"India has the sanction of her own past glory and future vision to remain strong," the prime minister stated. "The greatest meaning of the tests is that they have given Indian *shakhi* [power]; they have given India self-confidence." When the prime minister left his residence in Delhi, he was pelted with rose petals by a jubilant people.

Shortly after the Indian nuclear detonation, the prime minister of Pakistan made the announcement that Pakistan had also detonated five nuclear devices. The explosions took place on May 28 in a desert area in Baluchistan, near the border of Iran. "Today, we have settled the score with India," he asserted. In a lengthy statement in English he declared that "as a self-respecting nation, we had no choice left to us but to restore the strategic balance. Our decision to exercise the nuclear option has been made in the interest of national self-defense. These weapons are to deter aggression, be it nuclear or conventional." On May 30, another nuclear device was tested by Pakistan.

In Karachi and other cities, there was dancing in the streets and guns were fired in the air. Mosques were filled. The *New York Times* featured a photograph of a large square in Islamabad where a sea of white-clad men were at prayer. They were expressing their gratitude, their hands held open before them.

Whatever the outcome of the tests in a nuclearized subcontinent, the repetition of the "fearsome symmetry" between India and Pakistan, recalling the "fearsome symmetry" between East and West in the Cold War, did not bode well for the peace of the world. Perhaps a spiraling arms race would eventuate; perhaps there would be an actual nuclear exchange in which the border between the two nations would allow the irradiation of both peoples without discrimination.

What was tragic for the world was the realization that the possession of a "diabolical contrivance" was needed for the citizens of a nation to achieve self-confidence and self-respect. How many other nations of the earth would in the future give paramount importance to this aim, to the detriment of the poverty-stricken and powerless who, unaware of the implications of nuclear possession, might dance and chant in the streets? When could the aim of delegitimizing nuclear arms as a currency of power be achieved?

The Institutionalization of Nuclear Deterrence

The American community, in particular the peace-minded members of the Catholic community, has looked to its spiritual leaders for guidance in the matter of the possession of the nuclear deterrent. In October 1998, ninety-four members of the Catholic hierarchy addressed themselves to an evaluation of the morality of nuclear deterrence. They were all members of Pax Christi–USA and had gained insight from their two brother Pax Christi bishops who had visited the Livermore National Laboratories.

They pointed out that the Cold War weapons amassed over the Cold War years have survived the struggle. There is now a search for new justifications for their continued possession and for the program of updating and storing them.

The rationale for maintaining the nuclear stockpile was that it was not an end in itself but a step on the road "toward progressive disarmament." "Instead of progressive nuclear disarmament, we are witnessing the institutionalization of nuclear deterrence," the bishops concluded. They asked whether the huge investment called for by the Stockpile Steward Program does not represent a renewed commitment to nuclear deterrence. Nations without nuclear weapons might well see the institutionalization of nuclear deterrence as a threat to their societies.[49]

Concluding that our nation has no goal of eliminating nuclear weapons, and that deterrence is no longer an interim policy but a long-term investment, the bishops announced that morality demands an unequivocal commitment to the abolition of the weapons.

They pointed out that they spoke out of love for those suffering illnesses in areas where nuclear devices are tested; they spoke also of love for those whose poverty is increased because of the huge resources expended on the ongoing maintenance of nuclear weapons; and they spoke out of love for those involved in the continued possession of the nuclear deterrent, whether as persons carrying out ongoing programs or as the many victims of it. "We speak out of love for both victims and executioners," the bishops explained.

"On the eve of the third millennium," the bishops stated, "may our world rid itself of these terrible weapons of mass destruction and the constant threat they pose. We cannot delay any longer. Nuclear deterrence as a national policy must be condemned as morally abhorrent because it is the excuse and justification for the continued possession and further development of these horrendous weapons." The bishops asked for large numbers of the American community to move toward the elimination of nuclear weapons with all speed rather than to continue to rely on them indefinitely.

Whether or not the moral condemnation voiced by a group of bishops will be heeded by those responsible for the nuclear deterrent, the statement was supremely worth making.

From the pragmatic side, the bishops' statement was buttressed by a powerful statement on nuclear weapons by sixty-one generals and admirals from around the world. The generals and admirals, from sixteen countries, joined in asserting:

> We, military professionals who have devoted our lives to the national security of our countries and our peoples, are convinced that the continuing existence of nuclear weapons in the armories of nuclear powers . . . constitutes a peril to global peace and security and to the safety and survival of the people we are dedicated to protect.

In a long statement, issued in January 1998, the generals and admirals pointed to the threat of the acquisition of nuclear weapons by other nations — as was already happening in the case of India and Pakistan.

Among the signers were generals and admirals from all the nuclear nations except China. Most numerous were the generals and admirals from the Soviet Union and the United States. General John R. Gavin, supreme allied commander in Europe (1987–92), was joined by Rear Admiral Eugene J. Carroll (U.S. navy, retired), who had been the deputy director for defense information, and seventeen others as signatories from the United States. The seventeen signatories from Russia included Major General Youri V. Lebedev (the retired former deputy chief in the department of general staff of the armed forces of the USSR) and Lieutenant General Vladimir Medvedev (retired chief of the Center of Nuclear Threat Reduction).

The Gift of Time

When the representatives of the five faith-based groups who founded the new abolitionist movement met at a retreat center in November 1998, they found that the concept of the abolition of nuclear weaponry was gaining acceptability. Moral leaders were calling nuclear deterrence evil, and military leaders were calling nuclear weaponry not only useless but "a peril to global peace and security." Yet the nuclear establishment seemed immovable in its ownership of thousands of nuclear warheads and likely to prevail in its view of deterrence as somehow an aid to peace. At the conclusion of the meeting, Mary Evelyn Jegen, SND, a leader of the New Abolitionist Covenant, expressed the mind of the group. She stated: "As the foundation of national security, nuclear weapons are idolatrous.

As a method of defense, they are suicidal. . . . The development, testing, and reliance on weapons of mass destruction in the name of national security is an evil we do not accept. At stake is whether we trust in God or the bomb."

The covenanters pledged to continue their campaign of prayer and public witness. To their meeting came Jonathan Schell, whose book *The Gift of Time* was reaching an immense readership.[50] It featured interviews with notable political figures like Robert McNamara and well-known generals like George Lee Butler, who served as the last commander of the Strategic Air Command. He would have been the man to whom the U.S. president would have issued the order to launch America's nuclear arsenal if the country suffered a nuclear attack, or feared one. Schell found rationality for deterrence in high places. He had hopes that new methods of "vertical" and "horizontal" nuclear arms reduction might help people realize that eventual elimination of nuclear weaponry was feasible. Schell was convinced that this period presented an opening toward the elimination of nuclear weaponry.

General Butler told Schell: "I have arrived at the conclusion that it is simply wrong, morally speaking, for any mortal to be invested with the authority to call into question the survival of the planet. . . . Now as the world evolves rapidly, I think the vast majority of people on the face of the earth will endorse the proposition that such weapons have no place among us. There is no security to be found in nuclear weapons. It's a fool's game." To turn away from the nuclear weaponry that threatens creation would be one of the most important events in the history of humankind.

While covenanters pray that in the profound designs of God, the turning may come soon, men like Jonathan Schell, speaking at gathering after gathering, energizing community after community, see the necessity for utilizing the "gift of time" to save humanity from itself.

The Black Obelisk

The location of the first test of the atomic bomb in the desert is marked by a black stone obelisk. It stands on the site of the hundred-foot tower which housed the first atomic bomb test, the site named Trinity by Robert Oppenheimer. The immediate area is surrounded by a metal fence and is guarded by military police. Twice a year, in April and October, the gates are thrown open for a short period to allow entry to visitors.

Peace groups learned that on July 16, 1995, the fiftieth anniversary of the first test, the gates would be opened to allow visitors to memorialize on site the exact moment that the atomic bomb blast had occurred.

At three o'clock in the morning on July 16, large numbers of people were at the gates ready to stream in and surround the obelisk. Among them were people from surrounding communities, including Native Americans, Americans from farther afield, Europeans, and a survivor of Hiroshima, who carried a small lantern from the atom-bombed city.

A film was produced of the event showing Emmanuel Charles McCarthy, a native of Boston and Melkite priest, celebrating Mass on the site. In a homily to American and European groups who had accompanied him, McCarthy asserted:

> There is an indispensable requirement placed on those who believe, to unashamedly and unhesitatingly proclaim the Good News of the Nonviolent God of love, the Nonviolent Trinity of love: Somehow, in 1945, the name of the God of nonviolent and everlasting love is given as the code name for the testing of an instrument designed to produce unbounded carnage — the first atomic bomb. How is it possible that the Trinity of nonviolent love is the name given to a weapon's test whose purpose is to assure victory by mass human slaughter?

Repeating the words "Nonviolent Trinity," McCarthy intoned to the desert air, "Glory be to the Father, the Son, and the Holy Spirit forever."[51]

A Way of the Cross followed, with participants carrying wooden crosses they had brought to Trinity. In prayers they marked the various stations, the bearing of the cross by Jesus, the stripping of his garments, the scourging at the pillar, and the driving of the nails that impaled him on the cross.

It was not only the crucified Savior who was in the hearts of the participants but a humanity crucified by the "diabolical contrivance" born at this spot that could threaten its existence.

Part Two

THE WAY OF NONVIOLENCE

Chapter 8

The Works of Mercy and the Works of War

> *I was hungry and you gave me no food, I was thirsty and you gave me no drink, a stranger and you gave me no welcome, naked and you gave me no clothing, ill and in prison and you did not care for me. Then they will answer and say, "Lord, when did we see you hungry or thirsty or a stranger or naked or ill and in prison and not minister to your needs?"*
> —MATT. 25:42–44

The last act of healing performed by Jesus was to undo the work of Peter's sword, restoring the ear of the high priest's servant which it had severed. Could there be a more dramatic disavowal of defense by the sword than the command to Peter to put away his sword, even though it was raised to protect one about to be unjustly executed (Matt. 26:51–52)? In contradiction to the work of violence, Jesus substituted a work of mercy, the restoring of the severed ear. This act stands as a precedent for his followers who, while rarely empowered by God to perform miracles, can by human means contribute to the healing of those maimed and disfigured by violence. After ordering Peter to sheath his sword, Jesus enunciated the fateful warning that "all who take the sword will perish by the sword" (Matt. 26:52).[1] Many followers of Jesus accept his example and refuse to take up the sword of violence for any purpose. It has remained for the nuclear age to reveal the dread reality of Jesus' warning.

The Christian has no choice but to relate the actions of daily life, even those concerned with war, to the message brought by the Savior. This chapter focuses on the relationship of war, in particular the wars of the twentieth century, to Christ's words in the parable of the Last Judgment.

War: Interrupting and Reversing the Works of Mercy

The Interruption of Mercy

In earlier wars that have marked European history, the works of mercy were interrupted for the duration of the conflict. Since battles were fought by knights and, in the main, by other trained soldiery, there was some built-in immunity for noncombatants. Battles could rage in a limited area while farmers, tanners, armorers, weavers, and shepherds went about their daily tasks.

Even brutal siege warfare, in which there was no distinction between combatant and noncombatant during the siege, did not end in the massacre of noncombatants once the siege was lifted. There were in Europe no massacres to match that inflicted on the inhabitants of Isfahan, in Persia, in 1388, by Tamerlane. He put at least seventy thousand men, women, and children to the sword and made a hill of the skulls of the dead.

The Hundred Years War (1337–1453, notable for the judicial murder of Joan of Arc in 1431 during its course), while bringing great misery as it waxed and waned, did not leave extensive areas totally depopulated. The Thirty Years War (1618–48), in which dynastic and territorial issues were also fueled by religious passions, did lay waste the territory now Germany and gave an example of the complete destruction of a city, that of Magdeburg, in East Germany. That the city was set on fire is part of history, but who set the fires in various parts of the city has never been established. Some accounts state that the inhabitants set the fires to hold back the opposing imperial Catholic forces. In any case, the city stands out for the horror in which it became a pyre for twenty-five thousand inhabitants, 85 percent of the city's population.

In long-drawn-out conflicts, the warriors went about their works of war, while civilians were tending to their own daily needs. Exchanges across battle lines were interdicted. Hungry people across a border could not be provided with food; wool for winter garments could not reach shivering children of the opposing side; herbs grown in one region could not be shared with those who needed their healing action.

In the parable of the Last Judgment, Jesus described for his followers the identity he chose, that of the person in need of merciful action. "I am the hungry one," he told us. "I am the thirsty one, naked, shelterless, suffering one, the one in prison" (Matt. 25:35–36). Wars impose a moratorium on works of mercy for those called enemies. For the duration of a war, Christians, whatever their own inclinations might be, are called upon to observe this moratorium.

Mercy is only love under the aspect of need; thus, in effect, love is interrupted for the war's duration. This illuminates war's incompatibility with Christianity, in that it displaces love from its central place in the Christian dispensation. Love for Christians demands above all that they desire for those loved the highest good, salvation. On the journey to that highest good Christians are called upon to express love through service to the needs of others, in particular, through the works of mercy. Love is treated here with respect to war. One insight into the breadth of Christian love is found in Paul (1 Corinthians 13), who tells us that even speaking with the tongues of men and angels avails nothing without love.

The Reversal of Mercy

During the experience of years spent in the debris of modern war, I became aware of a dread reality, the fact that modern war does not merely interrupt, but reverses, every work of mercy. Each side in a war shares in this reversal, and one side cannot refuse to practice the reversal without losing the war. The question of just or unjust is irrelevant, since wars are fought for victory.

Jesus commends those who receive a prophet, because he is a prophet and promises a prophet's reward; he also praises those who receive a righteous man, because he is righteous and holds out the promise of a righteous man's reward. Jesus continues with the promise that whoever gives, because he is a disciple, a cup of cold water to one of the little ones will not lose his reward (Matt. 10:42). The simplest and most lowly of all works of mercy, one that might not cost anything more than the gesture of offering it, is given a place of honor for Christians in ages to come.

The reversal of the works of mercy is hardly noticed because it is the general and accepted practice of warfare. Some concrete examples taken from twentieth-century wars may highlight the tragedy of this reversal. These examples bring out the gory underside of war, not the side of glory and conquering. They picture war from the side of the victims, as Pope John Paul II asked us to do when he made himself the voice of victims of all wars.

I Was Thirsty

World War II presents an obscene example of the reversal of giving drink to the thirsty. Thirst, instead, was brought to many. One of Europe's largest reservoirs, located in the Frankfurt area of Germany, was bombarded. Allied forces trumpeted it as one of the most important strategic victories of the war. No effort was made to justify the destruction of the reservoir under the principles of the "double effect," as incidental or unintended. The bombardment was planned as a mortal blow to civilian life. The soldiers at the front did not suffer the effects of the attack. The reservoir was a prized target and its bombardment an acceptable mode of warfare. The daily life of German civilians was dealt an unimaginably frightful blow. A whole area was deprived of a regular water supply, not only homes, but hospitals and institutions for children, the aged, and the ill. The attack on the water supply was one of the tragic instances of the reversal of the most basic work of mercy.

During the Vietnam conflict, chemical defoliants were sprayed along canals and roads to destroy vegetation and thus prevent the setting up of ambushes against American troops. The water in the irrigation canals poisoned the vegetables grown alongside and brought sickness to the villagers.[2] In the bombing and evacuation of villages, the villagers were separated from the wells from which they drew their drinking water.

One of the most horrendous acts of modern war occurred in the Persian Gulf War, in 1991, during one of the shortest and most destructive of modern wars.[3] The drinking water of the people of Baghdad and the surrounding area was contaminated when the electric system was destroyed by Allied bombing. Iraq was a developed society with water supplies controlled electrically by water pumping and filtering plants. With the power system a target of bombing raids, water pumping, filtration, and purification plants ceased functioning, as did sewage disposal plants.

Untreated sewage had to be dumped into the Tigris River, a source of the water supply. Attempts to purify the polluted water by boiling were made difficult by lack of fuel. In some areas, the population was reduced to drawing its water from polluted streams or trenches. As a stopgap, the International Red Cross imported a few tons of chlorine to purify the water supply of Baghdad. Fifty tons of chlorine a day were needed to make some impact on the contaminated water supply. Nowhere near this amount was obtainable.

The health of Iraqi civilians, in particular the most vulnerable among them, the children, the old, the ill, underwent incalculable hurt. Intestinal infections were widespread. A doctor in an Iraqi hospital explained the cruel dilemma: "To get water, I need electricity. To get electricity, I need a generator. To work the generator, I need fuel. To get fuel, I need a tanker. There are no tankers."[4]

The study of the Persian Gulf War entitled *On Impact* asserted: "The targeting of the life support functions of the civilian population, even for military effect, disabled the very objects that are otherwise restricted from attack: medical care, safe water supplies, food."[5] *On Impact* explained: "What this hyper war demonstrated is that the idea of a war in which noncombatants are safeguarded is no longer credible."[6]

When Iraq retaliated against air assaults by setting fire to the oil wells in Kuwait, the action was described in a White House statement as "environmental terrorism." Comparing the destruction of the life support system of a modern society with the destruction of oil wells leads the authors of *On Impact* to conclude: "Such destruction is as much de facto terror bombing as destruction of oil wells is environmental terrorism."[7] The study also asked: "Is it acceptable to destroy a modern society's way of life through the destruction of electricity production, water purification and fuel distribution yet unacceptable to destroy oil wells?"[8]

After the outbreak of war in former Yugoslavia in 1991, photographs of civilian life in Sarajevo pictured the deprivation. People stood with canisters around water outlets. The destruction of the water system was a step in the escalation of the war. So precious was water to drink, to prepare food, and to attempt to keep clean, that persons braved the bullets of snipers and the attacks of artillery fire to obtain it. One amazing day, the water began to flow from the faucets in houses and in buildings still standing. Engineers had located the breaks in the water system and had cobbled it together to make civilian life in war conditions more bearable. The complete reconstruction of the water and sewage systems in Bosnia would be a long and daunting task, as the American soldiers sent to keep the peace in 1996 discovered. In the city of Tuzla, they found that water systems were so damaged that water usage had declined from 170 quarts a day per person to 32 quarts daily.

I Was Hungry

It is not enough in wars of this century to interrupt the flow of food to the hungry among the enemy population; technical means are at hand to scorch the earth that produces food. By planting mines, the armies of Nazi Germany were able to scorch the earth of areas of eastern Europe from which they were forced to retreat. Polish farmers and their children returning to the works of peace in a stricken land were often killed by buried mines. Many lost hands or feet or were blinded when the earth yielded a lethal harvest. One of the first relief officials to reach Poland after war's end (Aloysius J. Wycislo of Catholic Relief Services) related his meeting with a village boy who was leading a cow into a barn. A strap was entwined around what the relief leader thought were hands. In fact, they were stumps where hands had been. He had lost both hands to a mine planted on his family's farm.

Another example of the use of mines occurred in France. When peace came in 1945, three hundred thousand hectares (about seventy-five thousand acres) of crop-producing land were mined — a consequence of Nazi strategy that spread hunger among the population long after fighting had ceased.

Even civil wars produce long-term hunger. People around the world shuddered at the news films and photographs of African children skeletonized and at the point of starvation. Their hair had become yellowed by Kwashiorkor, a disease linked to prolonged hunger. The children were in Biafra, a province of Nigeria that had seceded from the rest of the country. Food supplies were cut off from the people of Biafra as part of the civil war. A continuous succession of airlifts undertaken at great personal danger by members of voluntary agencies was an expression of the outrage of many people at the enforced starvation. While starvation was at its height, a minister for the Nigerian government put the policy into words: "All's fair in love and war, and starvation is a method of warfare." This strategy, though generally unspoken, is an acceptable part of warfare. Citizens of several nations of Europe, as well as American nationals, defied the ongoing war, flying in food and medicine and unloading them in secret airports. It stands as one of the first planned efforts to confront war by refusing to limit the works of mercy. In the lead were religious organizations, notably Church World Services and Catholic Relief Services.

In Vietnam, crop destruction was achieved in "area denial" programs by spraying herbicides. Many basic crops were lost. A petition for compensation from the American military was presented by six Vietnamese farmers who explained that they were people who had fallen into a desperate situation. "American military performed this," they stated, "by planes spreading chemicals.... The effect has made various types of fruit trees lose their leaves and [the trees] were ruined. ... Crops such as green beans, white beans and black beans died."[9] Not only are the hungry not fed, but hunger is prepared for present, and perhaps future, generations.

A city can live and feed its inhabitants only with provisioning from outside its limits and with means of transporting those provisions. Baghdad, a city on both sides of the Tigris River, has more than three million inhabitants. In the Persian Gulf War it was described by the chairman of the Joint Chiefs of Staff of the United States as "a very target-rich environment."[10] The targets included the bridges which linked Baghdad together as a city and also those that linked it to its hinterlands. The Department of Defense announced, after two weeks of the Persian Gulf War, that thirty-three of thirty-six targeted railway and highway bridges had been bombed in over 790 air sorties.[11] Hunger stalked the population. In 1991, in the immediate aftermath of the Gulf War, the mortality rate of children under five had tripled according to the report of researchers for the Harvard University School of Public Health to the UN Food and Agriculture Organization.[12]

The hope that the end of the hostilities would ease the hunger of Iraq's people was doomed by the imposition of economic sanctions. The UN Security Council decided to maintain the sanctions until its members were satisfied that Iraq had complied with full reporting on technical matters, including the possession and development of biological or chemical weapons, or nuclear capability. The continued embargo raised the pertinent question of when a war is ended. For the Iraqi people, over eight years after the cessation of hostilities, the privations of war continued to be imposed. UN agencies, including UNICEF, warned about the deterioration of the health of the people of Iraq, especially the young. The Harvard researchers estimated that as many as 576,000 Iraqi children died in the four years after the ending of the Persian Gulf War as a result of the sanctions imposed by the UN Security Council.[13]

A moral response to the hunger imposed on the people of Iraq came in a petition to the government of Iceland from a group of its people. They asked their government to oppose economic sanctions since they amounted to collective punishment inflicted against the people of Iraq. They asserted that the imposition was a form of hidden violence and constituted punitive measures contrary to established law.[14] An organization called Voices in the Wilderness united American and British citizens in programs of supplying food, medicines, and water purification equipment to Iraq to ease the plight of civilians. The group persisted with works of mercy despite the ban on such actions, a ban voted by the UN Security Council with the strong support of the United States. Whatever the rights and wrongs of the Persian Gulf War, the bringing of the long hunger to the Iraqi people could hardly be justified. The infliction of privation by the victor on the civilians of a defeated nation has given rise to an addition to the conditions of the just war, namely, justice in the postwar period. The *jus post bellum* was violated in postwar Iraq.

I Was Shelterless

Giving shelter to Jesus as the shelterless stranger has always evoked the giving of hospitality by his followers. The early Christians who gathered around the house of God (*para oikos* in Greek) were the ones who gave the word "parish" to succeeding generations. The new Christians, at the breakup of what remained of the order of the Roman Empire, found hospitality with newfound communities of believers. For example, the monks of St. Benedict of Nursia (480–547) offered shelter to travelers.

It remained for modern warfare to develop means of destruction so that thousands, even hundreds of thousands, of human beings can be made homeless in a few atrocious hours.[15] In World War II, the rationale for the bombardments unleashed over enemy cities, including Dresden and Hamburg, was the destruction of military targets. Such targets as Krupp works of Essen, arms factories, and crucial rail junctions were supposed to be pinpointed as the focus of bombing raids, and at war's beginning they were.

Only later was there admission on the Allied side that bombing raids had also expressly been targeted at destroying civilian housing. Those who took part in such bombing raids, even when they realized that they were raining down death and destruction on noncombatants, accepted it as their duty to weaken the morale of an implacable enemy.

One member of the Royal Air Force Bomber Command was moved to write about his participation in "indiscriminate air attacks on German towns," long after his participation in them. Peter Johnson, as a group captain and RAF pilot, related in *The Withered Garland,* published in 1995, how his doubts surfaced regarding the participation in the bombing of civilian housing.[16] His doubts grew as the war progressed, with civilian deaths in Dresden estimated at over two hundred thousand, of whom many were refugees from other cities and foreign workers. Many Britons were outraged when the commander-in-chief of the RAF Bomber Command responsible for the fire bombings of Dresden asserted that it was better to bombard any target in Germany than to bomb nothing. To the commander-in-chief, General Sir Arthur Harris, no distinction existed between military and civilian targets: it was Germany that had to be destroyed.

Johnson describes himself as initially harboring a hatred for Germans because his father had been a victim of a German submarine attack in World War I. He rejoiced in the success of bombing missions over the cities of Germany as his contribution to bringing the war to a victorious conclusion. He began to have doubts about the targets of RAF bombing missions over Germany when he saw the photographs of the results of a raid over Dusseldorf. He saw "rows and rows of apparently empty boxes which had once been houses. They had no roofs or contents. This had been a crowded residential area, long streets of terraced houses in an orderly right-angled arrangement, ... a huge dead area where once thousands of people lived."

Johnson realized that the "de-housing" of the enemy was a concept promoted by Lord Cherwell, scientific adviser to the British prime minister. Estimates were made of how many houses could be destroyed by so many bombs. If heavy bombers were employed, in time, "a third of the entire German population would be turned out of house and home." An RAF commander issued a directive with the proviso that "operations should now be focused on the morale of the civil population. ... I suppose it is clear that the aiming points are to be the built-up areas, not, for instance, the dockyards or aircraft factories." The U.S. bombing command operated under similar concepts. As the months and years of his bombing career went on, Johnson noted that the purpose of the raids changed from maximum damage to military targets (or enemy industrial centers) to the destruction of the enemy city itself.

On March 16, 1945, when Germany was already at the point of collapse, Johnson learned that the day's target of bombing would be the town of Würzburg. He inquired as to the military targets in the town and was told that there was nothing much of importance there. When he asked why a raid was planned for a residential town with only some light industry, he was told that in addition to a rail center (which was actually of no importance) there were in Würzburg "thousands of houses totally undamaged, sheltering tens of thousands of Germans."

Johnson pondered his options. If he refused to fly, or refused to give the orders to his squadron to discharge their bombs, he could be accused of mutiny and court-martialed, even executed. He gave the orders to his bombing squadron. In twenty minutes, 85 percent of the city was destroyed. It was a mass of rubble, and where homes and buildings had been, there were smoldering, ruined walls open to the sky. The civilian dead, men, women, and children, reached a total of over three thousand. These were delivered to the Würzburg cemetery to be lowered into a mass grave.

Johnson confessed that he had "a fairly prominent part in the deliberate mass bombing of German towns." He asserted that those actions left a permanent stain on the record of British armies. "Nearing the end of my life," he stated, "I have to admit that even the post-war revelations of the greater and more ghastly crimes of the Nazis . . . do not, cannot, expunge the stain of my part in it." In the title of his book, Johnson echoed words of Shakespeare: "Withered is the Garland of War."[17] Those who support modern war, even for high moral causes, must see that the infliction of mass shelterlessness is an inevitable concomitant.

To refuse in wartime to carry out orders reversing the works of mercy can be a daring act — one that might even lead to death. Peter Johnson knew that the bombing mission he was commanded to carry out would deprive noncombatants of shelter. He considered refusing but knew that such refusal might mean his execution. He carried out the order.

In occupied Europe during World War II the Nazi command arrested Jewish citizens. In Poland, a dread edict went into effect: anyone giving shelter to a Jewish person was liable to the death penalty. Despite this, Polish families hid Jews, particularly Jewish children, throughout the war years. Some paid with their lives.

In peacetime, when embargoes are in force, as in the U.S. embargoes on Cuba and Iraq, American citizens are threatened with reprisals if they insist on supplying medical aid and food to the Cuban or Iraqi peoples. Americans who have insisted on carrying out the basic works of mercy, feeding and relieving the suffering of Cubans and Iraqis, have persisted in their efforts year in and year out. The U.S. government — possibly realizing that arresting or collecting heavy fines for acts of conscience could be an embarrassment — has refrained from taking action.

I Was a Homeless Refugee

The massive uprooting of peoples has been an accompaniment of every war of the twentieth century, called by some "the century of the refugee." In a study of immigration and refugees an author stated, "Domestic and international conflicts may produce real or only pyrrhic victories, but they always produce refugees."[18] Wherever there are refugees, there is, for longer or shorter periods, hunger.

Those who have any part in causing people to flee from their home places must realize that they are bringing hunger not to enemy soldiers, but to families, from aged grandparents to children and infants, and not only to the presently afflicted generation but to generations unborn.

In Former Yugoslavia. A heart-tearing news photograph from former Yugoslavia shows a group of men, women, and children clogging a road. One woman, her head wrapped in a scarf, is pushing a wheelbarrow in which is crouched an aged man with sunken cheeks. His eyes are closed as he enters the world of the refugee. Another photograph which sears the imagination shows a slender woman with her eyes closed forever. She is hanging from a tree, having taken her life near the refugee camp at Tuzla.

The Balkans, from whose grim history World War I had taken its beginning, sent over a million men, women, and children spinning from their homes and home places in the war that exploded in 1991. Refugees streamed out of towns and villages at the approach of the armies. Others took flight as a result of bombardment. They spilled over borders into Austria and Hungary, many making their way into Germany.

It was in the Balkans that tribalism, more recently termed nationalism, allied with ancient religious hatreds burst forth with lethal virulence. It was this virulence that made it the crucible of conflicts. A new conflict exploded in 1991 when Yugoslavia began to fragment. While all the parties in the conflict in former Yugoslavia were of the same ethnic stock, whether Croat, Serb, or Muslim, they had been divided by history and by the emergence and fall of history's empires. For the Serbs, history had dealt them an irretrievable blow in 1339 at the Battle of Kossovo Field when they passed under five hundred years of Ottoman rule.

The 1991 war expanded inexorably despite over a dozen cease-fires achieved with the aid of European and American diplomats. Despite the fact that the highest ranking religious leaders of the three groups joined in public prayer and spoke with the voice of peace, their adherents continued to choose war.[19]

At length the violence was concentrated in Bosnia, one of the republics of former Yugoslavia. There, the Muslims were close to half the population, with about 33 percent Serbian and 18 percent Croat. The Serbs in particular resisted peace, and the Bosnian Serb army fought without let-up, because of their fear of Muslim domination. Their abiding fear was that they were in danger of a sweep of Islam from the Middle East that would again force them to submit to Islamic rule. This fear was intensified by the presence of small contingents of Islamic fundamentalist Afghans and Iranians, fighting side-by-side with the Bosnian Muslim forces.

By the time a peace agreement had been achieved through bringing the contending parties to a faraway place in the center of the United States (Dayton, Ohio), many towns in Bosnia showed the effects of ethnic cleansing: they had become towns of chiefly one community. The ethnic cleansing carried on during the war by killing and forcible displacement continued after hostilities ended— by an ethnic cleansing that was self-inflicted.

Sarajevo, once a city where the three communities had lived in some harmony, at least in mutual forbearance, exhibited a sadly significant movement of peoples. In Ilidza, Vogosca, and Hadjici, Sarajevo's Serb suburbs, whole families stripped their homes and moved away. Not only did they take all movable possessions with them, but they took their dead. They exhumed the bodies of their dead, especially those killed in the most recent violence, and transported them to a cemetery near Pale, the headquarters of the Bosnian Serbs.

This was a fateful sign for the future of the refugee population scattered around former Yugoslavia and in neighboring countries. They would not feel welcome in the towns or areas under domination of the community against which their forces had been fighting. Many of the ruined towns had changed hands, and where habitable structures remained, new owners had taken them over. The refugees and their children could look forward to years of privation before they could count on normal living.

In the meantime, under the U.S.-engineered peace agreement, many thousands of soldiers from the United States and the North Atlantic Treaty Organization were stationed in Bosnia so that the borders between the Serb communities and those under the Muslim-Croat Federation would be clearly marked. A mini no-man's-land, a zone of separation, was needed to keep each of the three warring communities apart. The underlying fear was that violence would flare once more and give rise to another stream of refugees.

In Rwanda. The tribalism that spawned violence and cascades of hungry and deprived refugees played its lethal role in many areas of postcolonial Africa as well, but never with such horror as in Rwanda's civil war of 1994. The former Belgian colony, which became a republic in 1961, was riven from its beginnings. The minority Tutsi tribe, less than 20 percent of the population, had an elitist control over the majority Hutu. Coups and sporadic violence marked the life of the country. Well before the massive violence of 1994, I was aware of how perilous the peace was between the Hutu and the Tutsi. When I visited Tanzania in 1970 to spend some time with the Missionaries of Charity of Mother Teresa, I found that some members of the team were not at the center in Tabora, but in a camp in northwest Tanzania at the Rwanda border. After a bloody uprising of the Hutu, the Tutsi had fled and had received a haven in Tanzania.[20]

The provocation that ignited the war in 1994 was the death in an airplane accident of the president of Rwanda. He was a Hutu, and his people claimed that his death was not accidental. The Tutsi, though better armed, could not repel the attack of the Hutu whose army had limited access to some arms, but as an agricultural people, had an abundance of machetes. Ethnic cleansing was achieved by massacre when Hutus descended on the Tutsi with unexampled ferocity. All restraints disappeared; chaos reigned.

No one escaped. Catholic Relief Services, which had been supplying aid and had been cooperating with local development programs for over thirty years, lost most of its local staff. Not only were the staff members killed, but their families were sought out for death. The old and the young were slashed and hacked to death, even women with infants strapped to their backs.

Large numbers of the Hutu belonged to the Catholic Church and Anglican Church. The church buildings which had been revered as places of sanctuary became instead centers where people were herded for slaughter.

When the director of Catholic Relief Services, Kenneth Hackett, returned to Rwanda in 1995, he decided to visit one of the large campsites. The refugees, finding themselves targets for massacre, had fled. "I found nothing but empty fields and barren hillsides," Hackett reported. "The fields, once brimming with people, were now deserted. Other stretches of land about the size of five football fields were stripped of trees and vegetation. All that was left were the skeletons of huts that had fallen to the ground." Hackett explained: "We have to follow the people with feeding programs. We are able to reach some 400,000 people in various places with basic food supplies."[21]

The word "genocide" began to be used when an estimated five hundred thousand of the Tutsi, about half of their number of one million in Rwanda, had perished. Priests and bishops who spoke in favor of the hereditary enemy were hacked to pieces. Early in the time of slaughter, the Catholic bishop of Awka, Nigeria, pointed out that "Christians live their Christian life in the context of their tribal life. In a conflict, it is not the concept of the Christian family that prevails but rather the maxim that blood is thicker than water, even the water of baptism by which one is born into the church."[22]

The churches which had been places of refuge became targets in which defenseless people who flocked there could be attacked. Refugee camps also became targets where armies of one side or the other could wreak their vengeance. Though agencies like Catholic Relief Services were making every effort to channel food and other aid to the newly displaced, restoration of anything like normal life would be slow. The Rwandan refugees in the heart of Africa traced their steps into neighboring Zaire, and finding safety and then danger in massive encampments, they were pursued for a long period by homelessness and hunger until expelled by the Zairian government.

During the course of the century, no continent on the globe lacked refugees whose plight arose from violence; there has been no continent where the hunger of refugees did not point a finger at those who threatened violence and dragooned others into inflicting it. In Latin America and the Caribbean area, uprisings from long-term oppression maintained by violence drove Nicaraguans and Salvadorans from their homes and over borders for years on end. Argentina and Chile sent many exiles to havens across the continent. Even on the roof of the world, under the snow-blanketed mountaintops of the Hindu Kush range, Afghans displaced by war and by the subsequent revolution for power hungered in a cold and foreboding climate. Hundreds of thousands of refugees lingered in Pakistan and Iran, fearing the privation and ruin they would meet in returning to Afghanistan.

There are too many instances of refugees condemned to hunger in the wake of violence to be recounted here. I only point to a few, some of whom I knew firsthand. The travail of the Polish refugees given haven in Mexico during World War II is little known. Deported from eastern Poland to Siberia and Asiatic Russia by Soviet forces in 1941 and 1942, they were given freedom of movement after the rupture of the Hitler-Stalin Pact. In response to an appeal by the Allies, Mexico was one of the countries that offered refuge for these men, women, and children, who were brought around the world to the safety of two camps in the Mexican hinterland. This was the first program of Catholic Relief Services, the new agency started by the American Catholic community, and it served as my own introduction to the world of the refugee. After World War II, I was involved in programs to feed the hungry in Europe, not only the former enemies in Germany, but the expellees of German ethnic stock who had been driven from eastern Europe into a destroyed landscape. As a member of the staff of Catholic Relief Services, I had seen, across the world, the plight of the close to four million Koreans who had fled south across the thirty-eighth parallel in the Korean War. Some had escaped by piling into trains, even clinging to the roofs. In icy winter weather, their frozen bodies were lifted down from the roofs of trains as so many descents from the cross.

In Vietnam, it was the seventeenth parallel that was the dividing line in a war that brought over a million people into refuge in the south of the country. I saw them in 1955 as they started new farms in the well-watered earth of South Vietnam, their hands the only tools. Others who had been fishermen in the north built new boats, often bending the heavy planks with the age-old method of slowly applied fire. Twenty years later, with the 1975 end of active warfare, the saga of Vietnam's "boat people" began. They took off from the coastline in every craft, from fishing boats to frail rafts. Their plight caught the attention of the world. Nearly a million found refuge in the United States, Australia, and Canada. Many thousands languished for ten, fifteen, even twenty years in places of temporary refuge in Asia, above all in Hong Kong.

Hong Kong had been the port of refuge in 1947 and 1948 to millions of Chinese from the mainland who clung to the hillsides for dear life. Within five decades they had made it the workshop of the world — a prize territory by the time it was joined to Mainland China in 1997 in the agreement with England. The Indian subcontinent, at its violent sundering in 1947, gave birth to at least twelve million refugees. Among them were the four million who streamed into west Bengal. Of these, a million came to rest in Calcutta, causing it to be called the "slum of the world."

On the occasion of my visit to Varanasi (Banaras), I was charged with the mission to bring aid to Tibetan refugees, some of the hundred thousand who had fled their occupied homeland. Through a strange turn of fate it was the brutality of the Chinese occupation of Tibet that brought them to the place tied to their Buddhist belief. Near Varanasi is Deer Park, where the Buddha had received enlightenment. A stupa, or hillock of stones, marked the place where the Buddha preached the sermon that "turned the wheel of law." Nearby a lamasery of young Tibetans was flourishing.

Not less than fifteen million human beings, perhaps as many as twenty million, have found themselves on the roads of the world every year since 1945, the end of World War II. These figures are a measure of the violence that caused their plight. There is every indication that these figures may continue as humanity enters the third millennium.

Jesus in His Distressing Disguise

When I was in Calcutta for the first time, in 1955, Mother Teresa had been working on its scourged streets for just over five years. Born in former Yugoslavia, of Albanian stock, the Catholic nun had become an Indian citizen in 1950. She founded an order of Sisters to serve the city. I went with her along slum walkways, where at the center sewage ran in rivulets. Around us were huts where mats resting on sticks were the only protection against sun or rain. It was these huts that collapsed in the monsoon rain, leaving families literally mired in mud. Many thousands simply lived in the open street covered with the spittle and filth of the gutter. Their hunger was only partly assuaged by coins they received from begging.

It was for the very least of the brothers and sisters of the street that Mother Teresa opened, in 1952, a hostel for the destitute dying. The city fathers of Calcutta gave her the free use of a pilgrim hostel. It was called Kalighat because it was close to the temple of Kali, revered as the goddess of destruction and purification, and near to the burning ghat for the cremation of the dead. In its two wards, one for men and one for women, lay those literally dying of hunger, along with all the diseases that hunger brings. I remember from my visit in 1955 seeing a woman with a loose hanging goiter being fed, slowly and carefully; in the adjoining ward, a man was being given soup spoon by spoon. His cheeks were sunk so deep and the skin drawn so tight that his skull seemed struggling to burst through.

These were the people whom Mother Teresa, and her original band of Sisters, washed, cleaned, and fed day after day. I drew away from the rows of pallets with the wasted bodies, and the suppurating sores. Ashamed at my revulsion from human beings rescued from the gutter, I asked Mother Teresa how she could care for these forsaken people day in and day out. Her reply stayed with me. "Our work," she said, "calls us to see Jesus in everyone. He has told us that he is the hungry one. He is the thirsty one. He is the naked one. He is the one who is suffering. These are our treasures," she said, looking along the rows of pallets in the hostel for the dying. "They are Jesus. Each one is Jesus in his distressing disguise."[23]

Jesus was before us in so many places, condemned to the agony of hunger, driven out and abandoned as a refugee. He might say, in so many tongues, "I was hungry and you gave me no food." Now there are Kalighats on every continent.

And to how many of every generation, especially the generations of the twentieth century, could the Lord say, "Your brother's blood cries out to me from the soil" (Gen. 4:11)? Again and again, I was reminded of Cain, the first wanderer over the face of the earth — Cain who suffered his fate after violence, the murder of his brother, Abel. Too many human beings like Cain might try to excuse themselves and hide from the truth with the question, "Am I my brother's keeper?"

Dehumanized, Desacralized, Demonized

Every work of mercy suffers the same pitiless reversal as soon as a war is declared. Human beings are made to suffer the agonies of thirst, hunger, nakedness, homelessness, untreated disease, and captivity, when they are not killed outright. The same evils are inflicted on civilian noncombatants as on combatants.

How is it possible for Christians to take part in this reversal of the works of mercy, of love, for the duration of a conflict? They reach the final step of reversal by three stages. The first stage is that in a war the people of the enemy nation are dehumanized, so that we lose the sense of common humanity; the second stage expunges any sense of the sacredness of the enemy as a human person, an image of the Creator; in the third stage, under the relentless propaganda of war, the enemy becomes demonized. The leader of the enemy state envelops his people in the demonization that is conferred on him by those who oppose his policies.

When I saw the barely human figures in Calcutta's hostel for the dying, it revived within me the sense of their utter and inviolable sacredness as children of God. They had been snatched from sure death in the gutters and alleyways of a city with no resources to meet the influx of a million refugees. Mother Teresa's words, "Each one is Jesus in his distressing disguise," expressed a deep reality for all followers of Jesus.

There are many disguises in which divinity is hidden, many disguises which cause us to shrink from other human beings. The most impenetrable disguise of all, however, is that of Jesus in the enemy. This is the disguise that is almost impossible to penetrate during the massive fog of untruth and propaganda in which wars are waged.

The Most Impenetrable Disguise

There is astonishment in the parable of the Last Judgment. Those who had given drink to the thirsty, fed the hungry, clothed the naked, welcomed the shelterless, helped the suffering, and visited the prisoner are astonished to learn that in reaching out in merciful love to the lowliest and the least, they were serving the Lord. There is also astonishment on the part of those who are condemned. They question when it was that they could have failed to minister to the needs of the Lord (Matt. 31:46).

Could it be that the one who was starving, who gasped for water, who was naked, who suffered as a prisoner, or who was forsaken as a homeless refugee is an accusing Christ?

A thought intrudes. In an age when human beings have the power to wage war in such a way as to virtually destroy creation, the works of mercy may no longer be open to them. The works of mercy may no longer be possible or necessary if human beings have vaporized their kind.

Chapter 9

Toward a Theology of Peace

The consideration of actual wars, with their undeniable reversal of the works of mercy, is an apt place to introduce some reflections on a theology of peace. The concept of bringing mercy to an opponent comes to us in the gospel when Jesus praises the Samaritan who came to the rescue of a man left helpless by the roadside. While the priest and the Levite pass him by, the Samaritan not only gives immediate aid, but lifts him on his own animal and takes him to an inn. His care goes so far as paying for his keep and promising to return to settle further debt to the innkeeper (Luke 10:30–35).

The reception by hearers of Jesus' words must have been hostile, since Samaritans were considered the enemies of the Jews, heretics who accepted only part of the scriptures, and people who had intermarried with alien peoples. Jesus was praising someone from whom a Jew would not accept a cup of water. "There was no deeper breach in human relations in the contemporary world than the feud of Jews and Samaritans," according to John L. McKenzie. He remarks, "The breadth and depth of love could demand no greater act of a Jew than to accept a Samaritan as a brother."[1]

Yet "the breadth and depth of love" were what Jesus was demanding, not only of Jews toward the despised Samaritans, but of his disciples toward those in every age considered enemies. This chapter will deal further with the Beatitudes and will reflect on the relevance of the Lord's Supper to peace.

The Beatitudes

They Will Be Shown Mercy

The beatitude "Blessed are the merciful" is followed by the promise, "for they will be shown mercy." Followers of Jesus know that their dependence on the mercy of God is utter and complete. To risk not being enveloped in the mercy of God would be a dark terror: thus, being merciful to others should form the warp and weft of the daily life of the disciples of Christ. Mother Teresa's habit of seeing Christ in his most distressing disguise resonates with the spirituality of the Invocation of the Name of Jesus. This is a prayer widely practiced in the Orthodox communion but earlier used by many generations of Catholics. The name of Jesus, according to a monk of the Eastern church, is a concrete and powerful means of seeing human beings transfigured into their hidden, innermost reality. "We should approach all men," says the monk, "in the street, the shop, the office, the factory, the bus, the queue... with the Name of Jesus in our heart and on our lips. We should pronounce his Name over them all, for their real name is the Name of Jesus; [we should] name them with his Name, in a spirit of adoration, dedication, and service.... [We should] serve Christ in them. In many of these men and women — in the malicious and the criminal — Jesus is imprisoned. Seeing Jesus in every man, everybody will be transformed right before our eyes."[2]

It is not only "Blessed are the merciful" that has a close relationship to peace; the other beatitudes relate to peace and peacemaking in different ways:

> Blessed are the poor in spirit,
> for theirs is the kingdom of heaven.
> Blessed are they who mourn,
> for they shall be comforted.
> Blessed are the meek,
> for they will inherit the land.
> Blessed are they who hunger and thirst for righteousness,
> for they will be satisfied.
> Blessed are the merciful,
> for they will be shown mercy.
> Blessed are the clean of heart,
> for they shall see God.
> Blessed are the peacemakers,
> for they shall be called children of God.
> Blessed are they who are persecuted for the sake of righteousness,
> for theirs is the kingdom of heaven.
> Blessed are you when they insult you and persecute you and utter every
> kind of evil against you falsely because of me.
> Rejoice and be glad,
> for your reward will be great in heaven. (Matt. 5:3–12)

Scripture scholars point out that the word "blessed" came to carry a finalistic sense, assigning all happiness to a future beyond time, whereas other translations carry the implication of present happiness as well. The Hebrew word *baruch* could mean present welfare, seen as the reward for the faithful carrying out of the law; the Greek *makarios* has the connotation of happy and fortunate, of being free from care and suffering, and was thus applied to the gods of Greek mythology. The Latin *beatus* stresses being happy in the here and now.[3]

Christians, in emphasizing spiritual happiness, the happiness that flows from abiding in God's law, look forward to the hope of the final happiness of the kingdom. In true happiness, the future does impinge on the present, for the Christian lives both in time and eternity. It is good to be mindful of the reality that the term "blessed" or "happy" is not robbed of its impact on daily life. The translation using "happy" reflects this.

Those who accepted Jesus came to see happiness shining through the "present-future" screen. The holy paradox is repeated nine times in the gospel of Matthew. With eyes opened to a new vision of the "not yet-already" reign of God, the disciples, then and now, are enabled to see happiness entering the hearts of the poor in spirit, of those who mourn, of the meek, of those who hunger and thirst for righteousness, of the merciful, of the clean of heart, of the peacemaker, and even of those who are persecuted for the sake of righteousness (Matt. 5:3–10).

Those who feel themselves blessed or happy, even though they are oppressed and without possessions, are serene in their utter confidence in God. Even they are expected to share what little they may have with others, if only in service. The kingdom of heaven belongs to them, for they are already living under God's reign. It is possessions, and the power they confer, that frequently bring people into conflict. The remaining beatitudes depend on the first, since only those who are transformed in spirit, recognizing human inadequacies and the almighty power of God, can find the door to the kingdom.

Those who mourn find solace in the God of mercy. When mourning for a personal loss tends to drive us into isolation, Christ provides a remedy. The Eucharist draws us out of isolation into communion with the resurrected one and with the community of the faithful. Mourning then transcends the heartache for the death of someone close and beloved, and reaches out in compassion for others who mourn across the world in homelessness, poverty, and war. Identification with the mourning of others can inspire disciples to flee from anything that could cause others mourning and loss, in particular violence and war. From such mourning can spring mindfulness of those who suffer under oppressive social systems and governments. A fruit of this mourning might be the taking of steps to disengage themselves and others from unjust social practices. It is mindfulness that gives rise to the formation of counterpractices and countermovements, however small.

The promise of inheriting the land which accompanies "Blessed are the meek" was undoubtedly shocking to many of the first who heard it. It would not be by preserving meekness or gentleness that the land, then occupied by the Romans, would be returned to the children of Israel. Violent uprisings against the Romans were proof of the Israelites' determination to satisfy this longing. The Psalmist sang, "The meek shall possess the land," but only after the evildoers are cut off. Jesus, however, was opening the minds of his hearers to the kingdom of God, a kingdom won by a gentle heart. No war or violence could win or inherit this kingdom.

The Thirst for Righteousness: Ends and Means

"Those who hunger and thirst for righteousness" reflect the yearning for the fulfillment of God's justice on earth. The right relationship with our neighbor is the first duty of the Christian — a relationship based on love and on giving all persons their due. The promise is that the hungering and thirsting will be satisfied. In one translation, we find the words, "They shall have their fill" (Jerusalem Bible). Men and women on fire to lift the cross of suffering and injustice from the backs of the poorest of the poor can become wildly impatient at the enormity of the task and the heartbreaking slowness in lessening of the cross's weight.

In Latin America, age-old poverty and built-in structures of injustice moved many to action to speed the day of liberation — the only way seemingly being violence. Priests who were anguished by the unending misery and exploitation of their people were among those who chose the way of the gun.

In a remote mountain area in Colombia, a skirmish between the Colombian army and a group of guerrillas of the Army of National Liberation (ELN) resulted in the death of three guerrilla fighters. One of the dead guerrillas was a priest. It was 1974. The priest, Domingo Lain, was from Spain, and he had exercised his mission by working side-by-side with laborers at a brickyard. His bishop had asked him to lead a parish in the poorest section of Cartagena. He had obeyed and had lived in a thatched shack like the neediest of his parishioners.

Lain's championing of several protests, including one against the exploitation of the lands on which his parishioners lived and worked, resulted in his being moved to Bogota. There, with brother priests, he participated in presenting a social manifesto. He was deported to Spain in 1969, but reappeared on Colombian soil in 1970. He had decided to throw in his lot completely with the guerrillas.

He was not the only priest to make this decision. Camilo Torres made the same choice of a life defending the oppressed and suffered the same violent end. Lain's explanation of his decision resonates with the thinking of other religious persons, including priests, who took to the cause of liberation by violence. Lain stated:

> I am taking the course of armed struggle because there is no other way than by revolutionary violence to oppose the reactionary violence of the establishment and its social system in Colombia and throughout Latin America. Violence has no connection with religion. It is neither atheist nor Christian, it is the result of economic, historical, and sociological laws, of the growth and scope of societies, and of the relations among those societies, members and groups. Hence oppressed societies and exploited peoples have a right to resort to counter-violence in order to get free of their exploitation.[4]

From 1970 until a bullet ended his life in 1974, Domingo Lain fought and killed with the Army of National Liberation.

In "The Bondage of Liberation," Gordon Zahn, a deeply committed pacifist, discusses an aspect of liberation thinking of concern to believers in nonviolence. He says that this aspect has restored respectability to the "just war." He states that "by effectively reducing the traditional criteria for such a 'just war' to only two (just cause and right intention, both generously interpreted), [liberation thinking] has opened the way to virtually uncritical support not only for wars of national liberation but for a full range of guerrilla tactics, not excluding indiscriminate acts of terrorism. Thus active resistance to oppression, certainly a 'just cause,' and the 'right intention, . . . ' are deemed sufficient in themselves to cover the proverbial 'multitude of sins.' "[5] The question of means is often lost in the urgent concern for the cause.

Violence and Counterviolence. The concept of liberation, developed chiefly from the Latin American experience, has spoken to many hearts within the

Catholic Church as well as to other faith groups and to those of no faith alignment. It calls for reflections on the way society is structured, on who benefits and who is exploited.

The chief spokesperson for liberation thinking, Gustavo Gutiérrez, a priest of Peru, states: "The theology of liberation attempts to reflect on the experience and meaning of the faith based on the commitment to abolish injustice and to build a new society: this society must be verified by the practice of that commitment, by active, effective participation in the struggle which the exploited social classes have undertaken against their oppressors."[6]

According to Gutiérrez, all the theological theorizings "are not worth one act of genuine solidarity with the exploited social classes." Gutiérrez is emphatic in asserting that the response of the Christian cannot be passive in the face of exploitation. His message has inspired many lay Catholics as well as vowed persons, priests, Brothers, and Sisters, to address themselves to the remedying of tragic ills and injustices in the society around them.

John L. McKenzie has emphasized the role of the person. "Some liberation theologians," he states, "have found the gospel irrelevant to modern times because it is addressed to the individual person rather than to society. They seem to imply that society and not persons should be changed, or rather, that the persons will be changed if the society is changed. There seems to be a clear opposition here between the biblical approach to human problems and most modern approaches, which can be summed up, not unfairly, as political."[7] The transformed person as the agent of a transformed society has been the practice of Christianity since its founding.

"Why does the New Testament," asks McKenzie, "like the prophets, speak of a change of heart and not a reform of institutions?"[8] "Christianity," remarks Alfred Bour in a similar vein, "constantly oscillates between 'bodiless spirituality' and political messianism. The trumpet-call of some who want to transform the structures of a cruel society in their own lifetimes savors of messianism. At the same time, it underplays or even eliminates the crucial role of personal transformation."[9]

Persuaded Messenger of Nonviolence. The antithesis of "political messianism," to purge social ills and heal society's wounds, is exhibited in the life of the Argentinean artist and sculptor Adolfo Pérez Esquivel. His program was one of Christian patience and endurance. Pérez Esquivel had read and addressed in the abstract the accounts of the nonviolent struggles of Mahatma Gandhi and Martin Luther King. What had been abstract became concrete and living to him with the coming to Latin America of Jean Goss and Hildegard Goss-Mayr, traveling secretaries of the International Fellowship of Reconciliation.

At the age of forty, Pérez Esquivel's increased awareness moved him to join in the struggle for the nonviolent liberation of voiceless and oppressed peasants and workers. He was married and the father of three sons. In the *Peace and Justice Bulletin*, which he founded in 1973, he focused on examples of injustice in various countries of Latin America. He did not merely report on conditions, but went personally to stand by the side of the exploited, and when necessary, became the voice for the righting of wrongs. Pérez Esquivel traveled to Ecuador in 1974 to work with the Indians in Toctezinin in a struggle with a large landowner. The next year he spent time in Paraguay with the Christian communities of the Peasant Leagues, who needed help to survive under cruel repression. In his own country, he took part in the formation of the Permanent Human Rights Committee.

Meanwhile, he became secretary for Latin America of the Peace and Justice Service, affiliated with the Fellowship of Reconciliation. As a teacher of nonviolence, he described three steps for nonviolent action against injustice: identify the injustice and bring it to the attention of competent authorities; if dialogue with the authorities fails, involve the masses of people through appeals to public opinion and public fasts; put out a call for civil disobedience. As in the movement led by Mahatma Gandhi, he called for careful planning and organization. He discussed these steps with Brazilian laborers at the Perus cement factory in São Paulo, with the peasants of the Alagamar plantation in Brazil's Northeast, with miners' wives in Bolivia, and with mothers and fathers of the "disappeared" in Argentina.[10]

His work was interrupted in April 1977 when the military government of Argentina imprisoned him. He was never told the reason for his arrest. In the prison of La Plata, Buenos Aires, he was kept in solitary confinement in a tiny cell and beaten at intervals. He was forbidden all books, even the Bible. After his release from prison in June 1978, he was placed under house arrest for fourteen months. When he was liberated, he traveled to most of the countries of Latin America, encouraging the prophetic response of evangelical nonviolence in liberation from injustice. He brought separate struggling and protest groups into communication with one another. He supported the justice and peace programs of Latin American bishops, and they in turn gave him their encouragement and support.

In December 1980, Adolfo Pérez Esquivel received the Nobel Peace Prize in Oslo. His acceptance speech encapsulates his career and its spiritual undergirding. "I come from a continent that lives between anguish and hope," he told the audience at the ceremony. He explained how his work stemmed from a conviction of "the dignity of the human being, the sacred, transcendent, and irrevocable dignity that belongs to the human being by reason of being a child of God and a brother and sister in Christ, and therefore our own brother and sister." He went on:

> As I speak to you, I have before my eyes the vivid recollection of the faces of my brothers and sisters:
>
> - the faces of the workers and peasants living at subhuman levels, whose rights to organize are severely limited,
> - the faces of children suffering from malnutrition,
> - the faces of young people who see their hopes frustrated,
> - the faces of outcasts and marginal urban poor,
> - the faces of indigenous peoples,

> - the faces of mothers searching for their missing sons and daughters,
> - the faces of those who have disappeared, many of them children,
> - the faces of thousands of exiles,
> - the faces of people who lay claim to liberty and justice for all.
>
> Despite so much suffering and pain, I live in hope because I feel that Latin America has risen to its feet. Its liberation can be delayed but never denied.[11]

The recitation of the Beatitudes of Matthew was the conclusion of his address.

The president of the Nobel committee stated of Adolfo Pérez Esquivel: "For years he has dedicated himself to the cause of human rights in Argentina and in all Latin America. He is a persuaded messenger of the principle of nonviolence in the political and social battle for emancipation. He has lighted a light in the darkness. The committee is of the view that this light ought to be kept burning."[12]

One of the paintings of Pérez Esquivel that has been reproduced in the United States is that of Christ on the cross, wrapped in the garment of the Latin American poor, the poncho.

A Crucified People. Any war, from a war of liberation to a war between nations, calls for the annulling of mercy and the reversal of the works of mercy. A voice for liberation which never veered from the message of Jesus concerning mercy is that of Jon Sobrino. When Sobrino, a Jesuit priest from the Basque region of Spain, arrived to serve in El Salvador, he found a people misery-ridden and subjugated, victimized by year-in and year-out violence which left tens of thousands dead. A brother priest, Ignacio Ellacuría, called them a "crucified people," and the task was to attempt to "take them down from the cross." Sobrino's intensive studies of the latest theological thought had put him in what he called a "dogmatic slumber." In the face of the "gigantic cross for millions of people," there needed to be an awakening from the "sleep of inhumanity."[13]

In studying the situation of the Salvadoran people, Sobrino decided that it was necessary to enter deeply into the principle of mercy. "The term mercy," he decided, "must be correctly understood. The word has good and authentic connotations, even dangerous ones. It suggests a sense of compassion. The danger is that it may seem to denote sheer sentiment, without a praxis to accompany it."

Performing the "works of mercy," according to Sobrino, "may release the practitioner from the obligation of studying the causes of the misery that is being relieved, and of taking steps to transform structures." Sobrino points out that he uses the term "principle of mercy" to mean a "specific love which, while standing at the origin of a process, also remains present and active throughout the process, endowing it with a particular devotion and shaping the various elements which compose it. We hold that this principle of mercy is the basic principle of the activity of God and Jesus, and therefore ought to be that of the activity of the church."

In the Latin American context, Sobrino pointed out that the exercise of the principle of mercy may call for taking risks. He cited Oscar Romero, archbishop of San Salvador, as one who took such risks — to his life and to the institution of the church itself. Because Romero publicly attacked injustices, the archdiocesan radio and printing operations were destroyed and several of his priests were assassinated. On March 24, 1980, Archbishop Romero himself fell victim to a single bullet to the heart as he stood at the altar celebrating Mass.

Sobrino's commitment to gospel nonviolence was further tested when Ignacio Ellacuría, the rector of the Simeon Canas Central American University (referred to as UCA), was gunned down by elements of the military. On the lawn behind the pastoral center of the university, lying in pools of their own blood, were Ignacio Ellacuría, five brother Jesuits, and two women, a housekeeper and her daughter. The priests at UCA had committed the crime of calling into question societal structures, not only the structures of the economy, including landholding, but the ever-present military structure, which kept everything in place. For that, they suffered the vengeance of the military on November 16, 1989.

Sobrino memorialized his confreres as "signs of hope":

> My six Jesuit brothers now rest in the chapel of Monsignor Romero beneath a large portrait of him. All of them, and many others, would have given each other a firm embrace and would have been filled with joy. Our fervent desire is that the heavenly Father transmit that peace and that joy very soon to all the Salvadoran people. Rest in peace, Ignacio Ellacuría, Segundo Montez, Ignacio Martín-Baró, Amando López, Juan Ramón Moreno, Joaquín López y López, companions of Jesus. Rest in peace Elba and Celina, very beloved daughters of God. May their peace give us hope, and may their memory never let us rest in peace.

Not long after the murder of the priests and the two women, a Jesuit priest came to Maryhouse, the Catholic Worker Center in the East Side of New York, to celebrate Mass and give a talk to our guests and our volunteer staff.

He was young and vigorous and smiled easily. He was Dean Brackley, an American Jesuit. We said good-bye to him that evening, since he was leaving for San Salvador, one of those who would take the place of the priests whose lives had been taken. The chapel where the priests were buried became a place of pilgrimage, a site where nonviolent love speaks as it has through chapels and shrines dedicated to martyrs from the days of the catacombs in Rome.

Blessed Are the Clean of Heart

An astounding, even apocalyptic, promise accompanies the beatitude "Blessed are the clean of heart" — simply that "they will see God" (Matt. 5:8). It recalls the words of the Psalmist, "Who shall stand in the holy place? He whose hands are sinless, whose heart is clean" (Ps. 24:4). Such a one may be present in the temple of the Lord. The promise of seeing God must await the coming kingdom, but it is prepared for by seeing God in his image, the human beings who people our lives. Those among us who fail to see God's spirit in his children, who disfigure them, who dehumanize and desacralize them, who fragment or vaporize them in war — are we not risking our final union with the Creator? The vision of God in human form, of Jesus even in the "distressing disguise" of the enemy, should make us tremble before any participation in the works of war.

The heart, in scripture, includes not only the emotions but the mind and will, in fact, the whole person. Being clean of heart means turning the whole person toward God. Only those who strive to make this "turning" know how much has to be discarded as refuse, though the world might see it as good, before the heart is clean and unencumbered.

For the Psalmist, nearness to God was experienced in the holy place of the Temple. For the followers of Jesus, there is, incredibly, another holy place, the temple of the human body. The first-generation Christian community in Corinth was reminded of this spiritual reality by Paul: "Do you not know that you are the temple of God, and that the spirit of God dwells in you? If anyone destroys God's temple, God will destroy that person; for the temple of God, which you are, is holy" (1 Cor. 3:16–17).

The beatitude of the clean of heart can be seen as leading directly to the concept of the peacemaker: "Blessed are the peacemakers, for they shall be called children of God" (Matt. 5:9). The terrain is changed here, since the naming of peacemakers as children of God occurs on earth, in the midst of earth's turmoil.

A first necessity for peacemakers is to recognize their identity, and from this identity can be built the spiritual structure of peacemaking. As temples of the Most High, they are the highest point of his creation, infinitely sacred, irrevocably transcendent over all transient situations or events. It is when they have enfolded this conviction into their hearts, their whole persons, that peacemakers can look around them and see each human person as a holy temple.

The conviction of "the other," whoever he or she may be, as a holy temple is one of the pillars of a theology of peace. The Psalmist sings his gratitude to the Creator, "Truly you have formed my inmost being; you knit me in my mother's womb. I give you thanks that I am fearfully and wonderfully made" (Ps. 139:13–14).

It is now, only as science has probed the mysteries of the human body, that we are enabled to reach an inkling of how "fearfully and wonderfully" each child of God is made. Scientists make different estimates of the composition of the human brain, for example, but many agree that the few ounces of soft matter shielded by the carapace of the skull counts ten billion cells.

No computer could ever map or trace the intricacies or interweavings that make for the thoughts registered by these billions of cells. This should make us pause in wonder before the Creator's gift to his image, the pinnacle and crown of creation. The human person is a final end, an end destined for eternal life with the Creator.

Can any intermediate end justify the disfiguring, maiming, or killing of the human person? Yet, such intermediate ends as an increase in territory, rectification of a border, and retaliation for an offense to a national government are classified as sacred by regimes and national governments. Do these give them the right to force their citizens into the task of injuring and killing other human beings?

The time-bound necessities of national policies and the noxious disease of nationalism harness human beings, possessing an eternal destiny, to the "duty" of killing and being killed. This constitutes a supreme subversion of values and occurs in every act of violence toward a human being, and above all, in the mass activity of violence we call war.

The Persecuted

The gospel of nonviolence reaches a higher point in the eighth beatitude in Matthew. Each of the teachings contained in the Beatitudes starts with a phrase familiar to Jesus' hearers — "Blessed are you" — and then proceeds to overturn their world. The final beatitude tells them they will be blessed when they are persecuted for the sake of righteousness, for the kingdom of heaven awaits them (Matt. 5:10). But, in advance of the kingdom, they are to expect insults and every kind of evil on account of Jesus. In case they think it madness to rejoice in suffering and persecution, Jesus reiterates the reward of the heavenly kingdom and reminds them of the evils that befell their prophets. In Luke, those who are persecuted on account of the Son of Man are told that on that day they should "rejoice and leap for joy because their reward is sure" (Luke 6:23).

But I Say to You: The Sermon on the Mount

The new way of life of the community around Jesus is etched in strong, even stark, terms in the antitheses that make up part of the Sermon on the Mount. Each expresses a direct contrast to the commonly accepted mode of behavior of the children of Israel of that time. Before announcing them, however, Jesus explains, "Do not think I came to abolish the law or the prophets. I have come not to abolish but to fulfill" (Matt. 5:17). Jesus gives concrete examples with regard to this fulfillment. "You have heard it said, . . . but I say to you" opens a vision of living that would have peacemaking at its core.

According to McKenzie, "The 'six antitheses' of the Sermon on the Mount show Jesus as the new Moses, pronouncing a law of equal authority with the old." McKenzie points to the fact that the location of the "Sermon on a mountain . . . is a deliberate evocation of the first revelation of the law upon a mountain."[14]

"You have heard that it was said to your ancestors, you shall not kill; and whoever kills will be liable for judgment. But I say to you, whoever is angry with his brothers is liable for judgment." This antithesis contains the command to seek reconciliation with an offending brother or sister before leaving a gift at the altar.

Three antitheses dealing with adultery, divorce, and oath-taking reveal in dramatic examples what Jesus asks of his community, the new creation.

It is in the fifth and sixth antitheses that Jesus carries the Hebrew Scriptures into the new dispensation that he is bringing to his disciples. "You have heard that it was said, 'an eye for an eye and a tooth for a tooth.' But I say to you, offer no resistance to one who is evil. When someone strikes you on your right cheek, turn the other to him as well. If anyone wants to go to law with you over your tunic, hand him your cloak as well. Should anyone press you into service for one mile, go with him for two miles. Give to the one who asks of you, and do not turn your back on one who wants to borrow" (Matt. 5:38–42).

Retaliation

First of all, the principle of retaliation, the law of talion, taught that punishment should respond in kind and degree to the offense committed by the wrongdoer. The command in the Old Testament was intended to moderate vengeance so that the punishment should not exceed the injury. Many scripture scholars point out that the response of no resistance does not simply mean to submit to one who is evil, but rather not to resist in kind, in other words, not to descend to the same level as the one who is involved in an evil act. Much light has been thrown on the spiritual and social implications of this antithesis by Walter Wink.[15]

Turning the other cheek needs to be seen in the context of the culture of the Roman Empire, when a master could slap an inferior or slave with the back of his hand. If the person slapped offers his other cheek, it is instantly clear that the first slap has not had its effect. The response is not that of a coward. Something new and unheard of has entered the relationship. The person slapped has taken control of the situation. Even if the person continues to beat the person, the same reaction of nonviolence will continue. The person has dignity that emanates not from the perceived relationship of superior and inferior, but rather because the person slapped knows his infinite worth as a child of God.

The way of nonretaliation, of "never paying back evil for evil," was the sign of a new dispensation. "If anyone wants to go to law with you over your tunic, hand him your cloak as well" is a way of going beyond the demands of the law, beyond legalism. It is also a way of nonviolence. The effect might be ludicrous if the legalistic person pressing the charge would deprive a man of covering or reduce him to his undergarments.

The Roman custom of impressing a civilian to carry the *impedimenta* of a soldier to the next mile marker on the road was part of the law. For Jews under occupation it was especially galling and humiliating. The response of the civilian who without anger or sense of humiliation asks to continue carrying the heavy military gear for another mile would have caused complete mystification to the soldier. What could he make of it? First he would realize that the relationship of the Roman soldier to a member of a vanquished people, or inferior class, was now changed. The inferior showed his refusal to be humiliated by going beyond the letter of the law.

"Give to one who asks of you, and do not turn your back on one who wants to borrow" goes beyond the giving of alms with which every Jew was familiar from scripture. Being willing to help one who wants to borrow is something quite different. Turning one's back on the would-be borrower is to be expected, since one did not lend money to a bad risk or to one who would not be able to repay with the often usurious rates of interest then current. The generosity of the disciple would be shown by lending money to someone who might never repay it. The community of disciples would show their trust in each other and in the providence of God by acting in a way that seemed profligate.

Love of Enemies

"You have heard it said, you shall love your neighbor and hate your enemy, but I say to you, love your enemies and pray for those who persecute you, that you may be children of your heavenly Father, for he makes the sun rise on the bad and the good, and causes rain to fall on the just and the unjust."

In the Hebrew Scriptures, there is no specific command to hate one's enemy, but in the love command contained there, the "neighbor" was one's own countryman. The hatred of evil human beings is assumed to be acceptable behavior in the Old Testament.

Now Jesus breaks with the Hebrew Scriptures in an astounding new teaching which seems to be contrary to reason. How can you reach out in love to a person who hates you and who expresses that hatred in acts of persecution? It goes against what was accepted in the culture as sanity as well as self-protection. What argument could there be for such a break with the old dispensation and with the defense by which a normal person would prepare himself or herself against a known enemy? Loving your enemy is, in effect, to strip yourself naked against those who would wish ill to you and who might be emboldened to wreak evil. With that break with the old came the inbreaking of the new, the reign of a Creator whose love, and loving care, is as all-inclusive and indiscriminate as the sun and the rain: the Creator will hear all who call on him, and give good things to those who ask him. Jesus then enunciates what has become known as the Golden Rule: "Do unto others whatever you would have them do to you" (Matt. 7:11).

Toward the end of the Sermon on the Mount, Jesus teaches his followers how to pray in the words we call the Lord's Prayer. This will be taken up in a later chapter. Jesus emphasizes in many ways how difficult will be the way he has traced, and how narrow the gate for those who decide to walk his way. It is hardly to be wondered at that his hearers were amazed at his words and at the authority with which he uttered them.

The Lord's Supper

If Jesus had simply left his teaching in words, and in the example of his willingly accepted suffering and death, where could his disciples, then and later, find the strength and the nourishment to meet the demands of living according to his way?

The way of nonretaliation, of no revenge-taking, of forgiveness, and of a love that would enfold the enemy and persecutor, called for a discipleship that would transcend human nature in all its "fallenness." How would it be possible for this fallen human nature to imitate the limitless largess of the Creator? The trinitarian God did not leave us orphans; God left the Spirit to aid us in confronting the world's violence, the effect of the "fallenness," with total nonviolence grounded in love. God did not abandon us, but left as nourishment for the journey his resurrected body in the form of bread and wine.

The Lord's Supper is the very heart of a theology of nonviolence. As the memorial of the act of sacrificial, redemptive love on the cross, it is the central act which convenes the community of those who have been baptized in Christ. Those who approach the Table of the Lord know that they are coming to the heart of peace, the great peace between God and man. "God wanted all perfection to be found in him and all things to be reconciled through him and for him, everything in heaven and everything on earth, when he made peace by his death on the cross" (Col. 1:20; Jerusalem Bible).

According to Donald Senior, "Reconciliation is the byword of the kingdom."[16] The followers of Jesus came to the Lord's Supper from many national groups, as Gentiles of many roots joined with the original Jewish community; the differences of class, of old antagonisms, were submerged in one spirit. "For just as the body is one and has many members, and all the members of the body, though many, are one body, so it is with Christ. For in the Spirit we are all baptized into one body whether Jews or Greeks, slaves or free persons, and we were all given to drink of one spirit" (1 Cor. 12:12–13). By the shedding of his blood, Jesus healed the division between the people of Israel and the Gentiles. "But now in Christ Jesus, you who were once far off have become near by the blood of Christ. For he is our peace, he who made both [Jews and Gentiles] one and broke down the dividing wall of enmity" (Eph. 2:13–14). No wall divided those who met at the Lord's Table.

The message of Jesus was addressed to each person, but as each person received it and was transformed by it, he or she joined fellow believers at the Table of the Lord. Personal transformation led to the transformed community, the beloved community. The great mystery of the changing of the bread and wine into the body and blood of the Messiah occurred whenever a gathering took place under an ordained presider. The hands of a mortal man, perhaps stubby, work-worn hands, could draw down the Godhead so that the other mortals could partake of it.

Saints of the past describe how the Eucharist strengthens those who partake of Jesus in the lowly elements of bread and wine. "By communicating himself, therefore, as heavenly food," said St. Procopius of Gaza, "God nourishes souls in virtue, and inebriates and delights them with knowledge, spreading virtue and knowledge before them like the food of a spiritual banquet to which all are invited."[17]

Saints also help us recapture, if we tend to lose it, the awe of partaking of the Godhead in the form of bread and wine. "For neither was it enough for him to be made man," said St. John Chrysostom, "to be smitten and slaughtered, but he also commingles himself with us, and not by faith only, but also in the very deed, makes us his body.... That which when angels behold, [they] tremble, and dare not so much look up at it without awe on account of the brightness that comes thence, with this we are fed and with this we are commingled, and we are made one body and flesh with Christ."[18]

The expression "commingled" is used also by St. Cyril of Alexandria:

> All who eat Christ's holy flesh enter into bodily union with him, and not only with him who is in us through the flesh, but with each other. We are, then, clearly one with each other in Christ. He is the bond which unites, he who is both God and Man.
>
> In a similar way, we may say that all of us who have one and the same Holy Spirit are, as it were, commingled with each other and with God. For though we are many separate individuals and Christ makes his and the Father's Spirit dwell in each separate one, yet that Spirit is himself one and indivisible. He therefore makes those separately subsisting individuals a unity through his presence; he makes them all be, in a way, a single entity through union with himself. Just as the power of Christ's holy flesh makes those in whom he dwells to be co-corporeal with each other, so the indwelling Spirit of God, indivisibly, one and the same in all, forges all into spiritual unity.[19]

Cyril concludes that under the influence of the Spirit, human beings are no longer simply human beings but sons and daughters of God and beings of a heavenly mold.

The church, whose saints couch the mystery of the Eucharist in terms for the lettered, also had the task over the centuries of presenting the mysteries to the unlettered and barely lettered. The attempt to transmit this overwhelming reality to the generations has been the church's never-ending challenge. For every generation, for the lettered and unlettered, the teaching has been carried unbroken in two words, the "Real Presence," the "Real Presence" of Jesus in the new covenant.[20]

In the Upper Room on the night before he was betrayed and raised on the cross, Jesus gave the new covenant to the apostles: "Take and eat; this is my body.... Drink from it, all of you, for this is my blood of the covenant, which will be shed on behalf of many for the forgiveness of sins" (Matt. 26:26–28). The same covenant is repeated across the world by priests in parish churches on the mean streets of the crowded cities, in the tiny chapels of lonely missions, and in great cathedrals.

"Since we are united with Christ, who is our Peace," says St. Gregory of Nyssa, "let us put all enmity to death and thus prove by our lives that we believe in him. Let us reconcile not only those who oppose us from without but also those which cause inner disturbances within us; flesh and spirit."[21]

To be the mature Christian described by St. Gregory, one able to put all enmity to death, is the goal of the ordinary Christian, a goal won by constant prayer as well as surrender of the will to God. During many periods of history, the goal seemed far distant, when causing the death of the enemy was chosen over causing death to enmity. But in all periods, the teaching of the Eucharist, of the holy flesh and blood of Christ, has been maintained. The Eucharist, with its concept of the human person as vaulting to the heavens, is a secure ground for Christian peace and peacemaking.

Can we, believing that the Eucharist is the body and blood of Christ, the Messiah, ever destroy the bodies and shed the blood of human beings for whom he gave his life on the cross?

Chapter 10

Crime against God and Man

Any act of war aimed at the destruction of entire cities or extensive areas along with their population is a crime against God and man himself. It merits unequivocal and unhesitating condemnation.

—VATICAN II, *The Church in the Modern World*

Vatican II on War and Peace: Background

Only once in the twentieth century were the Catholic bishops of the entire world gathered in Rome. They came from all the peoples under heaven, called together by an aged pope, John XXIII, who had shown his concern for world peace by addressing a letter to all humanity, *Peace on Earth* (*Pacem in Terris*).

Some twenty-four hundred hierarchs, who poured into St. Peter's Square after each session, made a colorful spectacle, many of them cardinals garbed in billowing scarlet garments. Peace was a chief concern at the fourth and last session of the gathering, known as Vatican II.[1] It opened in October 1965, and the bishops were preparing to make fateful decisions on modern war and its threat of annihilation. There was great expectancy among the world's peoples, not only Catholics, regarding the decisions of the universal church on the most unredeemed aspect of human life, war. In the nuclear age, what could the universal church say about the social acceptance of war as a means of resolving conflict? Under scrutiny by moral leaders would be the socially organized infliction of injury and death by the children of God on each other.

Only at the third session of Vatican II, in 1964, was the debate on war and peace addressed, and then only for one hour to introduce the issue. The subject had not been discussed in the first two sessions held in 1962 and 1963. The peace debate was opened on October 5, 1965, the day on which Pope Paul VI returned to Rome directly after his memorable peace appeal at the United Nations: "No more war, war never again."

One of the most poignant addresses during the October debate was that of Pierre Boillon, bishop of Verdun, France. He reminded his fellow bishops that in his diocese alone, in World War I, there had been 1.3 million casualties, counting injured and dead. The end-result of the prolonged violence had been the winning of a few blood-soaked acres. The agony of so many human beings, blinded, maimed, disabled, and killed, moved him to assert: "The difference between war as envisioned by the theologians of old and war as it exists today is so great that even though the same term be used, the reality underlying the term is entirely different. Our mission," he added, "can be only one, to join with the plea of Pope Paul VI to the United Nations that mankind never again be faced with the specter of war."[2]

Boillon interrupted his speech to announce to the council fathers that a witness of laypeople was taking place in Rome, news of which had been withheld from the press. Nineteen women from six nations were spending ten days in prayer and fasting "in order to ask light for the Council Fathers in their deliberations on the banning of war and the safeguarding of world peace." (Dorothy Day was the only American member of the fasting group.) The fast had been organized by Lanza del Vasto and his wife, Chanterelle, of the Ark Community in Southern France, a community dedicated to simple living and peace in the spirit of the gospel of Jesus.

"These women know," said Bishop Boillon, "that the Assembly of the Council of the Fathers will assume before the world in danger of death, the fearful responsibility of making decisions on problems of war and peace. ... They will fast and pray, ... supplicating the Lord to inspire the Council Fathers with solutions according to the gospel, which the world works for."

A news notice was released to the press asserting that "the fast places a twofold question before the Christian conscience":

> Shall we, before God, and before future generations, assume the responsibility of tolerating any justification of the horrors of total war, and of arms of massive destruction ... ?
>
> Shall we accept the lesson given us by our poorest brethren, whether or not they be Christian? Does not the success in the nonviolent struggle throw new light on the Sermon on the Mount, revealing it not only as a way of personal perfection, but also as a power capable of transforming institutions and of giving a new meaning to history?

The notice carried a general invitation, which included the hierarchs at the council, to visit an exhibit highlighting in photographs as well as books and pamphlets the nonviolent experience of groups around the world.

Lanza del Vasto, called by Gandhi "Shantidas" (Servant of Peace), during his work with the Gandhian movement in India, utilized the exhibit as a center for the spread of nonviolent concepts in various languages.

Change the Course of History?

At the introduction to the question of war and peace, on November 10, 1964, one of the most prophetic addresses of the entire Vatican Council was delivered by the patriarch of Jerusalem and Antioch, His Beatitude, Maximos IV. The patriarch viewed the assembled bishops as "the defenders of the earthly city." He believed that the conclusions reached by the spiritual leaders of the universal church could change the course of history.

The patriarch, his long white beard stark against a black tentlike enveloping garment, announced:

> A threat of destruction hangs over humanity: nuclear armament. And this threat increases daily, because of the growing number of the possessors of these infernal machines.
>
> . . . A cry of alarm rises spontaneously from our heart, a cry of anguish, I was almost going to say, of despair. . . . We beg of you to do everything possible for whatever effect it may have, to avert such a disaster.
>
> The intervention of two thousand bishops from all parts of the world on behalf of peace could change the course of history and safeguard the fate of humanity.
>
> What reason could be sufficient to justify, on any sound moral principles, the kind of destruction that an actual world cataclysm would involve? Can you annihilate a civilization and entire peoples under the pretext of defending them?[3]

The fact that Maximos IV spoke in French, rather than Latin, was a reminder of the non-Roman origin of the Catholic Church. It stressed its link to the Middle East, the bridge between Europe and Asia. His title called to mind the founding of the church in Jerusalem and the actual naming of Christ's followers as "Christians" in Antioch.

A Solemn and Forceful Condemnation

Patriarch Maximos placed before the fathers of the council a duty and a hope, the duty of a courageous condemnation and the hope of its effect on the fate of humankind.[4] "In the love of Christ, who is friend of men and King of Peace, we earnestly supplicate you to pronounce a solemn and forceful condemnation of all nuclear, chemical and bacteriological war." For two thousand years, he reminded his fellow bishops, "history has seen the bishops as 'Defenders of the Earthly City.' "[5] He continued: "The world needs unselfish and courageous defenders today more than ever. Let us not disappoint the world, which is watching and which expects the church to remain forever a pillar of strength and faith."[6]

The "solemn and forceful condemnation" of the methods of modern warfare was a challenge that the bishops could hardly avoid. History showed that the last formal condemnation of a method of warfare by a church council had occurred just over eight hundred years earlier. As noted earlier, the condemnation had been issued in the Lateran Council of 1139, prohibiting the use of the crossbow. Through the crossbow, technology was added to the power of the bow and arrow. A mechanism released a bolt or dart with such power that it tore murderously into the entrails of the opponent. The fathers of the Lateran Council decided that this was too barbarous a weapon to be used against human beings and should be restricted to the killing of animals.

Peace on Earth

Lay action had been in evidence in Rome from the opening of the council in October 11, 1962. Peace activists, heartened by the encyclical letter *Peace on Earth*

(*Pacem in Terris*) of April 11, 1963, hoped that the opening to peace contained in that historic document would become a broad highway of peace at Vatican II. Pope John XXIII, in sending out the call for a council of the world's bishops, hoped for an *aggiornamento* of the church, an updating and renewal to meet the challenge of the modern world.

There were some ominous signs that as far as the issue of peace was concerned, the gains of *Peace on Earth* might well be lost. Catholic peacemakers had found in Pope John's letter a strong basis for renewed peace efforts. They were more than ready for such words as: "Justice, right reason and humanity demand that the arms race should cease. That the stockpiles which exist in various countries should be reduced equally and simultaneously by the parties concerned. That nuclear weapons should be banned." Prepared in the fall of 1962, during the height of the Cuban Missile Crisis, *Peace on Earth* stated that "it is alien to reason to imagine that in the atomic era war could be used as an instrument of justice."[7]

The revolution embodied in *Peace on Earth* was pointed out in an article in the *Nation* magazine by E. M. Borgese. "The truly revolutionary feature of the encyclical," he wrote, "was the abandonment of the age-hallowed distinction between just and unjust wars. 'It is alien to reason to suppose that in the atomic era war could be the instrument of justice.' Pope John XXIII did not say, 'war waged by atomic weapons'; but war as such is ruled out in the atomic era."[8]

The encyclical letter validated civil disobedience ("God has more right to be obeyed than man"). It argued against the Cold War rhetoric of the time, and disconcerted many "cold warriors" by urging coexistence between differing political systems.

The Easter Sunday 1963 issue of the *New York Times* mirrored the effect of *Pacem in Terris* in its lead editorial. It stated:

> The encyclical which Pope John XXIII addressed to all humanity this Easter week has struck a responsive note among millions. Protestants, Jews, Buddhists, Muslims, communists and atheists have joined Catholics in approval of the Pope's moving words.... The most striking demonstration of the Pope's desire for reconciliation of all mankind is the encyclical's veiled but unmistakable references to Communism.... What a blow this attitude deals to the ideological fanatics on both sides of the doctrinal dividing line who reject all ideas of a reasonable compromise and think only of burying their opponents.

Other reactions to Pope John's letter came from within the church. The pope's defiance of the political positions of the nations, and the blocs of nations, led Archbishop Lorenz Jaeger of Paderborn, Germany, to assert that in *Peace on Earth,* the church had reached "the end of the Constantine Era." Whether or not the church had broken with the era of alliances with temporal powers could be debated, but undoubtedly an important step had been taken in viewing communism, for all its evils, as a system which encompassed millions of human beings who were children of God.

Fifty Million Catholics in an Awesome Dilemma?

The indications that the gains of *Peace on Earth* might be threatened came from a committee of American Catholics belonging to the Catholic Association for International Peace (CAIP). Members of the committee had reached the attention of some of the council fathers in a statement urging them not to ban nuclear weapons. The statement was entitled "Morality, Nuclear War, and the Schema on the Church in the Modern World." It was written in October 1964, and a copy was sent to an American father of the council, Bishop John J. Wright of the diocese of Pittsburgh. It is not known how many of the fathers of the council had also received a copy when the council debate on peace was opened in November 1964. A copy eventually reached the leaders of Pax, a Catholic peace group much younger than CAIP, having been founded in New York in 1962 as a branch of Pax, England. Among the purposes of Pax, founded in England in 1936, was the education of Catholics in gospel-based peacemaking. The leaders of Pax in England and the United States expected momentous decisions on war and peace at the last session of the council in 1965.

Pax members were aware that the Vatican II document dealing with war and peace was in its drafting stage and was referred to as Schema XIII. The section on war and peace, entitled "On Making Lasting Peace," was listed as article 25. The decision arrived at would eventually appear in the *Pastoral Constitution on the Church in the Modern World.*

The proposal of the CAIP leaders was in direct opposition to the prophetic utterance of Patriarch Maximos IV. It was listed as the work of the Arms Control Subcommittee of the Catholic Association for International Peace. The subcommittee was part of the International Law and Juridical Order Committee of the Catholic Association for International Peace. The statement was crafted by six Catholic laymen and was not a formal CAIP statement but rather represented the views of the drafters.

The CAIP committee gave clear advice to the council fathers. The schema should not include, they asserted, "a blanket condemnation of nuclear weapons on the grounds that their effects are incalculable and cannot be reasonably controlled by men. This would be misleading."

A particular sentence galvanized the activities of Pax leaders: "Should the Council issue a blanket condemnation of nuclear weapons, it would place fifty million American Catholics in an awesome dilemma as to whether to listen to the solemn findings of a Vatican Council or to the hitherto accepted assurances of their government that America's nuclear deterrent is the foundation for international stability and the "sine qua non" of the defense of the United States."

The members of the Arms Control Subcommittee made other recommendations to the world's bishops. They urged them to consider the likely consequences of "a unilateral abandonment of nuclear deterrence." They pointed out that the compliance of the communist regimes, also possessors of a nuclear deterrent, could not be expected. They urged on the Vatican council the necessity to have recourse to qualified professional recommendations, such as those from persons who participated in shaping the CAIP statement. Later research revealed that while the CAIP was anxious to have its views known at the council, it anticipated that contrary views might also be aired. Patricia McNeal cites a letter by the president of CAIP to a member on the issue of nuclear warfare:

> No subject is of more importance today, as you well know, and I am pleased to hear that you are planning to try to get something out to the Council. The European pacifist influence in Rome may need to be offset. Here, too, we need to speak out because of the formation of Pax. Like all extremists, they have the most persistent, devoted and possibly persuasive adherents.[9]

The Catholic Association for International Peace had every right to make their position known. Those who opposed that position had the same right to express their point of view.

The Peace Lobby

As the president of CAIP had anticipated, a response came from the members of Pax. The Pax action formed part of the larger peace lobby, the most dramatic statement of which was the fasting of the women at the last session of the council.

The peace lobby, which functioned during some or all of the four sessions of the council, included Hildegard Goss-Mayr and Jean Goss, leaders of the International Fellowship of Reconciliation. The husband and wife team had a long history of contacts with church leaders on the issues of peace and war, and maintained their contacts during the council. Gordon Zahn, whose book *German Catholics and Hitler's Wars* had jolted the consciences of many believers, found a warm reception for the concepts of conscientious objection and gospel nonviolence among several council fathers, in particular English bishops.

James W. Douglass, the first of whose books, *The Nonviolent Cross,* had awakened many to the implications of church teaching on peace and war, took up residence in Rome during the first two council sessions. He discussed peace and nonviolence with many fathers of the council and their *periti* (specialists). His ideas were reflected in the statements of several hierarchs, in particular that of Bishop John Taylor, the American bishop of Stockholm, Sweden.

In October 1965, while the members of the "praying lobby" led by Lanza del Vasto were cloistered in the Cenacle Convent outside Rome, the peace lobby of activists was reduced to three persons. One of them, Barbara Wall of English Pax, was the fortunate possessor of a *tessera* (an official press pass). This allowed her to attend the Vatican II Masses in St. Peter's Basilica. The second member, Richard Carbray, was a university professor, a Latinist, and a peace activist who served as volunteer secretary to Archbishop Thomas D. Roberts during the four council sessions. Roberts, the former archbishop of Bombay, was considered one of the leading voices for peace at Vatican II. As a founder of American Pax and its publications editor, I was the third member of the peace lobby.

American Pax officers, having read the statement of the Arms Control Subcommittee of the Catholic Association for International Peace, were convinced that the council fathers should be apprised of the fact that opposing strands of thought existed among American Catholics. They also felt that it was necessary to make it clear that the CAIP statement represented only one strand.

Dorothy Day: Mentor and Sponsor

The activity of the peace lobby of three in Rome had been preceded by a summer of preparation. In May 1965 we met with Dorothy Day, a mentor and sponsor of Pax, to discuss the challenge before us. It consisted of informing the world's bishops that in the American Catholic community, citizens of a nuclear superpower, there existed opposition to the possession and use of nuclear weaponry, and to the whole concept of the nuclear deterrent. There were other peace issues of critical concern, as mentioned earlier — the explicit right of Catholics to be conscientious objectors to military service and war, the right to "selective objection" to a particular war or means of war, and the validity of the witness of gospel nonviolence.

Could we prepare a special edition of the *Catholic Worker* to share with the council fathers? I asked Dorothy Day. We had learned that about 150 bishops were charged with the final wording of article 25, "On Making Lasting Peace." Dorothy Day gave me the task of editing the special edition. We were given sufficient funds by a Pax member to share the special edition not only with the 150 framers of the peace/war section, but with all the fathers of the council.[10] The fund would provide for the airmailing of the *Catholic Worker* to all the members of the hierarchy listed in the *Annuario Pontificio*.

My task was rendered easy by the willingness of every person I approached to contribute an article or a message of peace. They included James W. Douglass, Gordon Zahn, George H. Dunne, SJ, Father John McKenzie, Dr. Benjamin Spock, Philip Scharper, Howard Everngam, chairman of Pax, and Dorothy Day herself. I contributed the editorial.

The issue of the *Catholic Worker* of July 1965 was accompanied by a covering letter from Howard Everngam. He began:

> We are addressing you as a spiritual Father since in the renewed sense of collegiality, the bishops are the Fathers of all the faithful. We are a group of American Catholics united against what Pope Paul VI has called "that murderous device, the nuclear bomb and all that it represents."

The Immorality of Warfare Based on Deterrence

The special edition of the *Catholic Worker* was a tabloid of eight pages. On its stark first page was the title, in heavy black letters, of James W. Douglass's article "The Council and the Bomb." Above it, in letters almost an inch high, were the words of Pope Paul VI: "We are responsible for our times, for the life of our brothers, and we are responsible before our Christian consciences."[11] The entire speech of Patriarch Maximos IV was featured in the *Catholic Worker*, as was the preliminary text of article 25, "On Making Lasting Peace." Besides going by airmail to the world bishops, the *Catholic Worker* went to its usual eighty thousand subscribers.

"In view of the Church's doctrinal understanding of Scripture and of the nation's deepening involvement in total war," wrote James W. Douglass, "there would be nothing imprudent in the Council's support of a Christian dedication to total peace, especially by a recommendation in the Schema that each Christian explore in conscience the nonviolent love and teaching of the church."

Father John McKenzie, whose *Dictionary of the Bible* had just appeared, stated: "If the warfare based on the deterrent is not immoral warfare, then there is no immoral warfare." George H. Dunne, SJ, of Georgetown University, asserted that "neither the manufacture nor the use of nuclear or other indiscriminate weaponry could be morally justified."

Gordon Zahn, in his article "American Experts and Schema XIII," discussed the CAIP proposals in detail. He made the point that whatever the assurances of the U.S. government regarding the deterrent, the Christian has another ruling consideration. This must be "always and everywhere . . . the law of God and the counsels and example of his Son."

Dorothy Day devoted her "On Pilgrimage" column to Pope John XXIII's *Journal of a Soul*, and to her thoughts on peace. She explained that a Catholic peacemaker had asked her for a clear, theoretical, logical pacifist manifesto. She had not, in the thirty-two years of the Catholic Worker movement, been able to formulate such a manifesto.

Weapons of the Spirit

Day wrote, "Unless we use the weapons of the spirit, denying ourselves and taking up our cross and following Jesus, dying with Him and rising with Him, men will go on fighting, and often from the highest motives, believing they are fighting defensive wars for justice and self-defense against present or future aggression."

In "Questions on Modern War," Howard Everngam pointed to the fact that millions of works for social justice were stillborn because of the outlay in that year (1965) of fifty billion dollars on weaponry. Among the questions he posed to the bishops was one that went to the heart of nuclear warfare: "Is not any act or policy of direct or indiscriminate killing of innocent and noncombatant civilian populations to be condemned?" The query evoked an echo in the council's final condemnation.

Everngam also posed the question, "Does not every individual have the right and the duty in conscience to abstain from any form of war or killing which he judges not to meet the requirements of reason and morality?" A response to this query was also clearly echoed in the council's final document.

Six key statements from *Peace on Earth* were boxed and printed in bold type.

In the editorial "We Are All under Judgment," I referred to the judgment of love, the love that the follower of Christ owes even to the one called "enemy."

Citing the thousands of nuclear weapons in the American arsenal, I pointed out that the Russians and Chinese knew that the death-dealing missiles were aimed at them. Somehow, Christians must dissociate themselves from this threat and convey a transcendent concept to the children of God in the Eurasian heartland. For the followers of Jesus, they are seen not as targets for destruction but as objects of that universal love brought by Jesus. I drew attention to Thomas Merton's heart-stopping assertion that "total nuclear war would be a sin of mankind second only to that of the crucifixion." "It is in order to avoid that great sin," I concluded, "that we beg clear words from the fathers of the council."

Patricia McNeal has remarked that "at the Second Vatican Council the positions of the just war and pacifism collided."[12]

Water Nourishes

The three peace lobbyists were able to follow on a daily basis the developments on peace and war in the council. Through friends in the Documents Section and other offices we received copies of the key statements. Richard Carbray proved invaluable as a co-worker since his knowledge of Latin allowed him to translate at sight all the council documents that came to us.

In addition to the *Catholic Worker,* we had with us in Rome the printed text of an editorial from *Peace,* the Pax quarterly magazine, entitled "Will Nationalism Spoil the Schema?" One of the *periti* felt that this title and the editorial itself were too provocative.

"Any slightest reversal of *Peace on Earth* will not go unnoticed," the editorial affirmed. "The Fathers of the universal Church can give the Schema a Christocentric vision of peace that may move Catholics to become a worldwide community of love, reconciling man to man as Christ reconciled man to God."

We wondered if there were a way to put the editorial directly in the hands of the 150 bishops and *periti* engaged in drafting the peace/war section. Abbot Christopher Butler, OSB, was one of the chief drafters of the section. When I approached him, I mentioned that we of Pax had formulated another peace statement in addition to the special issue of the *Catholic Worker.*

Abbot Butler, in his correct English way, asked what we had in mind. I explained that our position on war and peace was distinct from that of some American bishops and of some lay Catholics whose proposal had reached Rome. Our statement would supply a viewpoint that the drafters might not be likely to receive from other sources. The abbot agreed that this was a reasonable argument in favor of distributing the Pax editorial. Then he added, "Of course, it must be in Latin. We need it now, however."

Richard Carbray, happy at this breakthrough, rushed with me to the library of the North-American College on the Janiculum Hill. Latinist though he was, he worked to consult the dictionaries for the exact equivalent of some crucial words. For one word, "overkill," he had to coin a Latin equivalent, "superhomocidium."

The copies were delivered the following day to Abbot Butler at the Benedictine Monastery of San Anselmo.

The debate was taking place in St. Peter's, and we were waiting to learn when

Patriarch Maximos would be making his statement. When we learned that he might not speak, we decided to visit him. He received us in his quarters on the top floor of the Salvator Mundi Hospital. After some preliminary conversation, which involved thanking him for his unforgettable peace appeal, we came to the point. Would he rise again to speak in St. Peter's before the debate came to a close? Did he not think he needed to reemphasize to the council fathers his warnings about modern war's danger to the human family?

There was silence. Like an ancient of days, the aged patriarch, with his long, snow-white beard outlined against his black garment, sat before us and meditated. At last, he spoke, saying in French: "I have already spoken."

Our attempt to move him to change his mind centered on the example of Dorothy Day, who, though sixty-eight years of age, was on a water fast on behalf of the fathers of the council. He shook his head slowly from side to side.

There was another silence. Then in a far-off voice he said, "L'eau nourrit!" (Water nourishes!). This came from a patriarch of the Middle East, acquainted with parched lands. We had no other argument. We thanked him and took our leave.

Presumption of Justice?

On our rounds to the cardinals and bishops, we stressed several concerns besides the threat of indiscriminate war. One was the need to affirm the right to conscientious objection, including selective objection, to military service and war; another was to eliminate what Barbara Wall called "weasel clauses" in the draft, meaning language that allowed for an escape from a clear or unequivocal stand. One "weasel clause" endangered support for conscientious objection, and had survived several drafts. It asserted that, in questions of participation in war, "where the law of God is not manifestly clear, presumption of justice must be acknowledged to lawful authority and its commandments must be obeyed."

This tallied with a common conception of Christians at that time. It was expressed by Pope Pius XII in his Christmas message of 1948. The pope explained that among the goods of humanity, "some are of such importance for society that it is perfectly lawful to defend them against aggression. Their defense is an obligation for the [people of a] nation as a whole, who have a duty not to abandon a nation that is attacked. The certainty that this duty will not go unfulfilled will serve to discourage the aggressors."[13]

The "presumption of justice," we felt, showed an incredible misreading of the role of Christian citizens as they faced the war-making nations of the twentieth century. The drafted German soldiers in World War II were advised to carry out their duty "with faith in the cause of the people." The drafted American airmen obeyed orders to bomb European cities and incinerate their inhabitants, and eventually, to unleash the lethal fire over Hiroshima and Nagasaki.

In such thinking the actual participants in war were given to understand that in any conflict, ordinary citizens cannot make a moral judgment that diverges from that of the all-knowing "lawful authority." Thus, one could argue that if they were ever able to make a moral judgment, it would have to occur after the conflict, when all the facts were known. By that time all the corpses would be gathered and the dust of those incinerated would have settled.

The "weasel clause" containing "the presumption of justice" was expunged from the draft and did not appear in the pastoral constitution *The Church in the Modern World.* The significance of this omission was that the individual expected to take part in a war would need to satisfy his or her conscience as to its moral rightness. Citizens could no longer abdicate their consciences in favor of the all-knowing state. Conscience, not the presumed rightness of the warring nation, would have to be the person's guide.

Another "weasel clause" affirmed that it would not be illegitimate "to possess modern arms for the sole purpose of deterring an adversary similarly equipped." "Modern arms" and "scientific weapons" were the code words for nuclear weapons. This wording, supported chiefly by a few American bishops, opened the door to the possession, and therefore use, of nuclear weapons. In the Pax editorial "Will Nationalism Spoil the Schema?" we pointed to the danger of "nuclear nationalism" and urged the fathers of the council to promulgate a peace statement that would go far beyond the concerns of nationalism.

Lobbying in Rome

Our reception by cardinals, archbishops, and bishops was, on the whole, cordial, even, at times, enthusiastic. They listened as we referred to the positions expressed in the *Catholic Worker* and the Pax editorial. The banning of the bomb, the elimination of the "weasel clauses," and the witness of gospel nonviolence were accepted by many, but not all, of the hierarchs. Cardinal Leger of Montreal said that our views coincided with his own. We were not able to talk with Cardinal Joseph Ritter of St. Louis, Missouri, but his secretary shared with us the text of his intervention. It called for the "absolute condemnation of the possession of arms which involve the intention or grave peril of total war." The cardinal asserted: "The condemnation must be pronounced so clearly and distinctly that no one will be able to hide or twist its meaning."[14]

Archbishop George B. Flahiff of Toronto congratulated us on our efforts. He had earlier published thoughtful articles on the little-known Catholic conscientious objector to the Crusades, Ralph Niger, as discussed in chapter 4. Flahiff urged that the rights of conscience in wartime should be respected. He asserted, "To cooperate in preparation for total war is thus to commit oneself to its execution and must be condemned equally with the act of execution."[15]

Many bishops had concerns of their own that they shared with us. One who made his point with great emphasis was Archbishop Raymond-Marie Tchidimbo of Conakry, Guinea. "The church," he told us, "should speak out on the solidarity of all men. The solidarity implies more than charity, for as St. Thomas teaches, superabundant goods belong by natural law to the poor for their sustenance." That solidarity, he felt, was breached by preparation for war. Such expenditures brought misery to the poor, who lacked the necessities of life. We were to hear of Tchidimbo later when he had to endure eleven years of imprisonment in his country until freed by the action of Amnesty International.

Our visits to the bishops went on from morning until late evening. We sped around Rome using the list of Roman addresses of the council bishops left by James Douglass with Richard Carbray.

Peace at St. Peter's

Then came what Barbara Wall termed the *dies mirabilis* for peace, October 6, 1965. Although, like Pope John XXIII, he did not participate in the council debates, Pope Paul VI did make an address to the assembled bishops. Within a day and a half, he had left Rome, had addressed the General Assembly of the United Nations, and had flown back to the Vatican.

In his address to the bishops, he said: "We give thanks to the Lord, Venerable Brothers, to have had the good fortune to announce to all the world a message of peace. The gospel message has never before had such a large audience or — we can say — an audience more ready and willing to hear it.... We must be now more than ever workers for peace." He added that the contribution of the church will be more "efficacious and precious insofar as all of us are convinced that peace must have justice as its foundation."

The debate in St. Peter's continued with affirmations of peace.

Bishop Alfred Ancel of Lyons, France, pointed out that the good of the human race demanded an unqualified ban on war. "It is objected," he asserted, "that the Council's rejection of war would serve no good purpose because no one would pay any attention to it. The effectiveness of our statement is not expected to come from the strength of arms but from the fact that acting in the Person of Christ we bear witness to truth before the entire world."[16]

Abbot Butler asserted that the intention of waging war unjustly (using indiscriminate weaponry) was itself unjust. He asked that the section claiming that the possession of "modern arms was not illegitimate as long as they are used for deterring an enemy similarly equipped" be expunged from the schema. He supported the strengthening of the right of conscientious objection, pointing out that such objectors may in fact be "prophets of a truly Christian moral order." He dismissed out of hand the concept of the "presumption of justice" in favor of the state.

He carried forward Bishop Ancel's concept of Christian witness. "We are the mystical body," he said, "and Christ is our head. He refused to defend himself and his mission by the sword of his disciples, or even by legions of angels. The weapons of the gospel are not nuclear but spiritual. The gospel wins its victories not by war but by suffering."[17]

Bishop George Andrew Beck, of Liverpool, England, asked, "May a state even threaten, by way of deterrence, that indiscriminate destruction of cities and whole regions which our Schema is condemning as a crime against God and man?" He expressed strong support for conscientious objection, even selective objection, to certain methods of warfare. He likened the moral objection of a surgeon refusing to kill a child in the womb to that of an airman refusing to release a nuclear device which would obliterate a town or region.[18]

Cardinal Bernard Alfrink of Utrecht, the Netherlands, president of Pax Christi, the international Catholic movement for peace, urged the elimination of the clause legitimizing the nuclear deterrent. Bishop Charles Grant and Bishop Gordon Wheeler, both of England, followed Alfrink in attacking the schema's treatment of nuclear arms. Wheeler urged the council to reaffirm the right and duty of conscience in disobeying the unjust commands of authority.

If October 6 was a *dies mirabilis,* October 7 was no less so.

One of the hierarchs in the Vatican most identified with rigid conservatism was the secretary of the Congregation of the Holy Office, Cardinal Alfredo Ottaviani. He not only joined the peace train of his brother bishops but traveled several stations ahead of where they would be willing to go.

He supported the right and duty of people to reject their own legal government if it is leading them to a ruinous war. He left his brother bishops behind when he urged the council to call all nations to unite in a worldwide republic, transcending national barriers. This would end the threat of nuclear doom and open the way for the peace of Christ to reign throughout the world.

The fact that Cardinal Ottaviani's rigid image among the council fathers was revised by this address caused them to award him an ovation. It was reported as the longest and warmest ovation of the four sessions of the council.

The open debate was over, but the final wording of the peace section was still to come. The whole section, we learned, was entitled "The Fostering of Peace," and included "The Avoidance of War."

The final voting on the complete pastoral constitution *The Church in the Modern World* was to be delayed until almost the end of the council.

We watched the fathers of the council streaming out of the basilica to fill St. Peter's Square. Cardinal birds were well named, their plumage no more brilliant than the scarlet robes of the Catholic cardinals. Hardly less striking were hierarchs in darker crimson robes. Among them all were the gray- and brown-robed ecclesiastics, and an occasional one in black robes, these standing out like drab birds among those of more spectacular plumage. Would there be in our lifetime a similar scene, men of God from every people under heaven, gathered under the sign of the Prince of Peace? Would their message meet the need of the time and would it, in any case, be heeded and acted upon?

Two council fathers from the United States, one a cardinal and one just made an archbishop, were absent from Rome when the voting on the various sections of *The Church in the Modern World* took place.

On their return to Rome, the two ecclesiastics, Cardinal Francis Spellman of New York and Archbishop Philip M. Hannan of New Orleans, led a movement of dissent against the peace section. We obtained a copy of their statement, cosigned by eleven bishops, four of them American. The remaining signers came from Mexico, Argentina, Australia, Lebanon, and South Africa. In explaining their objections to the text of "The Avoidance of War," the dissenters pointed out that they were opposed to the whole tone of the section. In particular, they objected to its negative view of deterrence. They disagreed with the assertion that "scientific weapons increase the danger of war."

"These affirmations," said their statement, "do not recognize the fact that the possession of 'scientific weapons' has preserved freedom in a great part of the world. The defense of a great part of the world against aggression is not a crime but a service.... In the world of today, there is no adequate defense for the more powerful nations without the possession of scientific weapons."

The statement, circulated to the council members on December 3, 1965, proposed that the "errors" be corrected and that other wording be substituted. Failing this, the whole pastoral, *The Church in the Modern World,* should receive a *non placet,* in other words, be voted down. It should, at a later time, be turned over to a synod of bishops for further study and correction.[19]

The American attempt to "torpedo" the peace/war statement failed. *The Church in the Modern World* became an official document of Vatican II at its 168th and last general meeting on December 6, 1965.

Crime against God and Man

The condemnation by the bishops of the Catholic world united at Vatican II dealt with the threat to the human family from modern war. The council fathers cut through all discussions by so-called experts and realists on "scientific weapons" by concentrating on their effects on human beings and their earthly home. Patriarch Maximos IV of Antioch and Jerusalem was vindicated. The gains of *Peace on Earth* were not reversed.

Pointing to the need for an "evaluation of war with an entirely new attitude," and to the fact that the people of our time "will have to give a somber reckoning of their deeds of war," the bishops stated in *The Church in the Modern World:*

> Any act of war aimed at the destruction of entire cities or extensive areas along with their population is a crime against God and man himself. It merits unequivocal and unhesitating condemnation.[20]

The Church in the Modern World began by asserting: "The joys and the hopes, the griefs and anxieties of the men of this age, especially those who are poor and in any way afflicted, these are the joys and hopes, the griefs and anxieties of the followers of Christ." It then stated that Vatican II addresses itself "not only to the sons of the Church and to all who invoke the name of Christ, but to the whole of humanity."

The document ended with chapter 5, "The Fostering of Peace and the Promotion of a Community of Nations." It asserted that in this hour of "supreme crisis," a more human world order will not emerge "unless each person devotes himself to the cause of peace with renewed vigor." For members of the church, this involved an explicit obligation: peace was recognized as a constituent element in the living out of the gospel. Above all, the document pointed to peace as "the fruit of love, which goes beyond what justice can provide."

The Church in the Modern World made explicit the recognition by the universal church of the right of conscientious objection to military service and alternative civilian service. It included a validation of the witness of gospel nonviolence (discussed in chapter 2, above).

Vatican II resulted in palpable changes in the church itself, so that church policies and actions were described as "pre-Vatican II" or "post-Vatican II." It began to have its effect on the members of the church, particularly on the younger members. One example was the recognition of conscientious objection. It was welcomed by young people in the United States, liable as they were at that time to conscription to military service in Vietnam. The greatest increase in the number of conscientious objectors to service in the Vietnam War was registered among Catholics.

Whether or not the condemnation of indiscriminate warfare was recognized for what it was, the condemnation had been formally issued by the universal church. Peacemakers around the world saw it as a condemnation of all modern war, which is necessarily indiscriminate: its weapons, which can shatter the life of a city, are too powerful to make the distinction between combatants and civilians.

As some bishops pointed out, the preparation for indiscriminate warfare and the threat to carry it out also fall under the condemnation. It may be that governments will not heed this condemnation, but the aroused consciences of millions of citizens may in the end hold back the destructive power of their governments.

Vatican II gave the bishops of the world the opportunity to show themselves as "defenders of the earthly city" as well as of the City of God. It was a time of *kairos,* a graced time, when the fathers of the universal church demonstrated their love and concern for the fate of humankind.

Chapter 11

Violence Forestalled

The Philippines and the Iron Curtain

One can only live if one's life serves to change this situation [of violence] to diminish all that is contrary to the life of man. If I continue to live, I must give my life so that violence may be conquered. —HILDEGARD GOSS-MAYR

For me the church is the community of Christians, of people who try to love as Christ loved them. Before, I had loved my fellow co-workers and not all other human beings. But in Christ I encountered the man that I had desired from the depth of my being: the one who loves all human beings, the good and the bad, the workers and the managers, believers and bishops, exploited and exploiters, for they have all been created by the Father in one immense love.

—JEAN GOSS

The Philippines: The Revolution of Love

Asserting that he had become "the field-marshal of the unarmed forces of the Philippines," Cardinal Jaime Sin of Manila expressed wonderment that an entrenched dictatorship had been ended by the methods of nonviolence.

The cardinal was speaking at commencement exercises of Stonehill College in Boston on June 1, 1986, just four months after two million people had filled the streets of Manila, not in rioting, but in acts of nonviolence. So dramatic was their witness that photographs of it went around the world. "It was a revolution of love," the cardinal explained. The international press carried the scene of a bouquet of flowers being offered by a slender girl to the armed marines as the people around her refused to budge from their position; the scene of nuns reciting rosaries as they faced oncoming army tanks, which had come to a halt; the picture of a tiny statue of the Virgin Mary being raised before the soldiers in an armed personnel carrier.

Cardinal Sin needed to serve as a "field-marshal" of the forces of the nonviolent for only seventy-seven hours in the last four days of February 1986. He had helped to prepare for those memorable hours, issuing with his brother bishops the historic pastoral letter of February 14 urging "the active resistance of evil by nonviolent means."

Niall O'Brien remarks that for the first time in recent history a hierarchy had condemned a dictator before he fell.[1] The people of the Philippines had prevailed by steadfast nonviolence against an armed force of three hundred thousand backed by the lethal might of modern weaponry.

The Manner of Christ

Credit was given by the cardinal to a unique married couple, Hildegard Goss-Mayr and Jean Goss, whose training in nonviolence grew out of a life of witnessing to the nonviolence of Jesus. What follows is an account of two Catholic peacemakers who challenged their church's blessing on so-called just violence and who evolved a method of response which inspired countless men and women to find creative ways to meet injustice and oppression with the nonviolence of the gospel of Jesus.

Boston had a place in the Philippine story since it was Boston which gave refuge to an exiled political dissident, a hero to the Philippine people, Benigno Aquino. Before his exile, he had spent seven years in jail in the Philippines, part of the time in solitary confinement, for opposing Ferdinand Marcos and his regime. Near the end of his time of exile, Aquino made known to his wife, Corazon Aquino, and to his friends that he saw no hope in violence dislodging those in power in an established national security state. He was studying the methods of Gandhi and the implications of the gospel for social change. He informed his friends that he had decided to return to his homeland to take part in the struggle for the restitution of democracy and human rights. He was fully aware of the danger that faced him.

The plane that brought Benigno Aquino, affectionately called Ninoy, to the Manila airport was awaited by his family and followers and also by a contingent of the Philippine army. As he stepped out of the plane and onto the steps that led to the tarmac, his body crumpled. He had been shot, but no sound was heard by the onlookers, so it was assumed that a silencer had been used. At the bottom of the steps, Ninoy lay, inert, on the tarmac, a corpse. The hopes of many died with him on that day, August 2, 1983. If an assassination could be achieved so openly, what could dissidents expect? If ever there was a time when the Filipino people could have been justified in rising in rebellion against the dictatorship, it was at the death of Benigno Aquino. Hundreds of thousands of mourning Filipinos viewed his bloodied body in an open casket — the body just as it was when it had been claimed from the military hospital. The wake lasted ten days. Manila was thronged with people on the day of the funeral. They were ready to move against oppression. Yet violence was forestalled.

The account of the forestalling of violence by a swift and bloodless revolution begins with the decision of Cardinal Sin and other religious leaders to bring Hildegard Goss-Mayr and Jean Goss to the Philippines. The couple were asked to present training seminars in gospel nonviolence. The Philippine seminar followed the pattern of similar seminars conducted in corners of the globe where violence was likely to erupt or where structures of violence and oppression might provoke counterviolence. Jean Goss and Hildegard Goss-Mayr had been presenting training seminars in what they termed evangelical nonviolence for over thirty years in their capacity as traveling secretaries of the International

Fellowship of Reconciliation. Their witness had taken them into the countries of Latin America and the Caribbean, into Angola and Mozambique and South Africa, and into countries then behind the Iron Curtain, as well as the Soviet Union itself. They were in all likelihood the first persons to penetrate the Soviet Union with the message of nonviolent resistance and its methods. They even took their seminars into that crucible of violence, the Middle East.

Their work was done without publicity, often leaving transformative marks on souls and conduct, as with bishops and peace activists in Latin America. One of the persons whose heart was opened to the response of nonviolence to unjust structures was Archbishop Helder Camara, who became one of its most articulate spokesmen. Those so transformed often went on to receive peace awards. As mentioned earlier, one who received the Nobel Peace Award, Adolfo Pérez Esquivel, headed Paz y Justizia, the organization founded by Hildegard Goss-Mayr and Jean Goss. Hildegard and Jean went about the world in the manner of the first apostles, on fire with sowing the good seeds that would nurture the kingdom.

Cardinal Jaime Sin had no hesitation in inviting Hildegard and Jean to bring training seminars in evangelical nonviolence to his priests and people. Their work in many countries, and in particular their witnessing for peace with various hierarchs in Rome during Vatican II, had earned for them the trust of churchmen.

Hildegard Mayr, born in 1930, was influenced toward gospel nonviolence in her own household. Her father, Kaspar Mayr, was a peacemaker who became one of the founders of the International Fellowship of Reconciliation. After the *anschlusss* of Austria and Germany, his opposition to the policies of the regime cost him his position as a teacher. He supported his family of five children on the meager income of a translator. Hildegard related to me that when she was ordered, along with her classmates, to raise her arm in the National Socialist salute, she was the only child in her class to refuse. She recounted that she was subjected to verbal abuse as a child and young girl, because of her father's and her own dissidence under the Austrian regime.

Ten years after the end of World War II, she reached a conclusion that gave direction to her life. Faced with the violence, destruction, and chaos of postwar Vienna, she resolved, "One can only live if one's life serves to change this situation, to diminish all that is contrary to the life of man. If I continue to live, I must give my life so that violence may be conquered. . . . And so I decided to commit myself, through the Fellowship of Reconciliation, to justice and peace by the force of nonviolence."[2] Her doctorate, which she received from the University of Vienna, was in philosophy and history.

Those of us who knew Jean and Hildegard were aware that they were an extraordinary couple, utterly distinct in temperament. Jean Goss, ardent, outgoing, and sometimes explosive in his avowal of God's overpowering love, was the complement of Hildegard Mayr. She was collected, thoughtful, measured, the kind of person who could spend time in libraries pursuing philosophical and historical research.

The background of Jean Goss contrasted at every point with that of Hildegard. His father was an anarchist in the French working class. Born in 1912,

Jean began to work at the age of eleven. When he found employment with the French railways, he became a labor union activist. Conscripted in 1939 into the light artillery of the army of France, he was awarded the Croix de Guerre, the military medal for the defense of France. He wrote of his experience:

> During the invasion, by day and by night, I killed many Germans who entered France as if it were their own country. . . . My regiment was ordered to kill them at any cost: How was it that I — a member of a labor union who wanted to defend the dignity and life of men — could kill workers, peasants, young people like myself, of the German people?
>
> Although I was decorated and a French hero, I felt more and more crushed within myself.[3]

On Easter night 1940, after Jean had been taken as a prisoner of war, a vision took over his consciousness. It was the vision of a Creator whose love envelops all, who gave his life on the cross out of the love for each one of them. The message he was to tell the world was: "Teach them to love one another as I love them. Teach them quickly."[4]

Jean spent the next five years in a prisoner-of-war camp. In the camp, he worked side by side with a man whose speech was punctuated with hate and revulsion against all Germans. Jean remonstrated with him about universal love and the enlightenment that filled his soul on Easter night. The man revealed himself to be a priest and urged Jean to share his insights with fellow prisoners. Jean had discovered that the blazing love he experienced could not be lived alone; it could only be lived with the help of others, of a community.

One Sunday evening he shared with his comrades this love, which, he insisted, should also be shared with German soldiers. He asked them to alert him whenever, in their judgment, he failed to show love, any occasion where he seemed to betray love. Whenever it was necessary to support a prisoner who was falsely accused, they looked to Jean Goss to speak up. Whenever he did not, Marxists and others would ask, "Goss, is that your love?"

Once when his courage failed utterly in helping a fellow prisoner, Jean confessed his lack of courage. A Marxist commented, "So, it is impossible, this love, just a utopia?"[5] Jean persisted, gaining courage, even to address the camp commandant about his treatment of prisoners. People came to him for help, priests, pastors, workers, all attempting to glimpse the love that had done so much for Jean. About eight fellow prisoners joined him in the daily practice of loving response, and in so doing helped in some measure to transform the camp of war prisoners.

Jean Goss concludes, "It is this revolutionary force of nonviolence which continues to make me travel throughout the world, even now that I am seventy-seven years old. . . . "[6]

A highlight of Jean's life was his meeting with Cardinal Alfredo Ottaviani at the Vatican in 1950. He had heard that the cardinal, considered to be among the most conservative members of the Vatican curia, had condemned modern war. The translation of the cardinal's statement "Bellum Omnino Interdicendum" (War is to be altogether forbidden) moved Jean to travel to Rome. Evading the Swiss guard, he ran up the stairs and landed in the Office for the Doctrine of the Faith. He learned from the cardinal's lips that he had actually condemned modern war. This incident Jean related to me with delight.

Ready with their training seminar in evangelical nonviolence, Jean and Hildegard arrived in Manila in February 1984. Cardinal Sin had spread the word among other bishops of the Philippines, notably Bishop Francisco Claver, who had fought tirelessly to spread the concept of the nonviolent dimension of the gospel. Groups of priests, religious, and laypeople had been given times when they could assemble to pray and search the gospels for their message of nonviolent resistance to evil. Jean and Hildegard met first with a group in Manila which included Corazon Aquino, who had suffered during her husband's long imprisonment and had learned from his letters his conviction that Christ's message and Gandhi's commitment to nonviolent action presented the way to proceed. Other groups were prepared for Jean and Hildegard's presence, and when all the first groups had participated in the training seminar, Jean and Hildegard had to leave the Philippines.

They were persuaded to return for two more series of nonviolence training seminars. They spent six weeks of the summer vacation on the seminars and found that the leaders from the labor unions representing four hundred thousand members had become involved. Workers sat side by side with bishops, nuns, and priests. While Jean and Hildegard concentrated on the groups belonging to the Catholic Church, Richard Deats of the Fellowship of Reconciliation met with groups from Protestant communities. Deats, an ordained Methodist minister with the experience of thirteen years of work in the Philippines, had the credentials of nonviolent witness and training in every continent. Out of the seminars grew the nonviolent movement in the Philippines, Akkapka, a branch of the Fellowship of Reconciliation.[7] The local Catholic church came to life with the message of nonviolence under a priest of the Jesuit community, Father José Blanco, who became the leader of Akkapka. Forty seminars in thirty provinces were held in response to his inspiration and leadership.

Training Seminar in Evangelical Nonviolence

What follows encapsulates the methodology of the seminars. All seminars owed their format to what Jean and Hildegard had developed over the years: the scriptural basis for nonviolence; nonviolence in the history of the church; the various reactions against injustice: nonviolence, counterviolence, and active nonviolence; and methods of nonviolent action. The 1990 edition of the manual for the training seminars has a special section on humanist violence, with examples from Prague and Czechoslovakia.

Each seminar is rooted in the problems and experience of the local people and evokes from the participants their own hopes and sufferings, their disappointments and achievements, and their own remedies, violent or nonviolent.

In the Philippines, the hopes were for the ending of an oppressive regime and the actions of the regime that brought suffering, humiliation, deprivation, and even death. The rage and frustration of the people were part of the discussion.

For people who had hoped against hope for a lifting of oppression and a softening of the harshness of the regime, there might be a temptation to passivity. The people's reaction of passivity or counterviolence was analyzed. The multiple causative factors in passivity are detailed — for example, a complaint may result in losing one's job, or in bringing down punishment. Counterviolence has the effect of allowing the aggressor to dictate the means to be used in a struggle.

Together, seminar leaders and participants would examine the dimensions of active nonviolence and the methods that nonviolent action would take. "The nonviolent struggle is carried out on the level of the conscience," says the manual. "A deep and irreversible faith in the capacity for human beings to open the conscience is at the base of every nonviolent commitment." Basic to the seminars is the commitment articulated by Jean and Hildegard:

> We bring our sufferings and our joys before God in celebration of the word and the Eucharist, knowing that it is the revolutionary force of the love of God which upholds us and wishes to act through us.

Hildegard and Jean always emphasized that the methods of nonviolent action are unpredictable since they must respond to vastly different situations and to vastly different forms of violence — structural, psychological, as well as the visible violence of police and military squads.

Jean and Hildegard had the power to provide instances from the rich experience of their lives. They pointed to the farmers of Larzac, France, who were struggling to prevent their land from being requisitioned for the extension of a military base. To tell the story to the French electorate, the farmers brought their sheep 450 miles to the space under the Eiffel Tower in Paris. They thus dramatized the conflict for French people who had never heard of Larzac.

In the town of Medellín, Colombia, the women of a poor barrio whose campaign for running water brought no results decided to make their problem visible. They brought their babies to the central square and bathed them in the puddles of water that surrounded the large fountain whose abundant spray was never cut off. Middle-class women, moved by the tale of children deprived of a most basic need, clean drinking water and water for cleanliness, made common cause with the women of the barrio. Instead of hostility there was reconciliation and solidarity in meeting a crying human need.

But what methods of nonviolent action would apply in the Philippines? Not only was the Marcos regime one of naked power, but it was the protector of the system of unrestrained power in the hands of landowners and of chiefs of industrial enterprises, who often funded private armies. Those who called attention to the exploitation of poor laborers, or who tried to organize them, were treated as were two priests on the island of Negros where sugar was king. These priests were effectively silenced by being falsely accused of a crime, convicted, and silenced to a filthy jail. Then they were exiled from the Philippines.

In New York City I had dinner with the two priests, Niall O'Brien, an Irish missioner, and Brian Gore, a missioner from Australia. In the group was their friend, the widow of the mayor they were supposed to have murdered. Niall O'Brien wrote a powerful account of life among the peasant workers of Negros and the corruption of the system that jailed them, *Revolution from the Heart.*[8] It gave heart-stopping accounts of how the lives of the poor could be expunged almost at will and how they could be cheated of landholdings coveted by large landholders. On his return to Negros, O'Brien rejoined the struggle of his people. In *Island of Tears, Island of Hope,*[9] he recounted the experiences that highlighted the types of injustices perpetrated in too many parts of the Third World. He described the poor in Negros as powerless beings to be used by the wealthy sugar plantation owners. O'Brien had to encounter not only the often bloody violence of the landowners but the bloody counterviolence of those who pitted their weaponry against that of the Philippine armed forces. O'Brien and some fellow-believers persevered with the ideals and practical methods of gospel nonviolence. Their difficulties and their achievements in community-building have meaning for nonviolent struggles in the most turbulent areas of the Third World.

Cardinal Sin pointed out that before 1985 the church in the Philippines had never taught nonviolent resistance. The church was reluctant to enter upon a course which might end with a confrontation with the dictatorship. Pastoral leaders could not be indifferent to the increasing disregard for every human right and the crushing poverty that accompanied the ever-heavier militarization of the Philippines. The poor and the powerless had nowhere to turn except to the violence of the New People's Army or to the church. In this polarized situation, the church found itself drawn into the vortex of the social and political life of the nation. It shifted its position from simply ameliorating the conditions of the people to one of conscientization, of proclaiming the gospel of human dignity and human rights.

The Crisis

A "snap" election was suddenly announced by Ferdinand Marcos, an election he felt confident would come his way. Corazon Aquino was prevailed upon to run for the presidency only after she had spent a day of prayer at a convent of contemplative nuns. In her prayers, she asked the Lord to forgive Marcos. Her decision to run was based on belief in the power of nonviolence in the face of the armed forces of the Marcos regime. She had the support of the people. "The people believe me," she said. "I am a victim; we are victims of this regime. We have suffered a lot. I can say that I am with them in their sufferings. My sincerity is real and will make people free from fear."

The outpouring of votes in favor of Corazon Aquino showed where the popular choice lay. A committee of poll-watchers allowed thousands of citizens to verify the elections. Poll-watching called for heroism, and many were attacked or killed at the polling places.

Marcos announced that he was the winner of the election. A storm of opposition arose in all parts of the country. Aquino and her supporters devised a strategy that called for boycotts of the banks and businesses that were in league with the regime. Akkapka and Fasters for Peace and Justice set up tent cities around Manila and other cities where people could gather for mutual support, for worship, and for strength to resist any attempt of Marcos to continue in power.

Aquino met with 250 of her followers to work out detailed plans for the strikes, boycotts, and civil disobedience that were expected to last for months. She sent out a call for a rally in Rizal Park in the center of Manila. The response was enormous. The park was filled with thousands of people ready to protest Marcos's stealing of the election. A contingent of resisters, carrying Akkapka banners announcing their adherence to justice, democracy, and peace, marched to the presidential palace. Hitherto, the approach had been blocked at the Mendiola Bridge, where the army had stopped all anti-Marcos protesters. This time, the protesters crossed the bridge and stood at the very gates of the presidential palace. When the troops ordered them to move back, they stood their ground, holding hands and praying. They showed no hostility to the troops, and they simply remained through the night at the palace gates.

At the other end of Manila, the defense secretary of the Philippines, General Juan Ponce Enrile, defected from the Marcos army, along with General Fidel Ramos. They may not have shared the vision of Aquino or the nonviolence preached by the Catholic bishops, but at that critical juncture, their actions paralleled those of Aquino's followers.

The wide EDSA (Epifania de los Santos) Road leading to the army camps where Enrile and Ramos took refuge was the natural target of the regime's tanks and helicopters. When Cardinal Sin learned of defection in the army, he did two things: he asked four convents of contemplative nuns to pray for the outcome of nonviolence, and he broadcast through the Catholic radio station, Radio Veritas, a call for his people to place their unarmed bodies between the defectors and the armed forces of the Philippines.

Bishop Sin, speaking also for his brother bishops, proclaimed:

> The people have spoken. Or have tried to. Despite the obstacles that have been put in the way of their speaking freely, we, the bishops, believe that on the basis of our assessment, as pastors, of the recently concluded polls, what they attempted to say is clear enough.
>
> In our considered judgment, the polls were unparalleled in the fraudulence of their conduct....
>
> We are not going to effect the change we seek by doing nothing, we would be party to our own destruction as a people. We would be jointly guilty with the perpetrators of the wrong we want righted.
>
> Neither do we advocate a bloody, violent means of righting this wrong. If we did, we would be sanctioning the enormous sin of fratricidal strife. Killing to achieve justice is not within the purview of our Christian vision in our present context.
>
> The way indicated to us now is the way of nonviolent struggle for justice.
>
> This means active resistance of evil by peaceful means — in the manner of Christ. And its one end for now is that the will of the people be done through ways and means proper to the Gospel.
>
> We therefore ask every loyal member of the Church, every community of the faithful, to form their judgment about the February 7 polls. And if

> in faith they see things as we the bishops do, we must come together and discern what appropriate actions to take that will be according to the mind of Christ. In a creative, imaginative way, under the guidance of Christ's Spirit, let us pray together, reason together, decide together, act together, always to the end that the truth prevail, that the will of the people be fully respected.
>
> These last few days have given us shining examples of the nonviolent struggle for justice we advocate here. . . .
>
> Now is the time to speak up. Now is the time to repair the wrong. The wrong was systematically organized. So must its correction be. But as in the election itself, that depends fully on the people; on what they are willing and ready to do. We, the bishops, stand in solidarity with them in the common discernment for the good of the nation. But we insist: Our acting must always be according to the Gospel of Christ, that is, in a peaceful, nonviolent way.
>
> May He, the Lord of Justice, the Lord of Peace, be with us in our striving for that good.[10]

When armored tanks approached the camps of the defectors, the people were on the EDSA Road, massed and immovable. Many offered the soldiers gifts of food, and flowers. They placed crucifixes before the faces of the soldiers; nuns held up rosaries as they prayed. Richard Deats related that "after the thousands had gathered at Camp Crame" (where the defectors had taken refuge), "two planes were sent to bomb the camps. When the lead pilot saw the hundreds of thousands of people on two highways crisscrossing the boundaries, it looked like a mammoth cross. At that moment, he said, the hair on his head stood up and he felt it was a sign from God. Instead of dropping their bombs, both pilots flew to Clark Air Force Base and asked for asylum." Clark was the United States base in the Philippines.

The Filipino people were ready for a siege, perhaps a long one. Their spirits were high, their pride buoyed up by their resolute maintenance of nonviolence. Suddenly the struggle was over. Ferdinand Marcos, his immediate family, and his relations were lifted out of Manila by helicopter and deposited in Hawaii. Corazon Aquino, vowing to maintain the ideals of her martyred husband, took the oath as the seventh president of a nation burdened with the darkest economic and social problems, but one that shed a dictator without having shed blood. Hildegard remarked that few bishops in history had spoken so courageously against oppression and so unambiguously in support of nonviolence as had the Filipino bishops.

Many lessons come to mind from the Philippine experience, but two stand out. First, training in nonviolence was the crucial element in the resolute behavior of the Filipino people. Second, the followers of Christ need to continue and build on his infinite expression of forbearance and nonviolence.

Jean Goss and Hildegard Goss-Mayr continued their training seminars in evangelical nonviolence, taking them into Bacolod and other towns and areas of the island of Negros. There, Bishop Antonio Fortich became a champion of the struggle for justice through nonviolent resistance. With the energetic witness of such priests as Niall O'Brien, base communities of peasants were organized to strengthen each other as they held out for their rights for justice and respect. As happened with Dorothy Day, so with Jean and Hildegard — the movement toward gospel-based nonviolence inspires leaders within the hierarchy of the church to return to the original way of Jesus.

In August 1986, at the annual assembly of Pax Christi–USA, held in Boston, Hildegard Goss-Mayr and Jean Goss were given the Teachers of Peace Award.

On April 3, 1991, Jean Goss died suddenly as he was about to leave with Hildegard for their work for peace in Mozambique. The next month, Hildegard received in their name the medal and prize of the Niwano Peace Foundation. Hildegard announced that the main part of the prize money of $150,000 would go to establish a fund for training in the spirit and practice of active nonviolence. One of the programs made possible by the fund was a colloquium on the significance of the life and works of Jean Goss. It was held in Paris for three days beginning October 30, 1993.

Recalling that during history's bloodiest war, Jean had been given the mission to witness to the absolute inviolability of each human person, Hildegard urged all present to go forward, along with other witnesses, to the nonviolent struggle in Europe, the Americas, Africa, and Asia.

Present were representatives of all continents, people of various faiths, laypeople, priests, and a bishop, Dom A. Fragoso, from Crateus, a poor diocese of Brazil. The members of the Catholic Church who participated represented a church which no longer blessed violence. With the help of persons like Jean and Hildegard, a new church was arising and taking form, a peace church that, breaking with a tragic past, no longer sanctioned violence.

Richard Deats of the Fellowship of Reconciliation, who took part in the colloquium honoring his longtime friend Jean, described him as a "stimulator, prophet, facilitator, and transmitter." He summed up Jean Goss as a "combative mystic."

It is rare that a campaign of nonviolence like that of the Philippines draws world attention. Perhaps rarer still is a call from Catholic bishops for unarmed resistance to oppression and destruction of human rights.

The hope that nourishes those involved in evangelical nonviolence is that there will be more instances of unarmed men and women choosing to face injury and even death in the struggle to right wrongs and achieve justice.

The Iron Curtain: Historical Change Linked to Nonviolence

Very few people, even the wisest and best informed, considered that the Iron Curtain, which had become an entrenched fixture of life in the European continent, could be brought down by any other method than violence. Armies faced each other across the great man-made divide, and behind the armies were the nuclear devices of every description trained by each side on the other. Yet, in the profound designs of God, the Iron Curtain was not brought down but rather fell away. A significant influence in this historic change was religion and a newly emerging belief in the efficacy of nonviolence. As in the case of the Philippines, leaders who spoke for human values appeared at the opportune time. A few examples will serve to remind us of those days in 1989 and 1990 when history was changed forever.

Violence Forestalled: The Iron Curtain Dismantled

The fearful symmetry of nuclear weaponry by which the Soviet Union, leader of vast territories and satellite states, could annihilate the Western world, and threaten the earth itself, while the United States and its allies could reciprocate in kind, was a long nightmare. People lived with the knowledge that their earthly home was in jeopardy, since a nuclear exchange could be initiated by accident.

The Iron Curtain, the man-made barrier which cut through the heart of Europe, seemed to keep the superpowers at bay. Yet the fateful barrier was dismantled without violence. The peoples of the satellite nations, though long restive and often at white heat with rage at the Soviet Union's brutal overlordship, achieved their revolutions against it without weapons. Those of us who knew the Iron Curtain firsthand, traversing it and bringing help to those who crossed it through human ingenuity and because of human desperation, wondered how it could ever be dismantled without violence. Traversing the entire length of Europe's great divide, from Kiel to Trieste, I could not forget the giant steel watch towers at various intervals, towers from which border police could train their telescopes, and sometimes their machine-guns, on those who tried to escape. Some stretches were marked by "dragons' teeth," concrete pillars, four feet high, pointed like fangs, and designed to block off vehicular traffic at important intersections.

Hundreds of roads were gashed, and a ten-yard-wide slash of plowed earth formed a no-man's-land on which every footstep made a mark. These plowed areas were called "death strips." At some points there were barbed-wire barriers nine feet high. The Iron Curtain assumed an aura of permanence. Yet all the elaborate and cruel mechanisms to keep East and West Europe apart were destroyed, the watch towers, the dragons' teeth, and the barbed wire. The slashed roads were knitted together and rail links reestablished.

The fact that the dismantling was achieved without weapons was due in large part to the fact that Europeans began to put their reliance on the weapons of the spirit.

In *Revolutions in Eastern Europe,* Niels Nielsen comments, "Internally, from the point of view of the motives of citizens who participated in the revolutions in different settings, the revolutions were a victory of nonviolence over violence, of spirituality over the debasement of human beings."[11] There has been a tendency to take for granted that the freedom that came so peaceably to such countries as Poland, Hungary, East Germany, Czechoslovakia, and Romania was inevitable. The evidence of the crumbling of the Soviet Union, with the reality of *glasnost* (openness) and *perestroika* (restructuring), allowed hopes to rise that people could regain their human rights and religious freedoms. But the brutalization that had fallen on them, the enforced atomization from above, the imprisonments and humiliations, could have welled up in acts of revenge that might have spread to involve localities and entire national groups.

David Cartright contributed a reaction similar to that of Nielsen in a discussion of nonviolent mass action at the Albert Einstein Institution. "The people of Eastern Europe," he said, "have given a stunning example of the power and effectiveness of nonviolent struggle." He termed the wave of popular revolutions "perhaps the most sweeping wave of nonviolent revolution in history. Millions of workers, students, intellectuals and others have used strikes, mass rallies, sit-ins, parallel institutions and numerous other instruments of nonviolent struggle to overthrow more than 40 years of totalitarian rule."[12] Cartright quoted Martin Luther King as evidence that the nonviolent outcome was not inevitable: "Human progress never rolls on the wheels of inevitability; it comes through the tireless efforts of men willing to be co-workers with God."[13]

What had been imposed on the peoples of the countries of Eastern Europe was the materialist concept of the human person. To achieve this, the role of God in human life and the influence of religion had to be curtailed and removed from respect and power. The religious concept of the person as free to operate according to the demands of conscience was anathema to regimes demanding the obedience of the person to political orders which they had no part in fashioning.

It was a protest against this "debasement of human beings" that undergirded the peaceful revolution of Eastern Europe. It was at base a spiritual revolution which had at crucial points the motivation and actual help of religious bodies.

A glance at the situation of five of Eastern Europe's nations will illustrate the point that the move from occupation to the free exercise of conscience and civic rights was not a process that was inevitably peaceful.

Poland: Conquer Evil with Good. Poland is emphasized as a case in point since it seemed a prime candidate for a bloody rising against the Soviet Union. Poland's history demonstrated an affinity for uprisings against Russia. There was one in 1774, another in 1830, and one again in 1863. Poles knew that in addition to the horrors of the German occupation of World War II, and the planting of the worst concentration camps on Polish soil, the Soviet Union had not been behind in bringing destruction to the Polish people; some fifteen thousand of its intellectuals, members of the officers class, had been executed in cold blood in the Katyn Forest in 1940. It was some consolation to the Polish people that in 1993, on his first visit to Poland, the Russian president, Boris N. Yeltsin, made his way to the Powaski Cemetery in Warsaw. There, before the cross erected to memorialize the victims of the Katyn massacre, Yeltsin placed a large wreath. Among the flowers was a Russian flag. At least the mass murder was acknowledged; it followed on the publication of secret files validating the fact that Stalin had signed the order for this among other executions of Poles.

During the Hitler-Stalin Pact, a million and a half Poles from the eastern region had been swept into Siberia and Asiatic Russia. At the Yalta Conference, extensive areas inhabited by Poles were handed over to the Soviet Union. The free elections promised at Yalta never occurred; instead a reign of terror was instituted by a communist regime answerable to the Soviet Union. The spiritual leader of the nation, Cardinal Stefan Wyszyinski, was arrested and imprisoned for three years. When, in 1956, the workers in Poznan demonstrated to protest inhuman working conditions, seventy-four were killed and three hundred injured by the regime's troops.

A striking news photograph of the summer of 1980 was of a Mass celebrated in the Lenin Shipyard at Gdansk. Only the back of the robed priest is seen, and in front of him, their faces to the camera, are several thousand workers, all striking the shipyard. The church supported the strikers and stayed by their side as they opposed the regime's brutality. By November 1980, an electrician, Lech Walesa, had risen from the ranks to lead his fellow workers to a success undreamed of — the registration of the workers' movement, Solidarnosc, to become the first legal and independent trade union in a Soviet-controlled country. Ironically, it was a worker, leading other workers, who confronted what had been described as "the workers' paradise," to assert not only workers' rights but the basic human rights of an occupied people. The success of 1980 was regrettably no forerunner of better things to come.

In two years, Solidarnosc was outlawed. When Lech Walesa was awarded the Nobel Peace Prize in 1983, he feared that if he left Poland, he might not be allowed to return. In solidarity with those Poles still imprisoned, he declined to travel to Oslo for the Nobel ceremony. In 1984, a young priest, Father Jerzy Popieluszko, who called for fearlessness in the name of Christ and kept repeating, "Do not let evil defeat you; conquer evil with good" (Rom. 12:21), attracted thousands of worshipers to his Mass in a Warsaw church. Suddenly he disappeared. A witness reported that he had been kidnapped and that one of his kidnappers had worn a policeman's uniform. Five policemen attached to the security apparatus monitoring priests were questioned. One broke with the secrecy of the secret police and revealed that the priest had been killed and his body, weighed down with stones, thrown into a lake. For once, the police apparatus became public. The body of the slain priest was recovered, and close to a half-million Poles attended his funeral.

Despite every provocation, the Poles did not resort to violence. The Catholic Church was their anchor and for those years of occupation, their sanctuary. This was particularly the case when a pope drawn from their own nation was able to visit them.

The course of the spiritual, and often civil, resistance was maintained without a violent uprising until the fall of the Berlin Wall marked the demise of the Soviet empire.

When asked by a pressman in 1986 if choosing the course of nonviolence was not a form of weakness, Lech Walesa replied:

> It is the course chosen by the majority of the Poles, and the majority of people worldwide. And it's probably due to nonviolence that I am where I am now. I'm a man who believes in dialogue and agreement. I strongly believe that the twenty-first century will not be a century of violence. We've

> already tried and tested every form of violence, and not once in the entire course of human history has anything good or lasting come from it.[14]

Freedom without Violence. Nielsen points out that Eastern Europe, as part of the Soviet sphere, was never "the single whole that communist propaganda and Western ignorance had made it out to be." He adds, "There was a common 'Yalta experience.' Citizens of the satellite countries felt deeply that they had been betrayed to Stalin by Roosevelt and Churchill. The cost in human suffering had been very high."[15]

In Hungary, the suffering had included a failed uprising. The Hungarian Revolution of 1956 drew the response of five thousand Soviet tanks and the execution of its leaders. The Christians of Hungary were not as unified as those of Poland, since they were affiliated with Lutheran, Baptist, and Catholic churches. One effect of the occupation was a split within the church bodies themselves between those who struck a modus vivendi of sorts with the regime and those who resisted being swallowed up in the pragmatic demands of those in power. Bishops and church leaders in general found it possible to take an oath of loyalty to the regime. Active Christians who expressed dissent in any way were discriminated against in matters of education and jobs. Church officials were expected to keep their communicants "in line" as far as political dissent was concerned.

Impatient with the official church leaders, believers gathered in unofficial groups, calling them Bible study groups. Most were under one or the other faith group, but some were ecumenical, drawing Lutherans and Catholics together in an affirmation of their commitment to the lordship of Christ.

A Catholic movement, aiming to live along strict gospel lines, positing the Sermon on the Mount as the basis for daily living, was formed under the leadership of Father George Bulanyi. It soon branched out in a number of small base communities. Bulanyi made contact with the Catholic Worker movement and the Bruderhof movement in the United States. It developed that the groups had much in common — gospel nonviolence, conscientious objection to military service, and simplicity of life. There were those in the official Catholic Church who raised questions about Bulanyi, as they had regarding the Catholic Worker movement, but his movement was not condemned.

Repression of religion began to wane as Soviet power weakened. At the beginning of 1990, the regime announced that there would be full freedom of religion and separation of church and state. The role of religion was not explicit in the overthrow of the Hungarian regime, but the "debasement of human beings" through the coercion of totalitarianism and their experience with the violence of 1956 certainly inspired Hungarians to move toward freedom through means that called for no violence against their occupiers.

East Germany: The Revolution of Light. It was not surprising that an explicit call for nonviolent action against the occupiers in East Germany should come from Leipzig, a city still carrying the marks of its cruel destruction by war's violence. The appeal (framed by Dr. Peter Zimmerman, a Protestant theologian; Dr. Kurt Mazur, director of the Leipzig Symphony; along with four others known to the public) was issued on October 9, 1989. Increasingly men and women gathered to fill the churches. There, they could pray as free human persons and sing, "We Shall Overcome."

The prayer meetings became peace vigils, with the Lutheran Church as the focus. The idea that the success of a nonviolent revolution against the police state was inevitable is faulty. It became known that the East German leader, Erich Honeker, had ordered the army to be ready for an armed confrontation. Possibly the swelling number of protesters pouring out of churches with their lighted candles moved him to countermand the order. What came to be known as the "Revolution of Light" prevailed even though the dark netherworld of Stasi, the state secret police, was still in place. Only later, when secret files were made public, was it revealed how the tentacles of Stasi had reached into every aspect of East German life.

Czechoslovakia: The Velvet Revolution. Movements can even reach into jail cells to summon up the leaders who express their aspirations. How else explain the fact that Václav Havel, a writer who never sought power and spent five years imprisoned by the police state, became the leader of free Czechoslovakia? The regime knew that he had given his allegiance to Charter 77 and Civic Forum, both organizations championing human rights and democratic freedoms. Like Hungary, Czechoslovakia had known the ruthlessness of Soviet arms when Alexander Dubček, in 1968, had attempted a liberalizing of restrictions by a revolution from within; his Prague Spring was brief and aborted.

What Havel communicated to the people from the end of 1989 was a moral vision of a society which could unite believers and nonbelievers, the church, including its leading officials, and the unchurched. Though it seemed that the Czechs and Slovak peoples could put aside their differences under such a leader, it was not to be. Slovakia desired its own separate identity, and irenically, Havel led the Czechs into agreeing to two separate republics, Czech and Slovak.

Havel's eloquence moved many hearts. He drew a standing ovation when he spoke before the U.S. Congress in a joint session. He invited a spiritual leader to Prague whose words of peace might have embroiled Prague in trouble with communist China, namely, the Dalai Lama. After he had been awarded the Nobel Peace Prize, the Tibetan spiritual head gave a speech in Prague, thanking Havel for having given him the opportunity to meet him and visit his beautiful country. He said: "I am particularly happy about this because President Havel is in many ways so unlike other political or national leaders. He has been thrust into his present position quite reluctantly and is one of the few national leaders totally dedicated to peace, nonviolence and moral responsibility."[16]

It is likely that Havel had the Dalai Lama talk to his people in the knowledge that he would reinforce the messages already promulgated by Havel himself, in particular the concept of nonviolence which was implicit — but not explicit — in Havel's own speeches. This is one more example of why the Czechoslovakian turnover was called the "Velvet Revolution."

Romania: Religious Dissent Ignites Revolution. The only Soviet-occupied nation in which nonviolence did not hold was Romania. There was no Soviet army to dislodge, no symbols of Soviet occupation. Romania's occupation forces were its own natives; its state security apparatus, the Securitate, was staffed by Romanians. There came to be a folk saying that in any group of three Romanians, one was a Securitate. In Romania they killed each other. The Orthodox Church officials and a number of its pastors were paid by the regime, so that from that source, representing about 80 percent of the population, came relatively little dissent.

There seemed to exist a cozy relationship between church and state. Visitors were primed by the staff of the Ministry of Cults to experience the placidity of the Romanian state. Orthodox pastors were expected to follow the government line or be brought into line by the Metropolitan or other official leadership. There was seeming peace under the rule of a dictator, Nicolae Ceausescu. Underneath, however, there seethed a deep resentment arising from the depredations of the regime, including the closing of some three hundred Baptist churches in the 1960s, the outlawing of the Catholic Uniate Church of the Byzantine rite, and the persecution of some Orthodox metropolitans and priests who refused to bless the status quo.[17]

It was religious dissent that ignited the revolution that brought down the dictator. It came from the Reformed Church of the large Hungarian minority, some million and a half citizens. Under the dictator's onslaught, the schools, universities, and cultural institutes of the considerable minority were destroyed.

A pastor of the Reformed Church in Timosoara was outspoken against the deprivations of his people's rights. When he was ordered to leave his parish, the people knew it was because of the influence of the Ministry of Cults. They rose up in the church and in the streets to protest. Troops fired on the protesters in the streets. The date was December 16, 1989. When they returned the night after the massacre to ask for the bodies of their dead for burial, the troops repeated their volley against civilians, killing thousands of them. Word of the massacre spread through Romania.

On December 21, 1989, workers were brought in by bus to welcome Nicolae Ceausescu on his return from a trip abroad. It was his custom to appear on his balcony at noon to respond to the acclaim of his people. On this day, he was met by jeers and shouts of "Timosoara." Students took the lead in shouting insults and in trampling down the banners they were supposed to wave triumphantly to hail the leader.

Ceausescu, as caught by television cameras, shocked and bewildered, turned and left the balcony with his bodyguards. That evening the security forces opened fire on the crowd, targeting in particular the students. The following day, part of the army seems to have changed sides, aligning themselves with the opposition. Four days later Nicolae Ceausescu and his wife were cornered in a room as they were preparing to flee, and shot to death. It was Christmas Day. With limited violence and amid chaos, a replacement regime was installed. At least the violence had been halted.

The Wall Once Breached: The Greatest Victory

With the resignation of the entire leadership of the East German regime on November 7, 1989, East Germans rushed to the Berlin Wall. There was mounting pressure to cross to the other side. The crush of people became dangerous, and the mayor of Berlin ordered the border guards to allow the people to cross for that night only. It seems to have been a temporary expedient to ease pressure and forestall violence. A stream of people crossed the divide and kept coming. The border guards held their fire. Once the wall was breached on November 9, 1989, it was never closed again.

"The greatest victory is the war that is not fought" is a saying popularized by a leader of Pax Christi, and one of the founders of Global Peace Services, Mary Evelyn Jegen.[18] The destruction of the wall separating Eastern and Western Europe was the symbol of the greatest victory of the twentieth century. The two nuclear arsenals targeted to the east and to the west were no longer pointed at each other. More visibly, five million Russian troops stationed in Central Europe, facing some three hundred thousand American troops, no longer had any function. After the victory of November 9, 1989, the Soviet army of Central Europe began to move back into Russia, and its presence in Western Europe was eventually reduced to zero. The American forces were reduced to one-third of their strength but were held in readiness for defensive action or to assist in fulfilling UN obligations.

Pope John Paul II made a strong statement at the UN General Assembly on October 5, 1995, to mark the fiftieth anniversary of the United Nations.

After adverting to totalitarianism's assault on the rights and dignity of the human person, the pope asserted:

> The revolutions of 1989 were made possible by the commitment of brave men and women inspired by a different, and ultimately more profound and powerful, vision: the vision of man as a creature of intelligence and free will, immersed in the mystery which transcends his own being and endowed with the ability to reflect and the ability to choose — and thus capable of wisdom and virtues. A decisive factor in the success of those nonviolent revolutions was the experience of *social solidarity:* in the face of regimes backed by the power of propaganda and terror, that solidarity was the moral core of the "power of the powerless," a beacon of hope and an enduring reminder that it is possible for man's historical journey to follow a path which is true to the finest aspirations of the human spirit.[19]

The End of the Cold War

Certainly the dramatic ending of the long Cold War exhibited that a new reality had seized the minds of most Europeans, an aversion to war as a means of achieving political or social change. Questions arise regarding the future. Is humankind closer to believing in an alternative to war? Are members of the Catholic Church willing to lead in articulating and acting on alternatives to violence? Would they be ready to unveil for the third millennium the alternative of war and its violence, namely, the foundational teachings of love, forgiveness, and willingness to suffer innocently? Could their witness to nonviolence in the face of provocation help achieve the end hoped for by the pope in his UN speech, that the tears of the twentieth century might prepare the ground for a new springtime of the human spirit?

War avoidance might not seem the loftiest of aims, but besides the lives saved, the treasures of humankind and even the planet protected, the time without war might make way for the cultivation of a culture of peace and social development grounded in justice and human rights.

An earlier period of European history was as fragile and open to destruction as the end of the Soviet empire. It was the early thirteenth century, when the Mongols (also known as the Tartars) invaded Europe. Moving out of central Asia, their archers were unstoppable as they raced on their small horses across Turkestan and Afghanistan. They lived off the conquered lands and exacted tribute. In 1223, they invaded Russia, but dynastic problems connected with the death of their great Khan, Genghis (Chingis), delayed deeper incursions. After Ogatai's accession as the Great Khan, the Mongols again turned westward to conquer Moscow in 1238 and Kiev in 1240. Overwhelming victories were achieved by the Mongols in Leignitz in Silesia and Mohi in Hungary. When they penetrated to Vienna in 1241, they had behind them the great subjugated land mass of Asia. Tales of Mongol cruelty preceded them, and Europe's capitals were seized by dread. Armies of mounted archers, swift and mobile, might reach them next. Cumbersome regiments of knights and soldiery could not deal with the wave of warriors from Asia who disdained the rules of warfare.

Prayers were offered in churches, and people gathered in the streets to implore the help of God in turning back the Tatars. (While Tatars was the original name of the Mongol tribe, the Europeans corrupted it to Tartars, meaning men from hell or tartary.) Late in 1241, the Great Khan Ogatai died, and all high nobles were called back to Karakorum to participate in the choice of a successor. The warriors swept out of Europe as swiftly as they had arrived. Western Europe was saved. The Mongols never returned, living only in the folk memory of Europeans.

The people of Europe saw the hand of God in the fact that the "men from hell" turned around before bringing destruction to any more of Europe's cities. Whether by human help or a seeming act of providence, should not war avoidance always be welcomed? The question needs to be pondered as to whether there can ever be a good war or a bad peace.

Chapter 12

Prophets of Nonviolence

Nonviolence is not for power but for truth. It is not pragmatic but prophetic.
—THOMAS MERTON

History, according to most history books, is written in blood, the blood of those slain in the great victories that added glory to the nation and fed the pride of its nationals. When the blood of a nation's own people is shed, it becomes the blood of heroes. Not only battles between soldiers enter the pages of chronicles, but the blood poured on the altars of the revolutions that make history — the French Revolution and the Bolshevik Revolution, for example. The twentieth century surpasses all others in the amount of human blood spilled. I was so often reminded of the words, "Your brother's blood cries out to me from the soil" (Gen. 4:10).

The vision of a new type of history has been glimpsed amid the century's appalling destruction, a history of bloodless victories, of victories that not only leave no corpses, but even, on some blessed occasions, leave reconciliation in their wake. The human person's highest gifts, reason and the power of self-transcendence, even the power to love, are all brought into play in the achievement of such victories.

In two world wars, and in their aftermath, the breakup of colonial systems, with the subsequent clashes of cultural, racial, religious, political, or national values, leaders relied on violence to right wrongs or to quell uprisings. As often as not, both sides in a conflict may claim that a right has been violated, or that something, generally territory, has been expropriated by force. Many of the most intractable conflicts seem to have been rooted in religion, when old wrongs are recalled and relived in order to justify wreaking vengeance on the present.

The weapon in the new, bloodless struggle to right wrongs and reach justice is nonviolence, active nonviolence which takes various forms in response to differing situations. The concept of nonviolence has been kept alive and witnessed to for the emulation of others by persons who must be recognized as prophets for the twentieth century. A glimpse of a few of them is presented here: Mohandas Gandhi, Martin Luther King, Daniel and Philip Berrigan, and Thomas Merton.

Gandhi

The most notable example of positing unarmed nonviolence against armed power was that led by Gandhi (Mohandas Karamchand). Gandhi challenged a world-girdling empire with what he called the weapons of the strong —

noncooperation with unjust laws, resistance to them, civil disobedience, and voluntary suffering. The voluntary suffering might take the form of fasting or of the endurance of imprisonment. In the Gandhian code of conduct, imprisonment involved the willing obedience to the reasonable demands of jailers — but not to any humiliating orders.

Gandhi's method, growing out of his campaign for justice in South Africa, came with a name, satyagraha, literally, holding on to the truth, thus, truth-force or soul-force. He saw God as another name for truth, and truth for God.

"It was the New Testament which really awakened me to the rightness and value of passive resistance," Gandhi asserted. (Later he dropped the term "passive" in favor of a more active description of his methods.) "When I read in the Sermon on the Mount such passages as 'Resist not him that is evil, but whoever smite thee on the right cheek, turn to him the other,' and 'Love your enemies, and pray for them that persecute you, that you may be sons of your Father who is in heaven,' I was overjoyed."[1]

The Sermon on the Mount confirmed the teaching of a Gujarati hymn he had learned as a child:

> If a man gives you a drink of water
> and you give him a drink in return,
> that is nothing.
> Real beauty consists in doing good
> against all evil.

"To me," Gandhi said, "the ideas which underlie the Gujarati hymn and the Sermon on the Mount should revolutionize the whole of life."[2] "Jesus lived and died in vain," according to Gandhi, "if he did not teach us to regulate the whole of life by the eternal Law of Love."[3] That these concepts revolutionized the lives of Gandhi and his followers, who attempted to struggle against oppression without sacrificing love, is written in the history books.

While he relied on the age-old wisdom of India, and its ancient law of self-sacrifice and ahimsa,[4] Gandhi incorporated the wisdom of Christianity. He wrote, "Christ died on the cross with a crown of thorns on his head, defying the might of a whole empire. And if I raise resistance of a nonviolent character, I simply and humbly follow in the footsteps of a great teacher."[5]

The pitting of moral teaching against political power was Gandhi's gift to the twentieth century. He resurrected for Christians the reality of the method practiced by the first Christians during the persecutions by the Roman Empire. Through him, the Sermon on the Mount went public. "Religion," Gandhi wrote, "to be true must pervade every aspect of life.... Politics is an integral part of civic life."[6]

The tragic reality of the communal (Hindu-Muslim) violence that accompanied the partition of India and the British withdrawal did not invalidate the fact that millions of ordinary men and women in India had experienced the power of nonviolence and the practice of returning good for evil. Participants in the campaigns took a satyagraha pledge as devised by Gandhi.[7] The pledge was published in *Young India* of February 27, 1930:

1. A Satyagrahi, i.e., a civil resister, will harbour no anger.
2. He will suffer the anger of the opponent.
3. In doing so he will put up with assaults from the opponent, never retaliate; but he will not submit, out of fear of punishment or the like, to any order in anger.
4. When any person in authority seeks to arrest a civil resister, he will voluntarily submit to the arrest, and he will not resist the attachment or removal of his own property, if any, when it is sought to be confiscated by the authorities.
5. If a civil resister has any property in his possession as a trustee, he will refuse to surrender it, even though in defending it he might lose his life. He will, however, never retaliate.
6. Non-retaliation excludes swearing and cursing.
7. Therefore a civil resister will never insult his opponent, and therefore, also, he may not take part in many of the newly coined cries which are contrary to the spirit of Ahimsa (nonviolence).
8. A civil resister will not salute the Union Jack, nor will he insult it or officials, English or Indian.
9. In course of the struggle if one insults an official or commits an assault upon him, a civil resister will protect such official or officials from the insult or attack even at the risk of his life.[8]

Satyagraha

Outside of India, people learned of the courage of the practitioners of satyagraha through the Salt Satyagraha of 1930–31. The removal of the salt tax was a cause that inspired all Indians, in particular the poor, whose daily need for salt arose from their work in the fields in India's furnace heat. For the occupying power, the tax was a source of immense revenue. Men and women, young and old, joined in a march to the sea. A young marcher described his personal participation: "Realizing that millions of men and women were depending on our leader, we strained to do his bidding like greyhounds on a leash. All along the line of the march, we were sent to nearby villages to address farmer-laborer gatherings and to mobilize them for an attack on the government's salt monopoly."[9] The climax of this campaign of civil disobedience took place on reaching Dandi, a shore point near Bombay. The marchers joined in prayers and then in the symbolic act of panning salt from sea water.

Groups of marchers next moved to occupy the nearby Dharasana Salt Works, only to be beaten down by the police with metal-tipped clubs. As soon as they were beaten down, another group took their place. Volunteers carried away the wounded as new marchers came to place themselves at the salt works. An American journalist recorded the almost incredible spectacle of how relay after relay of satyagrahis marched on the salt works and responded to violence with total nonviolence. Over three hundred persons, bleeding from cracked skulls and limbs, were taken to the local hospital. In the nationwide resistance to the salt tax, over a hundred thousand Indians went to prison, each one vowing not to retaliate by violence to arrest or imprisonment.

Nonviolence: A Process of Conversion

Satyagraha was a method that could be used to eliminate social evils and was utilized in the movement not only to free India from British rule, but to free its people from the burden of "untouchability." Less known is the Gandhian method of nonviolence as utilized against the social evil of casteism. Low caste Hindus (referred to as "untouchables," but by Gandhi renamed harijans, "children of God") were banished not only from entering the Hindu temples but from the precincts surrounding the temples.

A telling example was the satyagraha campaign in 1924–25 in Travancore, at the southern tip of India. This is chronicled by Joan V. Bondurant in *Conquest of Violence.*[10] A group of fifty followers of Gandhi joined in a campaign to allow harijans to walk through the precincts of the Vykom Temple. The first effort was made by requesting the Brahmin priests to open the road through temple grounds to harijans, who were forced to reach their work by taking the long route that skirted the extensive temple grounds. This was refused.

Harijans and caste Hindus started processions to enter the forbidden road, stopping at the boundary of the temple precincts. They were beaten but refused to retaliate. When some were arrested, others took their places, as did the salt marchers six years later. Police placed barricades at the border of the temple grounds. At that point satyagrahis placed themselves at the barricades facing the police. Months went by while the Brahmin priests made no move to capitulate; meanwhile the harijans were subjected to the long, circuitous route to and from their work morning and evening. A satyagraha camp was set up in which satyagrahis prayed, discussed the principles of satyagraha, and took part in hand-spinning and other constructive activities.

With the monsoon floods, satyagrahis stationed themselves on the road even when they had to stand up to their shoulders in water. Rather than being seen as heroes, the satyagrahis were ostracized by many in the surrounding community. Eventually, Gandhi joined his followers, and, through his influence, the local government authorities were persuaded to remove the barricades at the entrance to the Vykom Temple. The harijans were now free to use the shorter, more direct route to their work. However, they did not enter the temple precincts.

The satyagrahis surprised the Brahmin priests by not taking advantage of the civic removal of the prohibition. After prayers, the satyagrahis announced that they would not cross the temple boundary before the Brahmins were persuaded that it was right to permit the entry of the harijans. The satyagrahis stood their ground, persisting in their prayers. Finally, the satyagrahis received word from the Brahmins that they could no longer resist the prayers that had been addressed to them. They were ready to receive the "untouchables," the harijans.

"Untouchability will not be removed by the force of the law," said Gandhi. "It can only be removed when the majority of the Hindus realize that it is a crime against God and man and are ashamed of it. In other words, it is a process of conversion, i.e., purification, of the Hindu heart."[11] This major decision had its effect in other parts of India and was the beginning of a larger effort against untouchability.

The sixteen-month campaign at the Vykom Temple was an instance of the patience needed to penetrate the minds of opponents whose lives are guided by totally different concepts and who must face the harsh reality of unlearning them. In Gandhi's lifetime, many temples were opened to harijans. I saw how in Calcutta the much-frequented temple of Kali, the goddess for whom Calcutta was named, was open to the lowest caste of Hindus. Its pilgrim hostel, the *dharmasala,* became a hospice where the dying of all castes could be brought to be cared for by Mother Teresa and her Missionaries of Charity.

It was the discipline of the satyagrahis and the peace discipleship led by Gandhi which supported the peoples of India as they moved into the final phase when they announced to the British, "Quit India."

Martin Luther King: Unearned Suffering as Redemptive

Many of those who read about the almost mythic period of the Gandhian struggle, or who see filmed depiction of it, would have wanted to be part of it.

A nonviolent movement of similar heroism took place in the United States under a charismatic leader. So much has been written about the movement that I include only a vignette from my own experience to illustrate the deep religious heart of the movement led by Martin Luther King.

In the spring of 1965, people from all over the United States made their way to Selma, Alabama, where King was leading nonviolent marches for voting rights for African American citizens of the United States. I was one of them. It was a pivotal day, March 15, 1965, when marchers from Selma made their way without hindrance to the Dallas County Courthouse and there asserted the unfettered right to vote.

Before attempting the day's symbolic march we attended a prayer service at Browns African Methodist Chapel in the heart of "black Selma." It was a memorial service for the Reverend James J. Reeb, who had come to Selma in support of Martin Luther King's struggle and had been bludgeoned to death. I was placed in the sanctuary and looked out at a church packed with people of many creeds. Religious Sisters of many Catholic societies stood out at a time when distinguishing habits were the norm. Even the balcony was packed with people.

We had an hour to wait before the service began. Archbishop Iakovos, primate of the Greek Orthodox Church in America, flanked Dr. King. Bishops of various Christian churches, priests, ministers, a rabbi, and representatives of labor led by Walter Reuther filled the sanctuary. There were readings from the Old and New Testaments by the various bishops, prayers, eulogies, and, in between, the singing of hymns. In the heat of the camera lights and pressed bodies, we swayed and sang, "Lord, hold my hand while I run this race. / Guide my feet while I run this race." When Martin Luther King rose to speak, the swaying stopped, and we could forget the rivulets of sweat pouring down foreheads and backs. His rich voice rang with moral passion: "Why must men die for doing good?" he asked. "His death," he said of James J. Reeb, "may cause the white South to come to terms with its conscience.... All unearned suffering is redemptive," he told us, englobing the massive sufferings and humiliations heaped on his people.

As he spoke of forgiveness and even love of enemies, I thought of what he had said in arguing with a federal judge for the permit to march on Montgomery: "Maybe there will be some blood let in the state of Alabama before we get through, but it will be our blood and not the blood of our white brothers." He was pledging his people to complete nonviolence no matter what the provocation.

I thought of all the other churches in the South where nonviolence had been preached. In a church like this in Birmingham, the minister had asked his flock during an earlier stage of the civil rights struggle, "If blood is shed, shall it be theirs or ours?" And back came the cry of the congregation, incredible and heart-stopping: "Let it be ours. Let it be ours!" There was more than an echo of the early Christians in that cry. The black Christians of Alabama were presenting themselves almost literally as "lambs of God" for redemptive suffering.

As Martin Luther King talked about Christian nonviolence and love, we knew that the impact of his God-ridden movement was reaching beyond his oppressed people to the wider community of Americans and the world.

The anthem of hope followed Dr. King's sermon: "Deep in my heart, I do believe, we shall overcome some day." All of us, the archbishop, bishops, of so many faith groups, priests, ministers, nuns, labor leaders, civil rights workers, grown-ups, and children, black and white, crossed arms, clasped hands, and swayed together. The people in the balcony perches swayed in opposite directions like tall grass bowing to contrary winds.

A minister asked us to continue "We Shall Overcome" in a hum, and the soaring voices subsided into a throbbing undertone. At that moment the rabbi rose and intoned the Kaddish, the ancient prayer for the dead. The Hebrew words rang out over the soft humming of five hundred throats. The resonance that filled Browns Chapel and made the walls reverberate was only an accompaniment for the magnificent deep-voiced lament in the tongue of the prophets of ancient Israel. How often the Negroes in their days of enslavement had likened themselves to the children of Israel awaiting a Moses to lead them out of the house of bondage. The words of the Kaddish rose and fell, and we could only guess at the meanings hidden in the rich guttural syllables. The humming seemed to become a little softer as many gulped back tears. That Kaddish and that most ecumenical hum, I felt, were not only for James J. Reeb, but for every victim, known and unknown, who had died by lynching, bludgeoning, shooting, in the struggle for freedom, and for the innocent victims of war's violence.

When the rabbi began the prayer in English, the humming trailed into silence, and we heard him beg mercy alike for the disinheritor and disinherited, for the arrogant and humiliated, for the giver and receiver of violence. That was the end of the memorial service.

The Dallas County Courthouse was the "fortified city" symbolizing the evils and restrictions heaped upon Selma's blacks. Here the voting registration schedule, the indignities, and the complicated questionnaires sabotaged all efforts by blacks to join the voting rolls. Here was the domain of Sheriff James G. Clark, who wore the lapel button marked "Never" as a response to "We Shall Overcome."

Twilight was beginning to fall as we emerged from Browns Chapel into Sylvan Street. We lined up three abreast. For long minutes we stood in silent vigil until all the marchers were in place. As the marchers assembled, the police fanned out and formed a wide arc around them. All we could see were the inverted soup-plates of their helmets arranged like an outdoor pottery display.

The announcement was made that everyone had been waiting for. The march to Dallas County Courthouse could take place without hindrance from the police. About four thousand of us lined up and began the march. Eight weeks of attempts to reach the courthouse had been blocked by men of the law. Hundreds had been arrested.

At the courthouse, Dr. King was handed a large wreath of purple and white flowers, a memento from Bucks County, Pennsylvania, for James J. Reeb. He placed it carefully against the main door of the courthouse. A hymn poured out led by the Reverend Fred Shuttlesworth, an aide to Dr. King:

> O God our help in ages past,
> Our hope for years to come,
> Our shelter from the stormy blast,
> And our eternal home.

Dr. King ended a short talk with the words, "We thank thee, Lord, even for the oppression that has brought us to this point . . . so that we may be better citizens of the eternal city." "We are the assembled here," he told us, "to rededicate ourselves, to assert our conviction that racial segregation is evil."

Then everyone crossed arms and locked hands with those nearest, and the air was rent by four thousand voices announcing in song:

> We shall overcome.
> Black and white together. . . .
> We shall overcome some day. . . .
> God will see us through some day.

That evening at the Good Samaritan Hospital in Selma, where many of us were given hospitality, we gathered to hear the president of the United States. He was scheduled to make an important address. President Johnson began:

> I speak tonight for the dignity of man and the destiny of democracy.
>
> At times history and fate meet in a single time and in a single place to shape a turning point in man's unending search for freedom. So it was at

> Lexington and Concord. So it was a century ago at Appomatox. So it was last week in Selma, Alabama.

The whole room broke into applause when the president announced, "We shall overcome." "In Selma," he said, and then he looked up smiling and seemed to be departing from his prepared text, "we had a good day there." He announced that the Voting Rights Bill of 1965 would "strike down restrictions to voting in all elections, federal, state, and local. It will provide for citizens to be registered by officials of the United States Government if state officials refuse to cooperate."

The voter registration action in Selma began on January 3, 1965. The March 15 protest came on a pivotal day in the campaign. On March 27, 1965, after a historic march, Dr. King announced at the state capital of Alabama, Montgomery, "The end that we seek is a society at peace with itself, . . . a society that can live with its conscience." Just as in India, generations would live before such a consummation became reality.

Crucial to winning support in the struggle against the evil of racial segregation was the commitment agreed to by the volunteers in the Birmingham movement. The commitment stated: "I hereby pledge myself — my person and my body — to the nonviolent movement." It then listed ten commandments that each volunteer was to keep:

1. Meditate daily on the teachings and life of Jesus.
2. Remember always that the nonviolent movement in Birmingham seeks justice and reconciliation — not victory.
3. Walk and talk in the manner of love, for God is love.
4. Pray daily to be used by God in order that all men might be free.
5. Sacrifice personal wishes in order that all men might be free.
6. Observe with both friend and foe the ordinary rules of courtesy.
7. Seek to perform regular service for others and for the world.
8. Refrain from the violence of fist, tongue, or heart.
9. Strive to be in good spiritual and bodily health.
10. Follow the directions of the movement and of the captain on a demonstration.[12]

Daniel and Philip Berrigan

War, seemingly the ultimate in drama, seizes headlines and time on television screens. During the Vietnam conflict, antiwar actions began to reach the attention of newspapers, magazines, and television screens. A famous action led by Daniel Berrigan, a member of the Society of Jesus, and his brother Philip, then a member of the Society of St. Joseph (an American congregation founded to serve and champion black Americans), consisted of capturing hundreds of the files of draft-eligible men from the local board of Selective Service at Catonsville, Maryland. The state of American society called forth a special heroism from the brothers, the heroism of protest and consequent suffering.

It was May 17, 1968, and half a world away, villages and the bodies of the Vietnamese were being consumed by a new weapon of war known as napalm. The media were advised of the public protest. Nine people in total took part in setting afire and destroying the files with homemade napalm.

The eventual trial, sentencing, and imprisonment of the Catonsville Nine, in particular the participation of two Catholic priests, caught the imagination, sometimes puzzled, of people throughout the nation and overseas. There was some hesitant questioning on the matter of criminal trespass and property destruction in the protest. In *Harder Than War*, Patricia McNeal remarks, "The old pacifist groups in the antiwar movement such as the Catholic Worker, the War Resisters League, the Fellowship of Reconciliation, and the various Quaker groups, though supportive and sympathetic of the motives of the Catonsville Nine, could not support the destruction of property as a valid nonviolent tactic even in the symbolic violence of draft file destruction."[13]

An earlier act of resistance, led by Philip Berrigan, the pouring of blood on the files of Selective Service, had given rise to only moderate media attention. From Catonsville forward, the Berrigan brothers and co-resisters were constantly in the news. A play, *The Trial of the Catonsville Nine*, by Daniel Berrigan was presented in New York City, by peace groups around the country, and as far away as Japan. For Daniel and Philip Berrigan, resistance to the war in Vietnam, to all war, and to the American stockpile of nuclear and so-called conventional weapons became their way of life. With peace-minded "accomplices," they led acts of resistance that moved from pouring blood on draft files to breaking and entering factories and arsenals, where they symbolically disabled weapons — often by damaging the nose cones of missiles. In the General Electric Plant at King of Prussia, Pennsylvania, Daniel and Philip, with six other resisters, hammered two nose cones destined for nuclear warheads, and poured blood on the documents of the plant. The actions, called "Swords into Plowshares," echoed the biblical prophecy of Isaiah to "beat swords into plowshares" (Isa. 2:4).

Daniel Berrigan's poem to the court after Catonsville became for many the symbol of resistance to the Vietnam War:

> Our apologies good friends
> for the fracture of good order the burning of paper
> instead of children the angering of the orderlies
>   in the front parlor of the charnel house
> We could not so help us God do otherwise
>   for we are sick at heart our hearts
> give us no rest for thinking of the Land of Burning
>   Children.[14]

"Swords into Plowshares" operations were repeated, and repeatedly, men and women accepted the consequences for them and went to jail. After his first prison term, Daniel learned that given the condition of his health, he might not survive another term behind bars. His witness continued to be as moving and as highly visible since, as a recognized poet, he published dozens of books and responded to invitations to give lectures and retreats in communities nationwide.

Philip Berrigan called his autobiography *Fighting the Lamb's War.* Its subtitle, *Skirmishes with the American Empire,* indicated his underlying quarrel with his country, in fact, his holy rage against its policies. His rage never abated. After he entered his seventieth year he went to prison for ten months for a Plowshares action. Philip's ferocity in fighting the "lamb's war" evoked Isaiah:

> Cry out full throated and unsparingly.
> Lift up your voice like a trumpet blast;
> Tell my people their wickedness. (Isa. 58:1)

The witness of Philip and his co-resisters, their blood dramas and their willingness to enter the doors of prison for "their fractures of good order," served as a burr to awaken and prick the American conscience. That conscience was only too ready to become benumbed before the stockpiles of monstrous weapons that were justified by a government ready to unleash them against the children of God who happened to live in the Soviet "empire of Satan." In one of the many works of Daniel Berrigan, the author takes the reader through the struggles of the apostles and of the Christians of the present time. He urges twentieth-century Christians to take up where the apostles left off by participating in their own loving acts for the coming of God's reign of peace.

"The Book of Acts remains open, unfinished, waiting in a sense; a sense both hopeful and horrific," he says. "It lurks there like a preternatural third eye, waiting. Or like a scribe's quill ready, eye alert. What acts are to follow the acts of the apostles?"[15]

Here Daniel becomes more specific, reminding readers, in particular believing Christians, that participating may not come easy: "Let us set them down; better let us enact them, repeat them, dwell on them, improve, invoke them, stand within them. Acts of Dorothy Day, of Archbishop Romero, of Thomas Merton, of the murdered Jesuits, of that vast unnamed 'cloud of witnesses' who live and labor and dream fidelity to the Word. For whom the Word made supreme and sole sense in a senseless time."[16]

In *Harder Than War,* Patricia McNeal asserts: "The combination of Philip and Daniel's unique strengths and talents enabled them to bring forth the message of peace to their church and country in a way never before experienced in the history of American Catholicism."[17]

Thomas Merton: The Choice of a Monk's Cell

Important as highly visible and media-covered acts of protest and resistance were in heightening the repugnance to war and the necessity of peace, there were other influences at work. These influences were crucial in helping people to unlearn the general acceptance of war as a means of resolving conflict and in supplying a spiritual basis for an alternative to violence.

In an extreme paradox, an influence as telling as that of Daniel and Philip Berrigan and their co-resisters came from a monk's cell, from a member of one of the strictest, enclosed orders of the Catholic Church, the Cistercians, or Trappists. The monk, Thomas Merton, in religion, Father Louis, whose autobiography, *The Seven Storey Mountain,* had reached millions of readers, became a spiritual force in supporting war objectors and in helping articulate a peace position emanating from the central core of one's union with Christ.

The Seven Storey Mountain, published in 1948, opened an exotic spiritual world to those outside the Catholic Church, the world of the Trappist monk where silence is so treasured that ordinary communication is conducted by hundreds of hand signs. (The edition published in England was entitled *Elected Silence.*)[18]

The spectacle of a young man of twenty-six taking leave of the world forever to enter an abbey named Gethsemani (named for the garden where Jesus fell prostrate, sorrowful unto death) fired the imagination, and provoked the curiosity, of the general public, especially the young. The book burst upon a postwar, post-Holocaust world, still living with the implications of nearly six years during which humankind had devoted its energies to mutual slaughter. Merton could not deliver a stronger criticism of so lethal a social order than by the act of turning his back on it. The story of the conversion of a young man from a life that included debauchery to a life centered on God drew many to learn the "why" of conversion. The autobiography might supply an anchor for those who felt spiritually adrift.

Did Merton have a secret he could share with those hungering for meaning in their lives? Merton described what had transpired inside his soul to lead him to baptism and eventually to lying face down on the stone floor of the Gethsemani abbey chapel to be ordained as a priest of God. Celibacy was to be his lot for the rest of his life, a life wedded to poverty and obedience.

Merton, the son of artistic parents, was born in Prades in the French Pyrenees. His father, Owen, a New Zealander, had met his mother, Ruth Jenkins, an American, while they were both studying art in Paris. Shunted between France, England, and the United States, Merton did not have a settled existence until he went to live with his maternal grandparents in Flushing and Douglaston, Long Island. Neither parent had any strong religious ties, although his mother did occasionally attend the Quaker meeting in Flushing. Merton's mother died, leaving Tom at eight in the care of his father, and his younger brother, John Paul, in the care of her parents.

Owen Merton's landscape paintings, shown at various galleries in London and New York, had begun to receive favorable notices when he died of a brain tumor in London at the age of forty-four. Thomas Merton was sixteen years of age. His grandparents provided funding for a year at Cambridge, where his intellectual life came second to his social life of liaisons and drinking. He transferred to Columbia University, where he became "a big man on campus," his literary powers expressed in writing for *The Jester* and other college publications. He went on to earn an M.A. in English and began studying for a Ph.D. when something intruded on his plans to become a writer or a teacher. The "something" was a wholehearted conversion from no religion to being a member of the Roman Catholic Church.

In March 1941, Merton received a notice from Selective Service to report to the draft board. During World War I, Owen Merton had been a pacifist; Thomas Merton followed his example. He replied to his draft board with a letter explaining his position. He summarized the letter in this way:

> I made out my reasons for being a partial conscientious objector, for asking for non-combatant service, so as not to have to kill men made in the image and likeness of God when it was possible to obey the law (as I must) by serving the wounded and saving lives or by the humiliation of digging latrines which is a far greater honor to God than killing men.[19]

When he appeared for his medical examination, he was rejected because of poor teeth. A notification reached him to report for another medical examination. When he telephoned the draft board, the secretary informed him that the ruling about his teeth had been changed. If there was no other health problem, he would in all likelihood be found combat-ready.

Before he was to appear for another medical examination, he presented himself at the abbey, where he had already been accepted, and the Gethsemani door closed behind him. Novices in religious orders were exempt from the draft. The date was December 10, 1941, three days after Pearl Harbor and the entry of the United States into World War II.

The Seven Storey Mountain was followed by several books of poetry and a series of works exploring contemplative life, the history of religious orders, and Catholic teaching.

Cold War Letters

At the dawn of the 1960s, Merton looked beyond Gethsemani and responded to, or initiated contact with, a large variety of persons involved with the struggles of the time, chief among them, the struggle for peace and for a spiritual basis for peacemaking.

He gathered a collection of his letters and mimeographed them. These, entitled *Cold War Letters,* he sent to peace activists around the country, among them myself. Though the recipients of the letters were identified only by their initials, they were recognized by those of us to whom he sent letters. A letter to Dorothy Day dealt, among other matters, with "enemies."

"As Christians," he wrote, "we have to keep insisting on the distinction between the man, the person, and the action and policies attributed to him and his group — we have to remember the terrible danger of projecting onto others all the evil we find in ourselves, so that we can justify our desire to hate that evil and to destroy it in them."[20]

An earlier letter to Dorothy praised her for refusing to take shelter in New York City's Civil Defense Drill. He wrote:

> It was a pleasure to get your letter. Of course I keep praying for you and for the Catholic Worker. . . . Again, I am touched deeply by your witness for peace. You are very right in going about it along the lines of satyagraha. I see no other way, though of course the angles of the problem are not all clear. I am certainly with you taking some kind of stand and acting accordingly. Nowadays it is no longer a question of who is right, but who is at least not criminal. If any of us can say that any more. So don't worry about whether or not in every point you are perfectly right according to everybody's book; you are right before God as far as you are fighting for a truth that is clear enough and important enough. What more can anybody do?[21]

Merton's letters to Dorothy reveal his reverence for her and the closeness of their views. He loved Dostoyevsky, as did Dorothy, and he appreciated why people would choose the Christian anarchist position. He confided to her that he could not in conscience continue to write about meditation and monastic studies without facing the life-and-death issues of the time. He, along with Dorothy, saw the need for a renewal of the social order, with its inbuilt structures of injustice and oppression, and its militarism, which devoured what was needed by the poor of the world. Their only difference was Merton's reluctance to proclaim himself an outright pacifist.

In a letter to Dorothy of September 22, 1961, Merton included an article, "The Root of War," which had already passed censoring by his superiors. "So if you want it," he wrote, "you can go right ahead with it." The front page of the October 1961 issue of the *Catholic Worker* carried the article, Merton's first contribution to the paper, on page 1. It began with a Jeremiad against the war-addiction of human beings:

> The present war crisis is something we have made entirely by ourselves. There is in reality not the slightest logical reason for war, and yet the whole world is plunging headlong into frightful destruction, and doing so with the purpose of avoiding war and preserving peace.
>
> This is a true war madness, an illness of the mind and the spirit that is spreading with a furious and subtle contagion all over the world, and of all the countries that are sick, America is perhaps the most grievously affected. On all sides we have people building bomb shelters, where, in case of nuclear war, they will simply bake slowly instead of burning quickly or being blown out of existence like a flash. And they are prepared to sit in these shelters with machine guns with which to prevent their neighbor from entering. This in a nation that claims to be fighting for religious truth and freedom.[22]

Merton revealed that he could not be published unless the material was precensored at the abbey. When I received a second mimeographed nonbook, I learned that Father James, abbot of Gethsemani, had allowed Merton to circulate his writings on peace in mimeographed form, since censorship would be required only for material that had the possibility of reaching the general public.

Another mimeographed communication I received was an article entitled "The Prison Meditations of Fr. Delp." Alfred Delp was a Jesuit priest executed in February 1945 for having joined the secret discussions of an anti-Nazi group concerned with reconstructing the social order along Christian lines after World War II. Merton asserted that those accustomed to the usual spiritual works, in reading Delp's meditations written when he was literally in chains, would need to adjust themselves to a new and disturbing outlook. The article contains Merton's view on the role of the mystic and prophet in society:

> The mystic and the spiritual man who in our day remain indifferent to the problems of their fellow men, who are not fully capable of facing those problems, will find themselves inevitably involved in the same ruin. They will suffer the same deceptions, be implicated in the same crimes. They will go down to ruin with the same blindness and the same insensitivity to the presence of evil. They will be deaf to the voice crying in the wilderness, for they will have listened to some other, more comforting voice, of their own making. This is the penalty of evasion and complacency.[23]

Merton points out that Western Christians might one day be in a situation similar to that of the condemned Delp.

"Peace in the Post-Christian Era"

The nonbook, entitled "Peace in the Post-Christian Era," had a chapter entitled "The Christian as Peacemaker." Merton stressed the importance of recognizing that we do not live in a Christian world and that while the "ages of faith" were not a time of earthly paradise, at least a Christian ethic of love was recognized, if only in the breach. The non-Christian world of the mid–twentieth century retains, in his view, some vestige of Christian morality in a few formulas and clichés. He sees Christians dismissing the gospel ethic of nonviolence and love as sentimentality. In an expanded version, he restated the Christian position unambiguously:

> Christians believe that Christ came into this world as the Prince of Peace. We believe that Christ himself is our peace (Eph. 2:14). We believe that God has chosen for himself, in the Mystical Body of Christ, an elect people, regenerated by the blood of the Savior, and committed by their baptismal promise to wage war upon the evil and hatred that are in man, and help to establish the Kingdom of God and of peace.

This short paragraph encapsulates his view on peacemaking and illumines all his convictions underlying conscientious abstention from war and the impossibility of justifying the killing of any human person, the image of God, for any motive. These conclusions simply describe the conduct demanded of a people "regenerated by the blood of the Savior." In this and in other statements, he posits the kingdom of God as the ultimate end of all peacemaking.

Merton hoped to have "Peace in the Post-Christian Era" published for general consumption. It was held back by an order he received in April 1962 to do no more writing on peace. Those involved in the peace movement had found in it treasures of insight and guidance. When, in two years time, the ban on peace writings was lifted, the text was published as *The Christian in World Crisis.*

Merton launched a series of essays that provided the ground for a Christian theology of peace and peacemaking. "Blessed Are the Meek" analyzed the false idea of meekness that might inhibit Christians from resisting what should be resisted. He cites words of Father Franz Metzger, executed by the Nazis for his resistance, and makes them his own: "War owes its existence to the Father of lies. War itself is a lie. War is the kingdom of Satan." He contributed "A Note on Civil Disobedience and Nonviolent Revolution" to the National Commission on the Causes and Prevention of Violence. In "Toward a Theology of Resistance," he asserted that "violence today is white collar violence, the systemically organized bureaucratic and technological destruction of man." His aim was to unmask the "unacceptable ambiguities of a theology of 'might makes right' masquerading as a Christian theology of love."

The Word "Pacifist"

In 1962, Thomas Merton agreed to become a sponsor of Pax, the American affiliate of the English pacifist organization Pax. We shared with him our dilemma about the use of the word "pacifist." He wrote in reply:

> About the word "pacifist." This is really a very important semantic problem which touches on theology, and I think it requires a little thought. I would even say that one of the more urgent tasks of Pax in America is education of Catholics on this point.
>
> For instance there has been so much talk to the effect that "A Catholic is not allowed to be a pacifist" and so on. What does it mean? If by pacifist we mean "peacemaker," then a Catholic is obliged to be one, or to tend toward being one.
>
> The ambiguity about the word "pacifist" is, I think, due to the fact that it has always, in America and England, been understood in the Protestant and liberal context of individualism and indifferentism. The idea that one could be a pacifist, even the approach that the U.S. draft legislation takes toward it, is based on the idea that there is no accounting for the urges of individual consciences and that each one is left inscrutably to his own whims on the matter. If a person feels strongly enough on religious grounds that there should be no wars, then that is his affair and society will respect his subjective convictions.
>
> Obviously this is not a concept of pacifism which can be too acceptable to Catholics, though even this is basically acceptable insofar as the individual conscience is always to be respected, and the individual is obliged to follow a conscience that is "even erroneous."[24]

Merton continued:

> Certainly we can continue to use the word, but we must make clear that it is to be understood in the light of nuclear warfare and *Pacem in Terris,* in

> other words that it is not just an individual and subjective revulsion of war, but a Catholic protest, based on the mind of the church, against the use of war as a way of settling international disputes when war is no longer a reasonable court of last appeal and when in the last analysis, all war today threatens to develop into nuclear war. As a matter of fact those who want to confuse the issue theologically (this includes most theologians in this country apparently) immediately try to get the argument on to this point and to maneuver the "pacifist" into a position where he seems to be a complete subjectivist and opposed to Catholic ethical thought even before Hiroshima.
>
> It seems to me that the important thing is not to let oneself get identified with a "pacifism" that is religiously marginal....
>
> Thus in résumé I would say it is most important to make sure that the word "pacifist" for Pax does not mean something purely marginal and subjective. Until the word is thoroughly cleared of such associations one must be rather circumspect in urging it, but I would say that with a little care and patience it could be rehabilitated.[25]

Merton had focused on the central point of our dilemma, and we were deeply grateful to him. In our rejection of war, we wanted our position to be seen as grounded in the mind of the church, and not as that of a religiously marginal group. We decided that besides the task of disentangling the ambiguities involved in the task of rehabilitating the word "pacifist," we would emphasize the expressions "gospel nonviolence" and "gospel peacemaking."

When Pax launched a campaign for the rights of conscience, Merton wrote us that we could use his name. By then he had been allowed to live in a hermitage a little apart from the abbey. Though he followed the monastic rule of worship and prayer, he was able to do much more writing and answering of correspondence. In 1964, he became sponsor of the Catholic Peace Fellowship, led by Father Daniel Berrigan, Jim Forest, and Tom Cornell. It was affiliated with the Fellowship of Reconciliation, which was avowedly "pacifist."

A meeting-retreat on the subject of "spiritual roots of protest" was led by Merton in November 1966, and brought a group of nine peacemakers to Gethsemani. Among them were Daniel and Philip Berrigan, Jim Forest and Tom Cornell of the Catholic Peace Fellowship, and John H. Yoder, a Mennonite theologian and writer. Merton had just read Gordon Zahn's *In Solitary Witness,* the story of a Catholic peasant who refused to take part in Hitler's war and went to the beheading block. During the conference he referred often to Franz Jagerstatter and the witness that led to his death.

He Shall Overcome

Merton was delighted to receive our Pax publication *The War That Is Forbidden,* dealing with Vatican II. Merton had been most concerned with the decisions of the council as they related to peace.

Since Pax was planning its annual meeting at the Catholic Worker farm in

Tivoli, New York, I asked him if he would be willing to send us a message. I told him that the subject of the gathering would be "peace and revolution." He replied with an article on the subject of the conference, "Peace and Revolution: A Footnote from *Ulysses*." The article and covering letter were dated July 16, 1968. Merton wrote me in the accompanying letter:

> I hope it is useful. It is a bit long, no doubt. Perhaps it will throw light on whatever Jim Douglass says — with a little variety and a new approach.
>
> Naturally, I leave you to edit, cut, adapt and so on to suit the needs of the conference. My own feeling is that we Catholics ought to stick pretty definitely to non-violence, and not for pragmatic reasons or for the sake of an image, but because it is the closest to the truth and to the gospel....
>
> I shall keep the conference in mind and prayer. Pray for me, too, and give my love to those present. And let us go forward in hope — though in these times our hope is necessarily sober.[26]

Marty Corbin, the editor of the *Catholic Worker,* read the article aloud to the group at Tivoli. We rejoiced as we discussed it. It was later published in our *Peace* magazine and reprinted in a collection of Merton's works.[27]

Pointing out that critics had given a favorable reading of Bloom (the central character in *Ulysses*) as a pacifist, Merton warns against Bloom as an authentic pacifist. His "pacifism," Merton explained,

> is in line with everything else in Bloom. It is the expression of pathetic weakness, confusion, frustration and ambivalence. It is not the product of any serious moral conviction.
>
> To say that Bloom is a pacifist and even to commend him for it certainly does no service whatever to pacifism or to the cause of peace. But the point I would like to make — and this I think fits in with Joyce's real intention — is that the Cyclops episode does in fact spell out the whole issue of Peace and Revolution in terms of popular contemporary cliché.

Bloom is contrasted with the "Citizen," a man of action who believed in revolutionary action, seen in the Dublin of 1904, as the revolutionary force that would remove the British occupiers of Ireland. The Citizen is a stand-in for the muscular Christian who disparages the weak-kneed pacifist. To him, "the pacifist, the nonviolent resister, ... is at once a crafty and ineffectual person, because his ideas about love and peace mark him as cut off from reality."

For Merton, one of the chief themes of *Ulysses* is "the breakdown of language and of communication as part of the disruption of Western culture.... Pacifism and nonviolence are fully involved in this question of language."

"Nonviolence, as Gandhi conceived it," Merton points out, "is in fact a kind of language."

"The real dynamic of nonviolence," in Merton's view, "can be considered as a purification of language, a restoration of true communication on a human level, when language has been emptied of meaning by misuse and corruption. Nonviolence is meant to communicate love not in word but in act. Above all, nonviolence is meant to convey and to defend truth which has been obscured and defiled by political double talk."

"Has nonviolence been found wanting?" Merton asserts, "Yes and no." His reply is contingent:

> It has been found wanting wherever it has been the nonviolence of the weak. It has not been found so when it has been the nonviolence of the strong. What is the difference? It is a difference of language. The language of spurious nonviolence is merely another, more equivocal, form of the language of power. It is a different method of expressing one's will to power.
>
> Nonviolence is not for power but for truth. It is not pragmatic but prophetic. It is not aimed at immediate political results, but at the manifestation of fundamental and crucially important truth. Nonviolence is not primarily the language of efficacy, but the language of kairos. It does not say "We shall overcome" so much as "This is the day of the Lord, and whatever may happen to us, He shall overcome."[28]

It is in this assertion, according to Gordon Zahn, that Merton makes one of his most significant contributions to the spirituality of peace. He raises nonviolent resistance to an "eschatological dimension," a dimension linked to the end of the human journey. In it, says Zahn, "Merton distances himself from the man he greatly admired, Martin Luther King, who said 'We shall overcome.' Merton points to the nonviolent victory of the Lord in the words, 'He Shall Overcome.'"[29]

It turned out that "Peace and Revolution: A Footnote from *Ulysses*" was Thomas Merton's legacy to Pax and the peace movement. It was one of the last, if not the last, of his writings on peace. In September 1968, he left on the journey to Asia where he died in what his friend Ed Rice called "a technological crucifixion," the falling of a standing electric fan across his body.

Merton warned that a spurious nonviolence, with any link to efficacy, would lead to the conclusion that only force is efficacious. Never was it more necessary, he warned, "to understand the importance of genuine nonviolence as a power for real change because it is aimed not so much at revolution as at conversion."[30] In this, he was at one with the century's greatest practitioner of nonviolence, Mahatma Gandhi.

Chapter 13

The Right to Refuse to Kill

For what purpose, then, did God endow every person with reason and free will, if we are nevertheless required to render blind obedience?

—Franz Jagerstatter

Must the citizen even for a moment, or in the least degree, resign his conscience to the legislator? Why has every man a conscience then?

—Henry David Thoreau

In the depths of his conscience, man detects a law which he does not impose upon himself, but which holds him to obedience. Always summoning him to love good and avoid evil, the voice of conscience can when necessary speak to his heart more specifically: do this, shun that. For man has in the heart a law written by God. To obey it is the very dignity of man; according to it will he be judged.

—Vatican II, *The Church in the Modern World*

The earliest followers of Jesus knew nothing of the *right* to refuse to kill. They knew from Jesus only that they had the *duty* to refuse to kill the children of God. To the amazed incredulity of the ancient world of Rome, they were ready to give their lives in compliance with this duty. They were prepared to give witness in blood to the duty of not shedding blood. Witness by the word was not wanting, as we learn from Clement of Alexandria, Athanasius, Origen, and Lactantius.

From ancient times, men have participated in the organized killing known as war. For some, becoming a soldier was ordained at birth, as in the garrison city-state of Sparta. In some cultures, as in India, there was, technically at least, a warrior caste. For large numbers, soldiering was considered an honorable career with lifetime benefits which might, on occasion, call for the shedding of blood, as in the Roman army. It fell to many to take part in warfare as a religious duty, as in the jihad. To many, joining the Crusades seemed a religious duty, though a voluntary one. Various European rulers impressed men into their armies as they were needed to prosecute a specific campaign. The naval and military forces were often kept up to adequate strength by "press gangs" which snatched men, chiefly the poor, from the ports and from the streets of cities.

In medieval times in Britain and on the continent of Europe, knights coming from the class that could command sleek mounts and costly armor cast the glow of chivalry and honor on warriors. Single combat made heroes of skilled contenders who exhibited special bravery. With the advent of gunpowder, and its availability to greater areas of Europe, the glow and romance of war-making began to fade. There was no glory in advancing toward an enemy who could blow your head off or split your body into bleeding fragments. (Yet in World War I the ordinary "Tommy" in British ranks somehow accepted this glory in defense of "home fires" and of a world-girdling empire.)

When Blood Is Their Argument

In a chronicle history play, *Henry V*, Shakespeare includes the famous battlefield conversation among three men, two of them ordinary soldiers and the third King Henry himself in disguise. King Henry pays homage to the just war tradition, telling the men that he would be content to die with the king, "his cause being just and his quarrel honourable." Will, one of the soldiers, interposes that as soldiers they do not know whether the cause is just. Bates, the second soldier, points out that such knowledge is not for ordinary soldiers. It is enough for them to know that they are the king's subjects. "If his cause be wrong, our obedience wipes the crime of it out of us."

Shakespeare then expounds, through Will, the accepted view regarding the person who would be responsible on judgment day for battle-deaths if the war were not just:

> But if the cause be not good, the King himself hath a heavy reckoning to make when all those legs and arms and heads, chopped off in battle, shall join together at the latter day and cry all, "We died at such a place; some swearing, some crying for a surgeon, some upon their wives left poor behind them, some upon the debts they owe, some upon their children rawly left. I am afeard that few die well that die in battle, for how can they charitably dispose of anything when blood is their argument? Now, if these men do not die well, it will be a black matter for the king that led them to it, who to disobey were against all proportion of subjection."[1]

Besides nodding in the direction of the just war, Shakespeare asserts the accepted view of the duty of unquestioning obedience on the part of the subject-soldier. Such obedience was absolute and released the soldier's conscience from all responsibility regarding the justice of the cause for which he was commanded to fight. His conscience was in the keeping of the king. In this, Shakespeare followed Augustine. There were no niceties regarding the doubtful conscience or conscience convinced of injustice such as had been raised by the decretists and decretalists (see chapter 5).

The Nation in Arms

It may come as a surprise to many that universal conscription into military service took its origin in fairly recent history. Conscription in the modern sense of the term dates only from the French Revolution. The concept of the "nation in arms," like so many of the theories and institutions on which modern France has been built, originated in the war-torn years of the revolution. "In August

1793, the Committee on Public Safety laid down both the theory and practice of what the twentieth century was to call 'total war': the young men were to go forth to battle, the married men would forge arms, the women were to make tents and clothing, and the aged were 'to preach hatred of kings and the unity of the Republic.' "[2]

Lynn Montross terms the decree announcing universal military conscription "one of the most memorable dates in the chronicles of war." The proclamation asserted:

> The young men shall fight; the married men shall forge weapons and transport supplies; the women will make tents and serve in the hospitals; the children will make up old linen into lint; the old men will have themselves carried into the public squares to rouse the courage of the fighting men, and to preach hatred of kings and the unity of the Republic. The public buildings shall be turned into barracks, the public squares into munitions factories; the earthen floors of cellars shall be treated with lye to extract saltpeter. All suitable firearms shall be turned over to the troops; the interior shall be policed with fowling pieces and with cold steel. All saddle horses shall be seized for cavalry; all draft horses not employed in cultivation will draw the artillery and supply wagons.[3]

Among the tragic effects of this concept have been the forcible involvement of whole peoples in total war through the enormous power given to the nation-state. The war-making state can coerce its own human and physical resources to plan and effectuate the destruction of other human beings: "The citizen turned soldier — indeed, conscript service as both the badge and moral consequence of citizenship — was one of the major results of the French Revolution. It was by no means the least important legacy bequeathed by the revolutionary age to the generations of the modern world."[4]

One of the first effects of universal military conscription was to make possible the Napoleonic policy of conquest. Once the French government had awarded itself the power to dragoon into army ranks the farmers from their lands, the tradesmen from their shops, the craftsmen from their workplaces, even students from their school, it had the power to embark on any military campaign. Napoleon, an obscure general from Corsica, springing to fame by putting down the 1795 uprising of Parisians against the Revolutionary Convention, went on to head the army of France. Taking advantage of the military power available to him from a ragged army, he forged a fighting force of unbeatable power (unbeatable, that is, until the face of Europe had been disfigured by the death of millions). "I can afford to expend thirty thousand men a month," Napoleon is reported to have boasted in 1805.[5] By then he was emperor of the French. The Battle of Borodino, before Moscow, in which there were well over a hundred thousand casualties, bloodies the pages of Leo Tolstoy's *War and Peace*.

The example of France in levying universal military conscription soon had imitators. By 1815 the state of Prussia had adopted universal military conscription in a move to strengthen its already strong army. In that year Prussian forces provided the turning-point in the last action of the Napoleonic wars at Waterloo.

In the United States a conscription law was adopted during the Civil War by the Confederates in 1862 and by the Northern side in 1863. Except for some exemptions, it was presumed to be a general draft of men between the ages of twenty and forty-five. Riots ensued in the North when it was discovered that the more affluent draftees could buy their way out of conscription with a payment to hire a substitute soldier. Japan passed a conscription law in 1873.

The encyclical of 1894 of Leo XIII addressed to all leaders attacking the principle of universal conscription went unnoticed. By 1914, conscript armies faced each other on the soil of Europe. Millions of men who were given no choice were drafted into military ranks. In France, where universal conscription was first imposed, even priests were conscripted. They managed to serve as stretcher-bearers and medical attendants, performing in the killing fields the works of mercy enjoined upon them by Jesus.

When the United States voted to enter what was called the "Great War," a central lottery was established in Washington in June 1917 by Selective Service. Close to three million men were conscripted into the army and navy, and about two million additional men, eligible for the draft, enlisted voluntarily.

What was the influence of the Catholic Church as a body in deciding on the justice of the Great War? What moral guidance could the church give, or its adherents accept, in deciding whether they would be carrying out the will of God in taking part in the killing? What was tragically clear was that church leaders were ignored in the declaration of the war. The words of popes counseling peace are poignant in their witness to the peace of Christ and their ineffectiveness in the war of men.

As the fires of war were being lit, Pope Pius X sent a prayer from the Holy See on August 2, 1914, "that God may be moved with pity and remove as soon as possible the disastrous torch of war and inspire the supreme rulers of nations with thoughts of peace and not of affliction."

On September 8, 1914, a few weeks after the torch of war had set the continent of Europe aflame, Benedict XV, successor of Pius X, spoke from the Vatican: "We are at once struck with horror and inexpressible sorrow at the awful spectacles of this war."[6] Addressing those who direct the affairs of nations, he "begged and implored them even now to turn their thoughts to the laying aside of their quarrels for the sake of human society. Let them reflect that there is already too much of misery and grief linked with this mortal life, so that it should not be made still more wretched and sorrowful. Let them agree that already enough of ruin has been caused, enough of human blood has been shed."[7]

The convoluted events that followed the fatal shooting of Duke Francis Ferdinand in Sarajevo on June 28, 1914, led to an orgy of bloodletting unprecedented in history. Rampant nationalism and the unhappy habit of peoples and ethnic groups of raking up historical wrongs to wreak vengeance on the present blotted out any reference to the conditions of the just war.

If the Catholic Church, either through Pope Pius X or Pope Benedict XV or the bishops of the various nations involved, could have no influence on the descent into organized violence, what would be the role of those who would do the fighting? Conscription obliged them to put on uniforms and carry out orders from above, orders handed down by officers formed in the military culture of standing armies. The conscript was forced, in effect, into an abdication of conscience once he was inducted into army ranks.

As a body, and in its members, the Catholic Church was distanced from playing a prophetic role in the greatest carnage that humanity had experienced until that time. The nation-states could not only dictate to the Catholic and other churches as to the declaration of war, but also dictate to the conscience of church members through universal military conscription. Any intrusion of the moral and prophetic dimension into war's violence was received by the warring nations as no more than that — an intrusion which could have no effect in halting the onrush of events.

The prophetic dimension of the Catholic Church and its members had never died over the centuries. The witness of nonretaliation in response to personal attack had never been recalled since Augustine emphasized it. Even in the breach the church kept alive the witness of the early followers of Jesus. The example of the nonviolence of Jesus was maintained throughout the history of the church by priests as they presided at the sacred liturgy and by monks occupied with the works of peace in their monasteries. The teaching and example of St. Francis and the Third Order Franciscans testified to gospel nonviolence as did the life and work of Desiderius Erasmus and those who preserved the Erasmian spirit.

A New Life

It was the Anabaptists who were the source for a prophetic witness against war. As Roland Bainton points out, the Anabaptists, after the Reform, were the only faith group who did not share the just war theory.[8] The name Anabaptist refers to the fact that new members, though already members of Christian churches, had to be rebaptized. Adult baptism, they asserted, was one sign of their return to the practices of the early church, in which, they claimed, infant baptism was not practiced. Not only adult baptism marked the new movement, but a radical redirection of life in accordance with the example of the earliest followers of Jesus. The Anabaptists turned to the scriptures as their source for a way of life.

The sign of the new creature of the gospel was a new life with love at its center, accompanied by the sharing of goods, simplicity of life, and a refusal, even when so commanded by the state, to kill another human being. On the continent of Europe, the Anabaptists rejected Catholicism as well as the Reform, as expressed in Lutheranism and Calvinism. Bainton comments, "The Anabaptists, being burned by the Catholics and drowned by the Protestants, saw no hope in man."[9]

From the Anabaptist root sprang groups which, though small in membership, profoundly influenced the larger church bodies by their fidelity to the blessed witness of gospel nonviolence. In Europe, the Mennonites, Hutterites, and Brethren grew despite atrocious persecution. In England, the Quakers arose in the period of the iron rule of the lord protector of England, Oliver Cromwell, who divided human beings into the elect of God and the nonelect.

Early founders of the Anabaptist movement were Conrad Grebel and Felix Manz, who in 1524 called themselves the Swiss Brethren. Grebel's pacifism derived from a conviction of the leadership of Christ and of Christ's captaincy over his nonviolent soldiers. These soldiers were called upon to conduct their lives in absolute nonviolence, their willingness to suffer in Christ's name serving as a sign and confirmation of their ultimate salvation.[10] Felix Manz suffered execution by drowning in Zurich by a revolutionary branch of the Anabaptists headed by Ulrich Zwingli.[11]

A brief accounting of the growth of Christian movements that taught and suffered for gospel nonviolence will indicate their importance to contemporary Christianity.

Should Christ's Disciples Carry a Sword?

In 1986, the Hutterian Brethren published the *Great Chronicle of the Hutterian Brethren.* It commemorated the 450th anniversary of Jacob Hutter's death at the stake, and served as a commemoration of the long list of Hutterite martyrs of conscience.

A decisive issue was nonviolence. "Should a disciple of Jesus carry a sword? Should he not carry just a staff?" This discussion had taken place at Nikolsburg in Moravia, now part of Czechoslovakia, where many Anabaptists had gathered, having fled from persecution in the Tyrol and elsewhere. In 1528 about two hundred adults, deciding in favor of unconditional nonviolence, left Nikolsburg to found a community at Austerlitz in Moravia. "On the way, the spirit of brotherly love impelled them to pool their few belongings on a cloak laid on the ground — the actual beginning of full community of goods among these people later to be known as the Hutterian Brethren."[12]

For over three hundred years, the witness of nonviolence was preserved unbroken by the Hutterians, through executions and repeated expulsions which brought them to Hungary, Romania, Austria, Poland, and the Ukraine. At various times the witness of "full community of goods" was limited or abandoned. As with other religious groups, their fidelity was sustained by Anabaptist hymns, one of which, cited by Bainton, is partially reprinted here:

> We creep for refuge under trees.
> They hunt us with the blood hound.
> Like lambs they take us as they please
> and hold us roped and strong-bound.
> They show us off to everyone
> As if the peace we'd broken,
> as sheep for slaughter looked upon as heretics bespoken.[13]

The migration of Hutterites to the United States began in 1874. Another move — from the United States to Canada — was occasioned by World War I, when the Hutterites refused conscription into the U.S. military. Two young Hutterites imprisoned for refusing war service died while in custody.

The account of the antiwar witness of the young Hutterites impressed a German citizen, Eberhard Arnold, who had become acquainted with the Anabaptists during his university studies. In 1920 Arnold with his family left the wealth and security of a respected academic career in Berlin to settle in a village in central Germany. Their aim was gospel-based community living. They called their community the Bruderhof, the place of brothers. Persecution by the Nazi regime, including the threat of conscription into the military, caused the group to leave Germany. The small community became incorporated into the Hutterian Church, and Arnold was ordained as a Servant of the Word. The Hutterian Bruderhof of Arnold's inspiration eventually planted communities in the United States (in Connecticut and Pennsylvania) and in England.

Arnold and Thomas Merton saw community as a harbinger of God's kingdom, a kingdom of love. The kingdom of loving nonviolence was that born in the victory of love over death on the cross. Another monk, Basil Penington, a Cistercian like Merton, in an introduction to a book by Arnold and Merton, points to the significance of the fact that "spiritual giants like Thomas Merton and Eberhard Arnold reach across what in times past seemed to be an unbridgeable gap."[14]

The question raised by the Hutterian communities, in their withdrawal from the life around them, is whether such withdrawal, with its rejection of holding public office, is a prerequisite to live out the message of Jesus, including nonviolence. The choice of a peacemaking livelihood, one that has clearly cut its ties to the war-making state, is a powerful witness. The close-knit community spirit strengthens those who have made crucial choices that involve an overall transformation of life.

Waging Peace While Others Waged War

Menno Simons (ca. 1496–1561), from whom the Mennonites take their name, stated that true Christians must "crucify the flesh and its desires and lusts, prune the heart, mouth and the whole body with the knife of the divine word of all unclear thoughts, unbecoming words and actions."[15]

Besides decreeing simplicity of life, in dress and in all aspects of living, the followers of Menno were to abide by the nonviolence of Jesus. Enemies should be loved, and all swords beaten into plowshares. Menno, a Dutch Catholic priest, became acquainted with the Bible only after his ordination. He chose rebaptism into a brotherhood that emphasized the inner relationship of the believer with God. He married, became a pastor, and then became the spiritual leader, or overseer, of the new community.

The American Mennonites had anticipated in January 1940 the move toward military conscription in the United States. They met with the president of the United States, in company with the leaders of the Quakers and Church of the Brethren, to propose the establishment of Civilian Public Service Camps where conscientious objectors to war could perform humanitarian service. The draft law, the Burke-Wadsworth Selective Service Act, was passed in September 1940. The historic peace churches covered the costs of their men in the camps.

Even when Catholics achieved the status of conscientious objection from their draft boards, there was no Catholic agency to contribute to their expenses (aside from small contributions by individual bishops). The Mennonites joined with the Quakers in paying the share of the Catholic objectors, thus buttressing the role of conscience in war.

During the years of World War II, Mennonite communities canned food and gathered clothing for civilian relief, waging peace while others waged war. As soon as the volunteer relief agencies were permitted to enter occupied Germany, they were ready with shipments of mercy.

The teaching of Menno is suffused with the spirit of the Lamb, that of suffering but not inflicting suffering on others. It is still the spirit of the Mennonites in the twentieth century.

Not Instruments to Shed Human Blood

The Church of the Brethren also stemmed originally from the Anabaptist root. In a peace statement of 1948, "The Church and War," we find the declaration:

> The Church of the Brethren, since its beginning in 1708, has repeatedly declared its position against war. Our understanding of the life and teaching of Christ, as revealed in the New Testament, led our annual conference to state in 1785 that we should not submit to the higher powers so as to make ourselves their instruments to shed human blood....
>
> The church cannot concede to the state the authority to conscript citizens for military training or military service against their conscience.[16]

The Brethren were not formally organized until 1708 when eight Brethren decided to be rebaptized in the Eder River in Schwarzenau, Germany. As they came later than other groups of this type, they did not, according to Dale W. Brown, develop a theology of suffering and martyrdom to the same degree as had the earlier Anabaptists.[17] The people were not subjected to the same atrocious persecution and death. In addition to the influence of the Anabaptists, there was the pietist strand, emanating from such spiritual writers as Jakob Böhme. From the beginning they were convinced that the call of Christ meant the refusal of military service even under conscription. One of the founders, Alexander Mack, alluding to Lutherans, Calvinists, and Catholics, wrote, "What is more horrible, they go publicly to war, and slaughter one another by the thousands."[18]

Persecution drove the Brethren to the United States, where under Quaker governance in Pennsylvania they found freedom of conscience. At the time of the American Revolution all males refusing to bear arms were required to contribute an amount of money judged equivalent to the time that conscripts spent in acquiring military discipline. "Substitute money" was paid by some who refused military service so that someone else would take their place. An annual meeting of the church forbade this, repeating the teaching that Brethren "should take no part in war or blood shedding, which might take place if we could pay for hiring men voluntarily."[19]

The practice of bounty-money (to supply a substitute) was continued in the Civil War, and Brethren were again forbidden to pay it. However, a changed commitment by the government, that the funds would be used for "sick and wounded soldiers," allowed Brethren to make the payment. Through the action by the Quakers, conscientious objectors from the "peace churches" would be considered noncombatants. They could then qualify to serve in hospitals or take responsibility for caring for freed slaves, rather than taking up arms.

An alternative to the violence of war was presented to the newly elected Provincial Assembly in November 1775 at the time of the American Revolution. Mennonite and Brethren elders presented a petition which described a pattern of alternative civilian service for those who "are persuaded in their conscience to love their enemies and to resist evil." Rather than take up arms, they should "be helpful to those who are in need and distressed circumstances;... it being our principle to feed the hungry and give the thirsty drink, — we have dedicated ourselves to serve all men in everything that can be helpful in the preservation of men's lives, but we find no freedom in giving, or doing, or assisting in anything by which men's lives are destroyed or hurt."[20]

Within what was soon to become the United States, this was the first program ever presented regarding civilian service as alternative to the military, and it was possibly the first such program ever devised for conscientious objectors. As with every act of testimony by members of the historic peace churches on behalf of abstention from war, it was taken not simply for the members of their own churches, but on behalf of the consciences of all.

The Lamb's War

"There is that of God in every man" is a statement of the Religious Society of Friends that has achieved familiarity in the wider society. It is acceptable to many in the Judeo-Christian tradition in its echo of the biblical teaching of God's image in human beings. What brought difficulty to the members of the Society of Friends was the conclusions that "there is that of God" in the enemy and that, therefore, to the followers of Jesus, the enemy should not be hurt or killed, but loved. From this seed arose the peace testimony of the Friends.

The Society of Friends, who were early called Quakers because of their quaking or trembling before the divine Spirit, arose during the bloody religious wars of seventeenth-century England. It was a time when Oliver Cromwell (1599–1658) was attempting to build a new Jerusalem in England through his armies. George Fox (1624–91), the son of a weaver, and apprenticed to a shoemaker, underwent a mystical experience. In that illumination "the Lord God opened to me by his invisible power that every man was enlightened by the Divine Light of Christ."[21]

Fox was well versed in scripture, and after his mystical experience he began to preach to others that the light of the Spirit could enter each soul. This dispensed with the mediation of ordained ministers or liturgy. He and his followers accepted neither the Church of England nor the Puritan theocracy. Fox suffered beatings and frequent imprisonment, but continued to bring followers to a position that rejected the society around them. After one imprisonment, in which he converted fellow prisoners to his beliefs, he was offered a captaincy in the army of Cromwell. When he was asked why he would not fight with Cromwell against the king, he replied, "I live in the virtue of that life and power which took away the occasion of all wars and I knew from whence all wars did rise, according to James' doctrine. . . . I told them [the Commonwealth Commissioners] I was come into the covenant of peace which was before wars and strifes were."[22]

That was in 1651, and twenty years later, Fox was to make a similar declaration against war service to Charles II, England's king:

> We utterly deny all outward wars and strife and fighting with outward weapons, for any end or under any pretense whatsoever. And this is our testimony to the whole world. The spirit of Christ, by which we are guided, is not changeable, so as once to command us from a thing as evil and again to move unto it, and we do certainly know, and so testify to the world, that the spirit of Christ, which leads to all truth, will never move us to fight and war against man with outward weapons, neither for the kingdom of Christ, nor for the kingdoms of the world.[23]

The only war which the Quakers would fight was the "Lamb's War," a war of those who have experienced the light of Christ's presence in their hearts. This would call for nonretaliation to evil; instead, the evildoer would receive from the Quaker love and the works of love. The term "Lamb's War" comes from the Book of Revelation. Brown describes the Lamb's War as "an external, prophetic, missionary, evangelistic, ideological, social, economical and political struggle to end evil in human history until God in mercy brings history to the peaceable kingdom promised by Isaiah and described in Revelation."[24]

The historic peace churches, though small in numbers, have given help out of all proportion to their numbers to those beggared and made homeless by the wars of others. The Quakers, more than the Hutterians or Brethren, take part in economic and political life, maintaining, for example, an office known as Friends Committee on Legislation in Washington, D.C. Quaker activity is important at the United Nations, with an office of the Quakers close to the UN headquarters in New York and an office near the Geneva UN Palais des Nations.

Wherever possible, Quakers bring their spirit of mediation, injecting their term "loving disagreement" into situations where men and women are seemingly locked in verbal combat.

The Person and the Nation-State

The question of refusal of war service inevitably results in an examination of the relationship of the person to the nation-state. This is the area where the individual confronts the power and ideology of the state and, equipped only with the peace message of Jesus, refuses the demand of the state. The maintenance of this refusal calls for a distancing of the person from his or her government. At a time when the nation-state gathers to itself, as in the United States, an inordinate amount of resources for weaponry and for a standing army that bestrides the globe, peacemakers have to look at the policies of their nation with deep suspicion. The country's readiness to go to war with little notice, and its capacity to wage indiscriminate warfare (even though nonnuclear), makes the pacifist Christian a critical or alienated citizen. The historic peace churches, to different degrees, separate themselves from the society in which they live, a society with more "overkill" capacity than any society since time began.

When the Catholic Worker community, led by Dorothy Day, rejected war and adopted gospel peacemaking as its way of life, it faced the same dilemma as the Hutterians, Brethren, Mennonites, and Quakers. As more members of the Catholic community moved toward a position of gospel nonviolence, the issue of a moral distance between the Christian and the war-making state would become more urgent and more in need of clarification.

The World's Catholic Bishops on Conscientious Objection

The affirmation of the right of conscientious objection to military service by the Catholic bishops of the entire world was made on December 7, 1965. It was contained in the pastoral constitution entitled *The Church in the Modern World,* a document from Vatican II. (See chapter 10, above, for a recounting of the actions of peace-minded Catholic laypeople from the United States to inform the bishops of their concerns and to communicate to them the urgency of updating the church pronouncements.) Those of us who were part of the "peace lobby" rejoiced at the words in chapter 5 of *The Church in the Modern World:* "It seems right that laws make humane provision for the care of those who for reasons of conscience refuse to bear arms provided that they agree to serve the human community in some other way."

Results in the United States

Many developments arose as a result of the affirmation in favor of conscientious objection by the Catholic Church. This narrative will describe two developments. One occurred in the United States, and one in the Human Rights Commission of the United Nations.

Background: Vietnam. In the United States, conscientious objection had become a burning issue in the late 1960s and early 1970s with the escalation of the Vietnam conflict. The glory had gone out of war when home television screens at dinnertime were peopled by Vietnamese villagers half a world away lying dead in grotesque poses. More and more the screens were peopled by young Americans comforting each other on the loss of comrades fragmented by bombs.

Opposition to the obligatory draft reached such a crescendo that young people demonstrated before the White House chanting such slogans as, "Hell No! We Won't Go." The Vietnam conflict, like World War II and the Korean War, was fought largely with conscript soldiers.

While more and more Catholics were becoming conscientious objectors, they were facing draft boards that held to the letter of the law, that only "all-war objectors" could be classified as conscientious objectors and become eligible for alternative civilian service. The 1965 opening to conscientious objection by the world's Catholic bishops had not been officially recognized by the American Catholic bishops. Such official recognition and promulgation were matters of great urgency.

The Selective Service System was to expire on July 1, 1967. The National Advisory Committee on Selective Service was set up by the U.S. government, a twenty-man citizens' panel to advise on the new law. We who were pushing for a greater opening to conscientious objection found out that the discrimination against "just war" objectors would be continued in the restructured Selective Service law. As reported in the press, the committee majority voted that "a determination of the justice or unjustice of any war could only be made within the context of that war itself. Forcing upon the individual the necessity of making that distinction — which would have the practical effect of taking away the *government's obligation* of making it for him — would put a burden heretofore unknown on the man in uniform."[25] This revealing statement highlights the fact that a conscience is a burden that is lifted from the individual soldier by his government.

The Rights of Conscience Campaign. The Catholic Peace group Pax, secure in the knowledge that the world's bishops had spoken, decided on a campaign to amend the U.S. draft law. Their aim was that "just war" as well as "all-war" objectors would be recognized as conscientious objectors. The action took the form of the Pax Rights of Conscience Campaign during May 1967. It was focused first on the American hierarchy to urge them to promulgate among American Catholics the statement on conscientious objection by Vatican II. It was also focused on the American Congress, where a change in the draft law would have to originate. Pax inserted a paid announcement recognizing selective conscientious objection in the center pages of *Commonweal* magazine.

The Pax Rights of Conscience Campaign asserted: "The consciences of those following the Just War Tradition should be respected." The text outlined the built-in discrimination of the U.S. draft law. We pointed out that the government was making a theological distinction which it had no right to make.

One of the first signers of the statement was Thomas Merton. Though he felt, as he explained in his letter, "that true Catholics ought to stick pretty definitely to the gospel," he hoped for a wider approach. He suggested that the bishops issue a "a statement that either the draft must be abolished or selective conscientious objection be permitted."[26]

In addition to the signature of Thomas Merton, there were signatures from seven Catholic bishops, the editors of three Catholic magazines exclusive of *Commonweal*, and prominent leaders of other religious groups. Reprints of the statement were distributed throughout the country, in particular to college campuses. Copies were sent to every member of Congress, both Senate and House. Congressman Don Edwards of California took the text to the floor of the Congress and inserted it into the *Congressional Record.* He stated that the Pax comments "deserve considerable weight in discussing the Selective Service System.... I call attention particularly to the inequity in the present legislation in not recognizing the moral and religious obligations of Christians to abstain from participation in an unjust war. The doctrine of the just war has a long tradition in Christian thought dating from the writings of St. Augustine in the fifth century C.E. The blanket requirement that conscientious objectors must disapprove of all war arbitrarily discriminates against those who prescribe to the just war tradition."[27]

The text of the statement was also sent to every American bishop. We wrote that it was an opportune time for the bishops to ask for a change in the Selective Service law before its restructuring of July 1967. The Pax Rights of Conscience Campaign had no effect on the U.S. Selective Service System. The Selective Service law continued with its in-built discrimination until the U.S. draft law expired in 1973 to be replaced by a volunteer army.

Pax continued to contact the Catholic bishops on the draft law. It was a time when thousands of young people, among them many Catholics, were fleeing the country for such havens as Canada and Sweden. This was proof, if any were needed, that they did not trust their draft boards to honor their choice of conscientious objection to the Vietnam War. When our communications went unanswered, we placed an open letter in three Catholic newspapers, including the *Catholic Worker*, asking the bishops "to make the Vatican Council's statement on conscientious objection a reality for American Catholics."

It happened on November 15, 1968. The American Catholic bishops acknowledged the Vatican II statement on conscientious objection in their pastoral statement, *Human Rights in Our Day.*

The bishops called for the abolition of the draft, reminding Catholics that Pope Benedict XV had seen compulsory military service as a contributing cause of the breeding of actual wars. They affirmed not only the right of conscientious objection to military service, but the right to selective conscientious objection, based on an actual war in which conscripts are asked to fight.

"We recommend," the bishops asserted,

> a modification of the Selective Service Act making it possible, although not easy, for so-called selective conscientious objectors to refuse — without fear of imprisonment or loss of citizenship — to serve in wars which they consider unjust or in branches of service (e.g., the strategic nuclear forces) which would subject them to the performance of actions contrary to deeply held moral conviction about indiscriminant killing.
>
> Whether or not such modifications in our laws are in fact made, we continue to hope that, in the all-important issue of war and peace, all men will follow their consciences.[28]

These statements were published nationwide on November 21, 1968. An immediate response came in the public press from the Selective Service director, General Lewis B. Hershey. "What kind of religious belief have you got that causes you to reject some wars and not others?" he asked in dismayed incredulity. Openly expressing his draft board theology, he commented, "If you say you object to all wars, I can't object to that.... But how can you have a religion that says it's okay to kill one time but not another?"[29]

While Hershey served as director, over fourteen million men had passed through the Selective Service System. Among them were millions of Catholics and others who believed in the tradition of the just war which taught exactly what General Hershey rejected.

Results at the United Nations

The Taboo on Conscientious Objection Broken. The validation of conscientious objection to military service by the world's Catholic bishops emboldened American Catholic peacemakers not only to engage in peace activism in their own countries, but also to take their activism to the United Nations.

In 1970, when the opposition of young people to the Vietnam conflict was at its height, the United Nations declared a World Youth Year. It was a year in which attention by member governments and UN agencies was to be directed to the "education of youth in respect for human rights." It seemed to many of us that the education of youth in respect for human rights was related to the reality of respect for the human rights of youth. The issue of conscientious objection to military service had been taboo at the United Nations since its founding.

A Pax statement on the matter — printed by the United Nations under the title *Education of Youth in Respect for Human Rights* — pointed out that one of the rights increasingly demanded by young people around the world was the right to refuse to kill fellow human beings. It adverted to the fact that the agenda of the Commission on Human Rights dealt with war crimes and crimes against humanity. "Many such crimes," it asserted, "would have remained dark dreams in the minds of a few pathological human beings if there were not a means to dragoon large numbers of ordinary people into carrying them out. That means is the conscription of conscience." It recalled that in the main it is youths who are conscripted for war and whose obedience has often been abused in the conduct of war.[30]

"The United Nations," it continued, "has recognized the right to life as a basic right in the Universal Declaration of Human Rights. It is hardly necessary to mention that great spiritual traditions, including Christianity, Hinduism and Buddhism, have fostered the right of conscience with regard to the 'taking of the sword.'" It concluded: "The right of life is a basic right of man as an ethical self-determined being."

The statement, translated into French, Spanish, and Russian, was duly distributed on March 17, 1970, among the members of the commission. The taboo preventing the airing of conscientious objection to military service at the United Nations had been broken.

I was invited to make an oral presentation of the case for conscientious objection before the commission. I knew that in addressing the commission I was to use other material than that contained in the statement. Nongovernmental organizations (NGOs) were informed that in their statements, they might not attack or disparage member states. It was possible for members of NGOs to submit statements and appear before the UN Commission on Human Rights with consultative status. I served as a UN representative of the International Confederation of Students (the student branch of Pax Romana) and received from the officers the permission to bring the matter to the next session of the commission. There was no rule against paying them compliments, and I paid tribute to Austria and to Soviet history. I alluded to the fact that Austria, after gaining control of its own political destiny, had incorporated into law the right of conscientious objection to military service. I referred to the case of a young Austrian, Franz Jagerstatter, who had refused military conscription into Hitler's army. He knew the penalty for conscientious objection, but persisted in his refusal to be part of the military in any capacity. The young man was taken from his wife and three young children to be beheaded in Berlin on August 9, 1943. He stood with many whose lives had been forfeited for conscientious objection to Hitler's war. Their names were unknown to history. I wondered if the delegates were aware of a petition lodged with the Commission on Human Rights by War Resisters International, a humanist organization with membership worldwide. The petition had been formulated during 1969, the centenary of the birth of Mahatma Gandhi. The thousands of signatories from twenty-seven countries, including three Nobel Peace Prize winners, stated:

> We the undersigned call upon the United Nations Commission on Human Rights to recognize conscientious objection to military service as a basic human right.

As 1970 was the centenary of the birth of Lenin, there had been references in the commission to the humanist tradition of Lenin as "a common patrimony of mankind." I adverted to this and reminded the delegates that the head of Lenin's office staff, many of his close co-workers, and his wife, Krupskaya, had been conscientious objectors. Even at the height of the Soviet Revolution, allowance had been made for conscientious objectors, who were given alternative service, generally in hospitals and in antiepidemic teams.[31]

Young people, I told the delegates, are rebelling in many parts of the world against being treated as replaceable parts of the machine of war. They seem ahead of their elders in respect for life. There was no need to mention that the Vietnam conflict had left a savage wound on American society, and that the youth had poured into the streets in massive demonstrations against war and the draft. Thousands of young men had chosen exile. For a generation of possible conscripts, war had been demythologized by televised scenes of broken bodies — soldiers, noncombatants, men, women, and children caught in the path of destruction.

No sooner had I finished speaking than the arm of the Soviet delegate shot up. He told the commission that he had listened with great interest to the NGO statement. As Krupskaya and Lenin had been mentioned, he wished to make a short reply. He wished to state that Lenin had, as was well known, drawn a distinction between just and unjust wars, such as wars of aggression waged by imperialist and colonialist powers. On the other hand, a fight waged against invaders in the defense or liberation of one's country was a just war and constituted a sacred duty of all citizens, irrespective of their religious or political conviction.

The Soviet delegate had barely finished his intervention when the Commission on Human Rights came in for a surprise. An observer-delegate of a nation not a member of the commission had entered the chamber as I began my remarks. He asked for the floor. He was Ambassador Jamil Baroody of Saudi Arabia, a prominent figure at the United Nations since its birth. With the chairman's concurrence, Baroody took a place at the commission table. "I salute the lady who spoke about the right of young people to object to taking part in war," he began. He went on to say that he felt compelled to speak on the question of the education of youth. His concern was not simply that youth held in their hands the future of humankind. In his view, the modern world had betrayed the confidence of youth. Under slogans cloaking imperialist and economic interests, youth had been led like sheep to the slaughter. Again and again, young people had been conscripted to fight in wars which brought nothing but more death. There were hostilities then in progress, he stated, which were examples of imperialist aggression. In such circumstances it was not surprising to him that young people in many countries were in revolt and demanding their rights. He asked the delegates if there would be any follow-up to the recommendation that would allow youth to object to war.

Their was neither reply nor comment to the intervention of Baroody, and the meeting adjourned.

At the next meeting it was clear that the question of conscientious objection had leaped across the table from the NGO to the government side. Felix Ermacora of Austria told the commission that the question of conscientious objection was a matter of special concern to youth and should become a matter of concern for the United Nations. Klaus Tornudd of Finland asserted, "My delegation would without hesitation consider conscientious objection to military service to be discussed as a human right." Further discussion of the question was postponed to the next session of the commission, to be held in Geneva in February 1971. The taboo on the words "conscientious objection" had been broken by four people in two days.

Conscientious Objection Reaches the UN General Assembly. Jamil Baroody, however, did not wait for the Geneva meeting of the Commission on Human Rights. He was known for unexpected actions. He had earned the title of "conscience of the United Nations" for his fearless, impassioned, and often impolitic interventions. Baroody decided to present a document on youth to the Third Committee of the General Assembly and in it brought up the right of conscientious objection.

His hand was strengthened by the fact that conscientious objection to military service had been endorsed by the delegates to the World Youth Assembly held during June 1970 at UN headquarters. The question had been brought up in the assembly's Commission on Education for Peace, Progress, and Cooperation. The young people asserted in their final report that "Conscientious Objection should be treated as a human right and should be on the agenda of the next UN Human Rights Commission."

The topic had been championed by a young man from a country which made no provision for conscientious objection, which indeed imposed repeated prison sentences on those refusing compulsory service. He was a young Spaniard, José Bravo, who represented the International Catholic Movement for Students. War Resisters League, led by Igal Roodenko, joined with the International Catholic Movement for Students in printing a leaflet on conscientious objection. It contained the War Resisters League's petition and the statement I had lodged with the Commission on Human Rights. Bravo circulated it among the 650 delegates from nearly one hundred countries. The leaflet featured a question asked by U Thant while serving as UN secretary general: "Is it surprising that the younger generation... should begin to refuse to kill their fellow men, whatever their race, their color, or their nationality?"

Baroody's statement to the Third Committee came right to the point: "It is suggested to member states not to punish any youth who refuses to join the armed forces of his country if such youth conscientiously objects to being involved in war, and to take into account his deep convictions." He suggested to member states "only to enlist in the armed forces those youths who volunteer strictly to defend their country from flagrant aggression." He proposed further that states call upon adults between thirty-five and fifty years of age to enlist in the armed forces in order to give the opportunity to the youth of the world to develop academically, culturally, and vocationally before being commanded to prematurely sacrifice their lives. "Baroody's Army," as a short description of the older army, was a phrase heard around the United Nations in those days.

Baroody's recommendations on youth, containing many more sections, constituted a statement of conscience rather than a formal resolution. Baroody, in his unorthodox way, was thus able to circulate his ideas without having them put to the vote for adoption. He was given the right, however, to take the podium of the General Assembly when the report on youth came to the vote. It was November 11, 1970, and I was present when Baroody presented his case for youth in impassioned terms. He then requested that the General Assembly agree to transmitting his document to the Commission on Human Rights and to all who had participated in the World Youth Assembly. The delegates voted in the affirmative on both counts without endorsing the contents of the statement. Baroody's unorthodox action was an important breakthrough. First, it broke the taboo on conscientious objection at the level of the General Assembly; second, it locked the question of conscientious objection into the agenda item on youth; and, third, it placed before the Commission on Human Rights a UN document which specifically alluded to conscientious objection.

Conscientious Objection: Persistent Agenda Item. For the succeeding seventeen years, conscientious objection to military service was an agenda item of the UN Commission on Human Rights. Having voted to request that the commission authorize a report on provisions for conscientious objection and alternative service among member states, the commission returned to the issue at later sessions as members supplied it with documentation. Conscientious objection became an East-West issue, with the Soviet Union, the Ukraine, and other commission members from the eastern bloc resisting any affirmation of the right of conscientious objection, while Western nations, notably the Netherlands and Austria, insisted on pursuing the subject and gathering further factual data. The fact that the conscientious objection agenda item was attached to the general item of youth saved it from being voted off the agenda altogether.

Nongovernmental organizations, notably the Quakers, supplied background material to the NGOs. The Quakers became, through their offices in New York and Geneva, the focus and animator of almost all the action on behalf of the right of conscientious objection.

Conscientious Objection Recognized. It was not until 1987 that the UN Commission on Human Rights voted favorably on a draft resolution on conscientious objection. It was presented by Austria with four nations as cosponsors. Twenty-six nations joined in the historic vote in favor of recognition of conscientious objection. The abstention of fourteen nations, led by the USSR and the Eastern European nations, made the favorable vote possible.

The change in alignment reflected the momentous historical shift following Mikhail Gorbachev's assumption of power. The policy of *glasnost* (openness) indicated the change in the way the Soviets and their allies dealt with the United Nations. It was a sign of the fact that the Cold War was fading into history.

The landmark resolution stated:

> Conscientious Objection to military service should be considered a legitimate exercise of the right to freedom of thought, conscience and religion recognized in the Universal Declaration of Human Rights and the International Covenant on Civil and Political Rights.

The resolution recommended that states provide for alternative civilian service for conscientious objectors, service which would be compatible with the reasons for which they became conscientious objectors. It recommended that states refrain from imprisoning conscientious objectors.

The linking of conscientious objection to the Universal Declaration of Human Rights was of deep significance. Each session of the Commission on Human Rights added provisions and nuanced interpretations to successive resolutions. Nongovernmental organizations like the Quakers, War Resisters League, and Pax Christi served often to spearhead expanded rights of asylum for objectors whose claim was not recognized.

In the session of 1998, marking the fiftieth anniversary of the Universal Declaration of Human Rights, Pax Christi broadened and expanded the issue in a statement entitled "The Right to Refuse to Kill."

The Dilemma of Conscientious Objection in an Unjust War

What happens to soldiers who discover that the war in which they are fighting does not meet just war conditions? An unprecedented event occurred at the fall meeting of the Catholic bishops of the United States on November 19, 1971:

the bishops declared that the Vietnam War as of that year was not meeting the criteria of the just war. The bishops declared:

> At this point in history, it seems clear to us that whatever good we hope to achieve through continued involvement in this war is now outweighed by the destruction of human life and of moral values which it inflicts. It is our firm conviction, therefore, that the speedy ending of this war is a moral imperative of the highest priority. Hence we feel a moral obligation to appeal urgently to our nation's leaders and indeed all the nations involved in this tragic conflict to bring the war to an end with no further delay.[32]

In addressing themselves to what they termed the "agonizing issue" of Vietnam, the American bishops decided that the criterion of proportionality (that the good to be achieved must outweigh even the probable evil effects of the war) was being transgressed. Since all the just war criteria must be met for a given war to be just, the Vietnam war was unjust. For the first time in the history of the United States, and possibly world history, a national hierarchy denounced as unjust a war being waged by its own nation. Bishop Thomas J. Gumbleton, auxiliary bishop of Detroit (who later served as president of Pax Christi–USA), pointed out that anyone who reaches the same moral judgment "may not participate in the war."[33]

Any Catholic in the armed services who became aware of this decision and who shared it could have asked to be relieved of army duty. Even if he were serving in Vietnam, he could approach his commanding officer and inform him that he had received word from a power higher than the army, namely, his spiritual leaders, that the conflict did not meet the standards of a just war. One can imagine the stupefaction of an officer when approached by a subordinate at the edge of a rice paddy in Vietnam. The soldier salutes, places his rifle on the ground, and states, "Sir, I wish to be excused from active duty since the conflict has been judged unjust by the religious leaders of my faith. I do not have the criteria of the just war on my person, but I can obtain a copy to prove my case." Of course, this did not happen.

The bishops expressed deep concern and sympathy for those who bore the hardest burden of the war, those who served conscientiously in the armed forces and their families. Many soldiers had lost their lives or limbs. Nothing could have proved more vividly the tragic irrelevance of the just war tradition to actual wars and to the men drafted to fight them. The U.S. government continued its prosecution of the war without reference to the negative decisions or warnings of the churches.

After persons liable for conscription had learned of their right to object in conscience and engage in civilian pursuits, the overarching need was to internalize the implications of this right. The external act of refusal would prove that they were following the dictates of conscience. All persons would need to answer for themselves the question of Franz Jagerstatter, written with his hands in chains and facing imminent death for following the dictates of his conscience: "For what purpose, then, did God endow every person with reason and free will if we are nevertheless required to render blind obedience?"[34]

At the 1998 session of the UN Commission on Human Rights, a comprehensive resolution was passed on conscientious objection that listed the rights of those who refuse military service. In order to follow reports on the treatment of objectors it was decided to include the right to conscientious objection as an agenda item in the session of the commission in the year 1999. The Pax Christi statement was submitted, as requested by the Commission on Human Rights, in October 1998 for the commission's session in Geneva in March 1999. Though the agenda item was conscientious objection to military service, Mary Evelyn Jegen and I, in preparing the statement, broadened it to stress the right to refuse to kill. We pointed out that in areas of the greatest violence, where young people, often no more than children, are impressed into service, as in Angola and Mozambique, there are no provisions for conscientious objection. Even the term is hardly known. Young people who feel revulsion against taking life could be informed of their right to refuse to kill as one that can be recognized and protected. Peace movements are broadcasting this right and giving it much attention in 1999, fifty years after the Universal Declaration of Human Rights had been voted on at the United Nations.

Chapter 14

Peace of Christ

Do all you can to preserve the unity of the Spirit by the peace that binds you together. —EPHESIANS 4:3; JERUSALEM BIBLE

It was not surprising that Europe, scene of war's slaughter from 1939 to 1945, should turn to peace as war ended. In the wake of World War II, a unique organization, which became the International Catholic Movement for Peace, was born. Its beginnings were in a prison camp in Compiègne, France, in the last year of the war. Bishop Pierre Theas of Montauban had been arrested by the Gestapo for his resistance to Nazi occupation policies. His fellow prisoners, chiefly men of the resistance who expected to be transferred to a German camp, asked if the bishop would give them a day of recollection.

Loving Enemies Like These?

"On July 14, 1944, I was preaching to about thirty political prisoners about love, and I had to include love of enemies," the bishop recalled. When he discussed the Lord's Prayer, which the men, whether practicing Catholics or not, would remember from their childhood, they received the words of the bishop without a murmur. When, however, he came to "Forgive us our trespasses as we forgive those who trespass against us," they balked. Did this mean forgiving the Germans? It was Bastille Day, the great French holiday, and these men had been captured as they tried to prevent German soldiers from further oppressing their families and despoiling their country.

Searing memories of atrocities were still raw. They knew that just a month before, on June 10, 1944, a detachment of German troops drove into Oradour, an ancient village of about fifteen hundred souls. It was situated near Limoges, in the heart of France. On a market day, many men, including farmers, were gathered in the center of the village. They were rounded up and shot. Women and children, about five hundred in all, were herded into the parish church. They perished when the church was set on fire. The number of the dead, counting men, women, and children, was 642. The whole village was then torched and left in ruins. The horror was perpetrated as a reprisal for the killing of German occupying troops.

Could the Lord's Prayer really mean forgiving and loving enemies like these? How can you ask this of us when they have butchered and burned noncombatants? was the question from the men.

Pierre Theas could not prevail. He said that these were the words of the prayer and the teachings of Christ. He could say no more and no less. The next day, the men were transferred from Compiègne to Buchenwald, where their survival was doubtful. The bishop was held at Compiègne.

The knowledge that he had to live up to the hardest lessons of the gospel moved the bishop to celebrate Mass for the Germans in Compiègne, chiefly camp guards.

Pax Christi

Pilgrimage of Prayer

A French woman, Martha Dortel-Claudot, a teacher in a lycée, was burdened by all the killing of World War II. What moved her with special horror was the spectacle of Catholic men from France and Germany, united in the Eucharist, yet continuing to maim and kill each other for years on end. Another burden on her spirit was that her country and society had suffered so much from Germany that she wondered if anything could promote reconciliation between the two peoples. She came to the conclusion that only by prayer could there be a beginning of Franco-German reconciliation. On November 11, 1944, the anniversary of the armistice ending World War I, she had an inspiration: a pilgrimage of prayer. The pilgrimage would focus on the fall of Nazism, which had unleashed so much evil on them, but more urgently on the German people, their needs, and their future. In the following March, a small group consulted Bishop Theas, now freed from prison. They found him a strong supporter of the concept. He suggested that the effort be christened "Pax Christi in Regno Christi" (The Peace of Christ in the Reign of Christ). The meeting, on March 10, 1945, was considered the foundation date of what came to be called Pax Christi.

A Peace Theology

The development of Pax Christi was unique in that it was the creative action of lay Catholics, the horizontal church, acting across frontiers and cultures rather than a movement of the hierarchical church. Even though a bishop headed each section as president (chairman), the movement was led by laypersons.

The theology of peace was midwife to the birth of Pax Christi, a theology arising from concrete situations. Besides the message of love, a love that englobed enemies as taught by a bishop in a prison camp, there was the assertion of a peace theology by a lay teacher in a lycée. To her, the scandal of believers sharing the body and blood of the Savior while intending to destroy each other had to be addressed, so also, the reconciliation of former enemies, followers of the Lord who reconciled humankind to God and God to humankind. Joint pilgrimage began to be made with French and German Catholics participating. Lourdes was a natural venue, since Pierre Theas became the archbishop of Lourdes.

Pierre Theas at Lourdes

In 1947 the national French pilgrimage came to Lourdes, and along with the French came delegates from Germany, England, and Italy. Since I was in Europe that year, moving between France and the DP and expellee camps of Germany, I learned of Pax Christi and the convergence at Lourdes. I made my way to Lourdes and asked to see Archbishop Theas. He expressed his joy that Lourdes was the locus of the reconciliation of former enemies and of begging God for the gift of peace.

I introduced the issue of nuclear weapons and asked the bishop if the church should not take the step of condemning such weapons. Hiroshima and Nagasaki had provided sufficient evidence that nuclear bombs were instruments of mass destruction, not weapons in the ordinary sense of the term. For a long interval, or so it seemed to me, the archbishop meditated. Then he spoke, deliberately. It would take a considerable time for the whole church to declare such a condemnation, he explained.

What could be done in the meantime? I wondered. There was another long pause. Perhaps, he suggested, speaking slowly and somewhat hesitatingly, the bishops in the United States, where the bomb originated, could accept it as their problem and make a strong statement against it. They might even warn against the use of the nuclear bomb in future wars.

The response of Archbishop Theas never left my mind. Whenever I heard of the actions of Pax Christi and its growth in various countries, I recalled the archbishop's reaction to the nuclear bomb.

In 1948 Lourdes was again the locus of a Pax Christi congress and peace pilgrimage. A French peace priest, P. Larson, gained acceptance for a statement containing an outright condemnation of war as immoral. It reflected the revulsion against war by those who had lived through it. The statement argued: "War is immoral because the instruments of destruction which it employs are blind in their working. Even those who use them lose control and foreknowledge of their action. Man as a free and responsible being has no right to initiate processes which he cannot control and limit." Among many other arguments, Father Larson, a member of the Society of Jesus, included in his reprimand the assertion that "if modern war is a collective sin, it also permits individual sins to multiply in a horrible manner." He concluded with the words, "War has spread a mountain of filth over the earth, which has infected humanity, debased souls, and even destroyed the sense of good and evil, together with the sense of God."[1]

The archbishop of Paris, Archbishop Feltin, agreed to serve as the international president of Pax Christi. In 1950, he called a constituent assembly, inviting delegates from thirteen European countries. From that year, Pax Christi took form as a movement, with publications and the planning of joint programs. These included an annual assembly, one of which brought together participants from Eastern and Western Europe to study the implications of gospel nonviolence. Representatives from Asia, Africa, and Latin America also attended the assemblies. Pax Christi International sponsored routes of peace on a yearly basis. A national section, for example, England, France, or Poland, would map out a route, with stops on the way for rest and hospitality by its nationals. Young people from the various national sections would assemble and march along the routes, holding sessions on peacemaking while they came to know each other and experienced the culture of the host country. Headquarters were initially established at The Hague.

The Chalice of Oradour

A moving chapter in the reconciliation of the French and German peoples was linked to the horror of Oradour, still not rebuilt, left in ruins as a reminder of the horror of the massacre of its inhabitants.

A German who was to head Pax Christi–Germany decided to visit Oradour in 1955 to kneel in penance in front of the ruined church. He was a priest, Manfred Hohrhammer, but he was not wearing the marks of priesthood. When he returned to Germany, he wrote an article for a newspaper about his pilgrimage. A woman who read it was moved to offer a chalice in atonement to the ruined parish.

A chalice was presented by the president of Pax Christi–Germany to Bishop Theas of Lourdes, as president of Pax Christi–France. Bishop Theas said, "This crime can only be redeemed through the blood of Christ." He then gave the chalice to the bishop of Limoges, in whose diocese Oradour was situated. The parishioners refused to accept it; the wounds left by the massacre were too painful in the hearts of the survivors.

When a group from Pax Christi–Germany went to Oradour in 1980, the chalice was still in a neighboring parish church. Another group of pilgrims from Pax Christi–Germany came to Oradour in 1994. After walking in silent prayer through the streets, they came to the parish where a German priest concelebrated with a French priest in a Mass of atonement and reconciliation.

When Pax Christi marked its fiftieth anniversary in Assisi in 1995, Cardinal Danneels celebrated at the Table of the Lord with the chalice of Oradour. It had been sent from the parish with the message that the parishioners were praying for the gathering. In his letter of thanks, the cardinal stated that "through the blood of Christ, the impossible became a reality." He expressed the prayer that "Through the communion in the same body and blood of Christ we may find the joy of living as brothers and sisters together across all frontiers."

At a time when small and bigger wars still give rise to killings and massacres, Cardinal Danneels reminds Pax Christi members in all continents that "reconciliation" is at the heart of the mission of Pax Christi for the world.

Assisi

In the time that had passed between the rejection of the chalice offered to the parish of Oradour and the reconciliation marked by the Oradour parishioners in sending it to Assisi, Pax Christi had moved around the world with the message of Christ's peace.

The raw wound of Oradour had reached the point of healing, as had countless raw wounds and agonies left by war in different parts of the planet.

A large international contingent — 650 persons from thirty-seven countries — gathered at Assisi, the town of St. Francis and St. Clare. Greetings were received from many international leaders, including the mayors of Hiroshima and Nagasaki, and the secretary-general of the United Nations. Representatives of national sections of Pax Christi joined in the deliberation of such topics as "The Theology of Peace" and "Reconciliation with the Rest of Creation." The participants joined in a proposal to the World Court to declare nuclear weapons illegal. At a poignant session, a priest member of the Assisi underground related how, during World War II, hundreds of Jews were saved by the Assisi secret network. An evening of drama brought the participants together for *Haunted by God,* a one-woman presentation of the life of Dorothy Day.

The climactic event of the anniversary was the celebration of Mass in the Basilica of St. Francis. The Table of the Lord as a unifying factor for humankind was palpable and visible at the Mass, with priests and bishops from Europe, Latin America, the United States, the Philippines, and Africa concelebrating for men and women of most of the peoples under heaven.

The Pax Christi vision statement was delivered by Cardinal Gottfried Danneels, cardinal of Malines-Brussels and president of Pax Christi International. It said in part:

> As a movement of reconciliation based on Christ's gospel of peace, Pax Christi International will continue its prophetic role within the Church and the world.... We will uphold the belief that waging war, aggression and destruction can never be acceptable.
>
> Pax Christi International seeks to be a leaven within the Catholic Church and to help transform it into a church which gives unambiguous witness to the nonviolent Jesus.

A Leaven within the Catholic Church

Pax Christi International found various opportunities to keep peace and peacemaking at the center of the world church ministry, as did the national sections. In its statement of purpose, Pax Christi–USA announced:

> Pax Christi intends to contribute to the building of peace by exploring and articulating the ideal of Christian nonviolence and by striving to apply it to personal life and to the structures of society.

While most of the leaders of the movement were committed to gospel nonviolence, the movement itself was broadly based to include those who believed that some use of violence might be unavoidable.

The national section in the United States grew out of the group known as Pax, which, in turn, had been inspired to start its activity by Pax, England. When the latter, originally founded in 1936, was accepted as the English branch of Pax

Christi in 1970, American Pax followed suit. First, a request had to be addressed to the council of Pax Christi. I was deputed to present it. Carrying with me reports of our programs and examples of our *Peace Quarterly* magazine, I made my way in April 1972 to an executive council meeting in Luxembourg.

All the printed material of American Pax were examined, and an extensive hearing, presided over by Cardinal Alfrink, international president, followed. Questions were posed about the Pax officers, headed by Gordon Zahn, and about the acceptance of the program in the U.S. church. The Pax program at Vatican II seemed an expression of apostolic activity acceptable to Pax Christi. On March 14, 1972, Pax was accepted as the matrix of the U.S. national section of Pax Christi International. Regular contact was maintained with the headquarters of Pax Christi in Brussels and with the international secretary, Etienne de Jonghe.

An instance of Pax Christi's functioning as a leaven in the church was its publication of *Studying War No More? From Just War to Just Peace.*[2] Its publication in 1993 commemorated the thirtieth anniversary of Pope John XXIII's *Pacem en Terris.* Two years earlier, Cardinal Danneels had consulted widely with Catholic universities, theological institutes, and Pax Christi sections regarding a book on modern war, its reality, and its alternatives.

The result was a publication presenting nineteen articles by theologians, scholars, and philosophers from three continents. It was published simultaneously in Europe and the United States in both French and English. A prominent feature of the book was the summary and analysis by William H. Shannon, theologian and educator, of an editorial that had appeared in *Civiltà Cattolica.*[3] It had been written by three members of the Jesuit order and was published in the Rome-based publication of the Jesuit fathers.

In the article, entitled "Christian Conscience and Modern Warfare," the Jesuit authors' overall aim was to unmask the face of modern war and to demonstrate that "it is always the poor and the weak who pay for war, whether they wear military uniform or belong to the civilian population." They asserted that "the theory of the just war is untenable and needs to be abandoned." The editorial points out that modern war has become total in a threefold sense. First, it involves whole nations, entire populations, all the cultural, artistic, and religious wealth and economic riches of a country. Second, it involves not only two nations or a single group of nations, but many nations and ultimately the whole world. And, finally, it uses weapons of "'total and indiscriminate destruction' which elude the possibility of human control." The authors of the editorial can only conclude that modern warfare is always immoral.[4]

Pax Christi was fulfilling its role as a leader in the church (this is but one example among many) by putting before the world Catholic community an editorial mirroring a controversial view on modern war held by an increasing number of followers of Jesus.

Civiltà Cattolica often reflected Vatican opinion. It would not publish anything that the Vatican would disown or contradict. The editorial gave moral and ideological support to those who, either pacifist or not, viewed participation in modern war as unacceptable for Catholic Christians.

The National Sections as Leaven

Each national section of Pax Christi was expected to react to the challenges and needs of its society while responding to the pacific aims of Pax Christi and the vision expressed by Cardinal Gottfried Danneels. The three chief concerns of Pax Christi–USA are demonstrated by its actions. One of its early national assemblies was held in Memphis, Tennessee, a choice that permitted participants to make a march-pilgrimage to the Lorraine Motel, where Martin Luther King was assassinated. On that site, penitential prayers were recited and the continuing challenge of racism was discussed. The national assembly held in the state of Washington near an Indian reservation was a reminder of an unhealed wound of the treatment of the original inhabitants of the United States. It was the home of the Swinonish people. A man and wife of the Swinonish tribe addressed the participants who visited and picnicked with the Swinonish people.

The visit of the bishop-leaders of Pax Christi to the National Livermore Laboratories, center of nuclear activities of the U.S. Department of Defense (related in chapter 7), was an indication of the concern of Pax Christi members for the nuclear potential of their country. The five-hour visit to the facility and the opportunity to talk to the directors did not serve to allay their fears about the morality of the whole nuclear program.

A cause of confusion was the difference between the conclusions regarding nuclear and other matters of war and defense reached by Pax Christi members and those reached by other citizens. In a way it marked a dividing line between Christian and citizen. How could any sane person, many wondered, not want to live under the security of the nuclear umbrella? Others, who rejected all violence, wanted to see the abolition of all nuclear weaponry, in fact all weaponry. A terse statement of belief in the form of a vow might be a means of reaching a measure of understanding. The question of a vow of nonviolence had been broached several times in discussions with the national council. Two persons brought versions of a vow to the national council at the same time, John Dear, later Father John Dear, SJ, and myself. The text presented by John Dear was a rather long version and emanated from a spiritual retreat. The version I presented was a shorter one, which was edited by the council and accepted.

Printed on a card that could be carried in a pocket or wallet, it was an item at hand to explain Pax Christi and the implications of nonviolence. Communities of religious Sisters jointly took the vow, stressing the application it had to personal relationship as well as to relationships with society and with the powers of the world. A member of Pax Christi–USA, in prison for a forbidden peace demonstration, found a ready audience among fellow prisoners. Many men, even some convicted of crimes of violence, formally pledged themselves to a new way of life through recitation of the vow. Members of monastic communities, men and women, recited the vow as individuals or as small groups in religious settings. A feature of the annual assemblies of Pax Christi–USA was the public recitation of the vow by officers and members.

The text of the vow:

Before God the Creator and the Sanctifying Spirit, I vow to carry out in my life the love and example of Jesus...

- by striving for peace within myself and seeking to be a peacemaker in my daily life;
- by accepting suffering rather than inflicting it;
- by refusing to retaliate in the face of provocation and violence;
- by persevering in nonviolence of tongue and heart;
- by living conscientiously and simply so that I do not deprive others of the means to live;
- by actively resisting evil and working nonviolently to abolish war and the causes of war from my own heart and from the face of the earth.

God, I trust your sustaining love and believe that just as You gave me the grace and desire to offer this, so You will also bestow abundant grace to fulfill it.

Cross-Bearing Witnesses: Ash-Marked Protesters

Pax Christi was convinced that its antiwar, antinuclear message was of such urgency that besides the usual methods, it needed to be dramatized, to reach governments as well as ordinary citizens. Members chose the Lenten season, the season of repentance, for their annual street drama. The protests that receive most attention in newspapers and other media are those that take place in New York City and London.

The annual protest in New York City always begins at the Isaiah Wall opposite the United Nations, and near the East River. The sixteenth annual way of the cross on Good Friday, April 10, 1998, drew over a thousand persons to the Isaiah Wall. Fifteen stations in all took the marchers from the United Nations at Forty-second Street and the East River across Manhattan to Forty-fourth Street and the Hudson River on the West Side. At each station, a leader led a prayer and read a Bible lesson from a flatbed truck; hymns were sung, and the participants joined in prayers. A graphic representing the station was displayed at each stop. In marking the first station, Jesus condemned to death, at the United Nations, the leader announced from the truck: "Here at this place we remember the condemnation of Jesus by state authority. We remember also the near million Iraqi children who have died as a result of the UN sanctions against Iraq."

At each station the marchers called to mind all the ways in which Christ continues to be crucified in the world today. Joining Pax Christi were Lutherans, Episcopalians, Mennonites, and members of other groups. Many carried handmade crosses. At other stations, Jesus falls three times under his cross, is stripped of his garments, meets his mother, Mary, receives help from Veronica, is helped by Simon to carry his cross, and is crucified and entombed. "Like Simon," the leader intoned, "we take Christ's cross with faith we do not understand." At the station where Veronica wipes the face of Jesus, the leader recited: "Veronica musters the courage to break from the crowd to tenderly care for the suffering one and to touch the untouchable."

At the station where Jesus dies on the cross, the leader prayed: "Help us to understand this terrible yet victorious death for our lives." When Jesus is laid in the tomb, the participants were reminded to "reverence earth, to protect it and heal it."

It is now common practice to add a fifteenth station, the resurrection of Jesus Christ. When Christ's death is announced, a chant arises after the prayers, "Christ has died. Christ has risen. Christ will come again!"

When the marchers in the 1998 procession reached the *Intrepid,* a warship converted into an air and space museum on the Hudson River, some of the marchers formed a human chain across the museum entrance. Fifteen persons, including the Reverend Daniel Berrigan, chose to court arrest. They were picked up by the police and taken away in two police vans. The rest of the marchers chanted their support before returning to their daily concerns. Each year the Good Friday peace witness, drawing between a thousand and fifteen hundred participants, is the largest continuing act of peace witness in New York City.[5]

In London, the Pax Christi community chose Ash Wednesday to mount its annual action entitled "Repent and Resist Nuclear War Preparation." In 1997, the thirteenth annual protest, Pax Christi, joined by Catholic Peace Action and the Christian Committee for Nuclear Disarmament, gathered in the Thames Embankment Gardens for a liturgy of prayer and repentance. Their foreheads marked with ashes as a reminder of the "ashes to ashes" of the Ash Wednesday service, they made their way, intoning Psalms, to the Ministry of Defense. Some protesters used charcoal, which had been blessed, to write their messages on columns in front of the building. They had time to trace in foot-high letters, "Repent," "Trident Equals Death," "Forgive Us," and "Choose Life" before police interrupted the action to take the writers to Charing Cross police station.[6]

I was reminded of the story of Jonah, who, though reluctant, finally brought the message of repentance to the people of Nineveh, the traditional enemy of Israel. As a result of his message, the people of Nineveh "proclaimed a fast, and all of them, great and small, put on sack cloth." Even the king "rose from his throne, laid aside his robe, and sat in the ashes" (Jonah 3:5–7). He and his people were to fast and to call loudly to God: "Every man shall turn from his evil way and from the violence he has in hand." The Lord relented from the judgment that he was going to mete out to the wicked people of Nineveh. The Lord then had to remonstrate with an angry Jonah, who had a vindictive streak and wanted punishment to be carried out:

> And should I not be concerned over Nineveh, the great city, in which there are more than a hundred and twenty thousand persons who cannot distinguish their right hand from their left, not to mention the many cattle? (Jonah 4:11)

May not the tenderness of God toward the erring people of Nineveh be expressed toward Londoners, New Yorkers, and the people of many cities —

including their leaders — who go about their lives blindly, hardly able "to distinguish their right hand from their left"? Members of Pax Christi, mindful of the stockpiles of horrendous weapons assembled from their taxes, are also mindful of the need to show repentance for their complicity in the evils of the time.

Chapter 15

The Lord's Prayer and the Reign of God

The first step in the creation of a new world order may be learning to forgive and to seek forgiveness. —LAWRENCE MARTIN JENCO, OSM, HOSTAGE

Our Father in heaven,
hallowed be thy name.
Your kingdom come,
your will be done, on earth as in heaven.
Give us today your daily bread;
and forgive us our debts,
as we forgive our debtors,
and do not subject us to the final test,
but deliver us from the evil one.
—MATT. 6:9–13

Dare to Say "Our Father"

In the gospel of Matthew, after warning his followers against ostentatious prayer, or babbling with many words, Jesus tells them how to pray. The Lord's Prayer is simple and powerful, consisting in Matthew of fifty-two words (Matt. 6:9–13).[1]

Commentators point out that Jesus uses an intimate word in addressing the Creator: "Abba." It is close to "Dad." The Creator is brought close to those who can say "our" Abba, not "my" Abba. Each of us who can say "our" is a member of a community, a community that cuts across all barriers of race, culture, tribe, or nation. The "our" binds us into the community of the baptized and serves as a reminder of that inescapable link, whatever may be the forces that weaken—or ever seem to destroy—it. The poorest of the poor as well as the mighty of the earth have the same access to the Creator and to his grace; they can all dare to say "our Father."

The almost unutterable tragedy in the Christian community is the spectacle of Christians daring to say "our" while lining up against each other under separate banners of unrelenting nationalisms. The murderous tribalisms and ethnic hatreds of earlier ages emerge in the twentieth century as cloaks proudly worn by followers of Jesus originating in France, Spain, Germany, Ireland, Poland, Italy, and other nations.

The petitions in Matthew have a different resonance, a different poignancy in accord with the needs and crises of each age. That the name of the Creator should be hallowed in the hearts of all Christians is the very ground of faith. In the nuclear age, however, the linking of some of the petitions in the Lord's Prayer with peace is close to the hearts of many Christians.

"Your kingdom come" has deep significance for peace. We remember that even after his resurrection, Jesus had to emphasize to his disciples that they were not to launch into a program of "restoring the kingdom of Israel." The earthly kingdom was of the greatest importance to members of the Jewish community humiliated under the rule of a pagan empire. The Messiah was to come and redeem them from this loathsome state of lowliness. This was what they expected, and when the one from Nazareth cruelly dashed their hopes of liberation, many Jews no longer walked with him.

The ones who did walk with him gradually realized that the kingdom preached by Jesus had a far different meaning from the one to which they were accustomed. It was a kingdom which, while already present in mystery when the poor are fed, the suffering are healed, and the enemies are loved, will only come to its fullness in eternal life.

When he cured the demon-possessed man, Jesus announced that the act had been performed by the power of the Almighty, telling the onlookers that "the kingdom of God has come upon you" (Matt. 12:28). When a scribe answers Jesus with understanding, echoing the teaching of love of God and neighbor, Jesus tells the man that he is not far from the kingdom of God.

The Lord's Prayer is part of Jesus' Sermon on the Mount, in which we learn that among those blessed are the "poor in spirit, for theirs is the kingdom of heaven." Also blessed are those who mourn, those who try to be peacemakers (a thankless and often dangerous task), and those who are persecuted, insulted, and the object of attack. These would have their place in the kingdom of God. How vastly different are the values of the kingdoms of the world and the kingdom of God.

The consequences of following a Messiah who preached the otherworldly type of kingdom were grave. The allegiance to this kingdom would shatter any allegiance to the kingdoms of this world, however powerful or demanding.

Kingdom of Wills Turned to God

What follows next in the Lord's Prayer, "Your will be done, on earth as in heaven," is the key to the kingdom and the key to peace. Heaven is a state of being where strife is no more. If our allegiance is to the kingdom of God, to his eventual reign, then our kingdom is a kingdom of wills turned to God. We are allowed to see a glimpse of the kingdom when an act of holiness, of charity, or of forgiveness is unveiled before our eyes. We call these moments the experience of seeing the "not yet" of the kingdom as "already" in our midst. Forgiveness as the central act of peacemaking makes for reconciliation. Only by reconciliation can human beings restore peace with each other and reflect the peace between human beings and their Creator.

And what of the kingdoms of this world? What power should they wield over us? How much obedience? Should we not examine every demand made by powers, by government, to find if it impinges on the prior duty to carry out God's will? Is there any teaching that has reached us through Jesus that calls for us to maim, disfigure, or kill human beings, images of God? It cannot be found. When followers of Jesus cooperate in the works of war, they cannot be said to be carrying out God's will.

The Grotesque Process That Began with Constantine

Even Christians who have vowed to serve God in celibacy, simplicity, and obedience join in giving obedience to earthly kingdoms over the kingdom of God. Like millions of other Christians, they allow their consciences to be bound in wartime to the worldly kingdoms. In World War II, George Zabelka, a priest with a great love of his country, became a chaplain with the rank of lieutenant colonel in the U.S. air force. He did not question the aims or the methods of the war. An outgoing, patriotic man, he shepherded and helped keep up the morale of airmen who risked their lives in bombing missions over enemy territory. Stationed on Tinian Island in the South Pacific, he became Catholic chaplain for the airmen who dropped the nuclear bombs on Hiroshima and Nagasaki.

Three months after the end of the war, Zabelka walked with two other chaplains through the ruins of Hiroshima and Nagasaki. They moved among the suffering people, above all, the children, and went through the hospital wards, where they saw people with empty eye sockets, disfiguring wounds, and mysterious illnesses which would sap their life-force. George Zabelka began to be aware of what he called "a worm that squirmed through my stomach." It was doubt, doubt about the use of violence against noncombatants, doubt about any use of violence for any purpose, even a good one.

On his return to civilian life, Father Zabelka resumed the task of spiritual shepherding in a parish. One of the churches he served as pastor was an inner-city parish, the Church of the Sacred Heart in Flint, Michigan. He entered into the struggles of his mostly black flock, adverting to the messages of Martin Luther King. In his sermons, he found he had to deal with the Sermon on the Mount and other teachings of Jesus on love and nonviolence.

In 1972, the priests of the diocese of Lansing, Michigan, were invited to a workshop on gospel nonviolence given by the Reverend Emmanuel Charles McCarthy. Father Emmanuel, a founding director of the Institute for the Study of Nonviolence at the University of Notre Dame in Indiana, was a lawyer, a theologian, an ordained priest of the Melkite rite, and an apostle of gospel nonviolence. Father Emmanuel had chosen as his life's work the preaching of nonviolence through retreats, workshops, and lectures to Catholics and other Christians in schools, parishes, and universities.[2] Among the priests at the workshop was George Zabelka. Gospel nonviolence was challenging to him as were the topics posed by Father Emmanuel, "Who is your king? Who is your God?"

Zabelka appeared at another of Father Emmanuel's workshops a few years later. Chatting with a young man who was reading a book about the *Enola Gay,* the plane which carried the atomic bomb to Nagasaki, Zabelka revealed that he had been the Catholic chaplain for the men of the 509th Composite Group, the atomic bomb group, on Tinian Island. Meantime Zabelka made a point of listening to cassette tapes made available for study by Father Emmanuel. One was entitled *The Kingdom of the Lamb.* When he appeared at a third workshop, Zabelka announced to Father Emmanuel, "I must do an about-face." This meant that he wholeheartedly accepted gospel nonviolence and saw violence as a total distortion and rejection of Christ's teaching. What had been growing as an inchoate response to his experiences now became more explicit. His spiritual journey brought him to forgiving the enemy, the "Japs," to forgiving himself for taking part in destruction so that he could ask for the forgiveness of God.

I Was Told It Was Necessary

"As a Catholic priest, my task was to keep my people, wherever they were, close to the heart and mind of Christ. As a military chaplain I was to try to see that the boys conducted themselves according to the teachings of the Catholic church on war,"[3] Zabelka told Father Emmanuel in an interview for *Sojourners* magazine. He continued:

> The destruction of civilians was always forbidden by the church, and if a soldier came to me and asked if he could put a bullet through a child's head, I would have told him absolutely not. That would be morally sinful. I and most chaplains were quite clear and outspoken on such matters as killing and torturing prisoners.
>
> In 1945 Tinian Island was the largest airport in the world. Three planes a minute could take off from it around the clock. Many of these planes went to Japan with the express purpose of killing not one child or one civilian but the slaughtering of hundreds of thousands and tens of thousands of children and civilians — and I said nothing.

To Father Emmanuel's query as to why he kept silent, Zabelka replied:

> Because I was brainwashed! It never entered my mind to publicly protest the consequences of these massive air raids. I was told it was necessary: told openly by the military and told implicitly by my Church's leadership. ... The whole structure of the secular, religious, and military society told me it was all right to "let the Japs have it!" God was on the side of my country.

Zabelka pointed to the seventy-five thousand people burned to death in an evening of fire-bombing over Tokyo and the hundreds of thousands destroyed by aerial bombing over Dresden, Hamburg, and Coventry. He added, "The fact that forty-five thousand human beings were killed by one bomb over Nagasaki was new only to the extent that it was one bomb that did it. . . . To fail to speak to the utter moral corruption of the mass destruction of civilians was to fail as a Christian and a priest as I see it."

Zabelka went on:

> When I walked through the ruins of Nagasaki right after the war and visited the place where once stood the Urakami Cathedral, I picked up a piece of a censer [the vessel for burning incense at the liturgy] from the rubble. When I look at it today I pray that God forgives us for how we have distorted Christ's teaching and destroyed his world by the distortion of that teaching. I was the Catholic chaplain who was there when the grotesque process that began with Constantine reached its lowest point—so far.

In response to Father Emmanuel's query as to what he meant by "so far," Zabelka explained that he saw no change in the moral climate regarding war either inside or outside the church since 1945: "The mainline Christian churches still teach something that Christ never taught or hinted at, namely, the just war theory, a theory that has been completely discredited theologically, historically, and psychologically."

One step proposed by Zabelka, though he knew he would be thought out of touch with the real world, was the calling of an ecumenical conference of Christians to deal with war and show its incompatibility with Christ's teaching. If the pope, the patriarch of Constantinople, and the president of the World Council of Churches could speak with one voice, teach unequivocally what Christ taught unequivocally on the subject of violence, they might be seen as "the divine leaven within the human dough."

George Zabelka: Repent

George Zabelka turned into a prophet of gospel nonviolence, a reluctant prophet at first. His message was "Repent." "Until the various churches within Christianity repent," he stated in the interview with Father Emmanuel, "and begin to proclaim by word and deed what Jesus proclaimed in relation to violence and enemies, there is no hope for anything other than ever-escalating violence and destruction. . . . Communion with Christ cannot be established on disobedience to his clearest teaching."

Emulating Father Emmanuel, Zabelka accepted invitations to give talks about his experience and his about-face on war. We would hear him from the podium at gatherings of Pax Christi and at peace demonstrations before the UN secretariat in New York City.

In 1982, Zabelka decided that in a penitential spirit he would walk the entire distance from the naval submarine base at Bangor, Washington, across the United States and Europe, to Bethlehem. He was generally accompanied by about twenty persons. The Ground Zero center for nonviolence of James and Shelley Douglass was in Bangor. It was situated alongside a Trident nuclear submarine base. Christmas Eve 1983 found Zabelka in Bethlehem, ready to celebrate Mass where Christ was born.

In one of his conversations with Father Emmanuel, Zabelka recalled that before they took off for their bombing flights, the airmen at Tinian Island would join with their Catholic chaplain in reciting the Our Father.

"We said the 'Our Father' as though God was the father of the American troops, of the Americans," Zabelka related. "There was no thought that God was the father of the Japanese, of all the human race."

On the morning of August 6, 1987, having participated in a fast of penance with Father Emmanuel, Zabelka joined with him in silent prayer for the victims of Hiroshima and of all wars. At 8:15 that morning, at the time of the nuclear explosions, Father Emmanuel relates, "George gave a sort of strangled sob. He began to weep. 'How could I have missed it?' he asked, and repeated 'How could I have missed it?' "

Lord, We Have Betrayed You

A film, *Zabelka, the Reluctant Prophet,* was available to tell the priest's story after his death in 1992. In the hour-long film, Zabelka's unequivocal stand against violence is buttressed by statements from Father Emmanuel, Gordon Zahn, Father Richard McSorley, SJ, and Bishop Leroy Matthiesen of Amarillo, Texas.

One of the most telling scenes is that of Bishop Matthiesen and Father Zabelka celebrating an open-air Mass at an improvised altar before the entrance to the Pantex nuclear facility. The eyes of a platoon of uniformed guards are on them. The activity of Pantex, behind the well-guarded entrance, was the assembly of nuclear weapons for transport elsewhere. Though many of the parishioners earned their livelihood at Pantex, their bishop for many years had no idea of their work. In the film, James Douglass, also present at Pantex, reminds viewers that it is from this point that warheads leave on the "White Train" to the place he then made his home, Ground Zero, at Bangor, Washington. The warheads are then loaded on Trident nuclear submarines.

At the Mass, there were prayers for forgiveness. "Lord we have betrayed you," intoned the celebrants, while the small congregation replied, "Lord have mercy."

For his own part, Zabelka explained that he could hardly call on the justice of God, but only on his mercy. He would do all in his power so that his church would again "teach unequivocally what Christ taught unequivocally on the subject of violence."

How Defend a Weaponless Kingdom?

If the words of the Lord's Prayer are taken seriously, then threatening or causing the death of one person is a rebellion against "Your will be done." How defend a weaponless kingdom of wills?

The daily bread for which we pray is bread for all, those named enemy or those named friend. According to Jean-Marie Lustiger, cardinal of Paris, in this, our prayer is joined to the sacrament of the Eucharist.[4] St. Cyprian tells us:

"These words may be taken either spiritually or literally, because, in the divine plan, both readings are helpful for salvation.... And we ask that this bread be given us daily, lest we, who live in Christ and receive the Eucharist every day as the food of salvation, be separated from His Body by some grave sin that keeps us from communion and so deprives us of our heavenly bread."[5] St. Augustine refers to the Table of the Lord as "the great table," describing it as "the one where the Lord himself is the food. No one feeds his guests with himself — except Christ, who sends the invitations and is himself the food and drink."[6]

Hardest Command of the Gospel: To Forgive

"Asking God to forgive us means asking him to re-create us, because sin is the destruction of ourselves," says Lustiger. Is it possible for us to forgive our brothers and sisters as God forgives us? Lustiger says, "Yes, because forgiveness is at the heart of redemption which gives us life and makes it possible for us to give life to others."[7]

What connection has this with war? It is not enough to forgive. The peacemakers must move others to forgive, men and women, groups, clans, tribes, citizens of many nations, who down the centuries recall old wrongs in order to wreak vengeance on the present. Jesus made it clear that only after we have forgiven others their debts (trespasses) can we expect from God forgiveness of our debts (Matt. 6:14–15).

The bishop of Banja-Luka, Franjo Komarica, asserted during a visit to the United States that his people were making a great effort to live "the hardest command of the gospel: to forgive."

Ordained bishop of Banja-Luka, Bosnia-Herzegovina, in 1989, Komarica found himself shepherd of about 180,000 Catholics. About 350,000 Muslims lived in his diocese. During the war which erupted in 1992, the diocese was the scene of what was called "ethnic cleansing." When the violence subsided, he was left with approximately 30,000 Catholics and only five out of seventy-five churches still open for worship. Muslims were reduced to 37,000, with all their mosques destroyed. Despite the violence, the bishop remained in Banja-Luka, pleading for human rights and publicly praying with the Muslim leader and the Orthodox bishop. In 1995, while under house arrest by the armed forces occupying the area, he began a fast to bring attention to the situation. During this period, refugees found shelter in his house, and he saw to it that Caritas Banja-Luka took care of Muslim war victims as well as Catholics. Together with Hadzi Ibrahim Efendi Halikovic (mufti of Banja-Luka), the bishop called for a multiethnic and multicultural environment for everyone. The mufti also decided to remain in Banja-Luka throughout the violence, which included attacks on his life.

In November 1996, Bishop Franjo Komarica received the Pax Christi Peace Award. The award was also given to Mufti Halikovic, who was described as a "deeply spiritual and compassionate person with an extraordinary commitment to forgiveness."

Christians cannot forget that they have the basic ground for forgiveness in the life of Jesus. They need to remind themselves, as Franjo Komarica must often have done, that forgiveness must proceed indefinitely.

Lustiger says:

> In order to forgive one another we must enter into the mystery of the cross; we are sinners who have ourselves been forgiven, thanks to the sacrament of reconciliation.
>
> So, the first step toward forgiving is to go, in a sincere gesture of humility, and ask God to forgive you. Confess your resistance to the act of forgiving, acknowledge your unwillingness to follow Christ in his Passion. Once you do, God will give you, as a grace, the capacity required to fulfill your mission, which is to forgive, in union with Christ crucified and resurrected.

For the Christian true forgiveness is inextricably tied to faith. "Christian forgiveness," Lustiger points out, "can only take place when a person has complete faith in God."[8]

To seek forgiveness from another person can be excruciating when one has to inform that person of a wrong concerning which he or she may be unaware. One has told a lie or has distorted the truth in such a way that the other person's life or livelihood might be harmed. The temptation might be to let well enough alone, but to be right with the Lord one must humiliate oneself by bringing into the open a hidden, perhaps deeply hidden, wrong.

To move others to forgive, at least to forgo the vengeance of retaliation and war, when they may be lukewarm followers of Jesus, or unbelievers, or followers of another teaching, seems an impossible task. Sometimes "presence" is the only approach that may effect change, that presence that does not judge, that listens, that puts in love where love is absent.

God Forgive Me: I Can't Forgive Them

At a peacemaking gathering in Derry, Northern Ireland, where on Bloody Sunday 1972 paratroopers had shot dead thirteen unarmed marchers for civil rights, I had hoped to meet some of the militant partisans of union with Britain, chiefly Protestant. It was the corpses of Bloody Sunday that reactivated a near-dormant Irish Republican Army, which then trumpeted its new life by a series of bombings in Northern Ireland, England, and the continent of Europe.

Because the three-day meeting was sponsored by Pax Christi, none of the Ulster Unionist group would put in an appearance, even at the purely protocol opening session. I did meet, however, with a person who had been quietly serving as a sign of peace to both sides. He was an immovable presence, immovable from his position of benignity to all. Will Warren, a Quaker, settled in Derry with the purpose of living by the nonviolent approach of seeing "the human dimension of the men of violence." Without preaching, he would listen to the complaints of the Loyalist partisans of Ulster as a British province and to their reasons for not wanting to be separated from their mother country. They were only being loyal to their ancestors, who had been settled in Northern Ireland three hundred years earlier.

The complaints from their opponents were often about specific, oppressive practices by the regiments of British paratroopers: arrests of suspects who were held without trial; refusals of requests by mothers and fathers to visit sons in jail. Warren would give each side the gift of his ear; he would listen and hear them out. Whatever he could do, he would do.

One of the most convoluted approaches to forgiving I heard was in a tiny house in the Bogside, a poverty-stricken part of Derry where traditionally the Catholic population had been moved to accommodate the settlers from England and Scotland.[9]

A mother recounted to me that her son had been picked up early one morning. He had done nothing, she said, and was only a suspect. The paratroopers had kicked in the door and torn through the tiny "two-up, two-down" row cottage, upsetting or breaking anything in their way. After showing me signs of the wreckage, she said, "God forgive me: I can't forgive them." A theologian would have to disentangle her conclusion: "them" were all those on the other side who believed in an Ulster tied to Britain.

Following the meeting in Derry, I was driven to Belfast by the writer Ciaran McKeown. He had been a participant in the civil rights march to Derry and stressed that its aims had been the nonviolent achievement of jobs and fair housing allocation. The marchers, he said, had emulated the methods of Martin Luther King. McKeown stressed that the claim that the paratroopers had replied to fire from the marchers was false. The fact that innocent people had been murdered was never acknowledged. We discussed the "Declaration of the Peace People," which he had composed in 1976 when Mairead Corrigan and Betty Williams led a peace demonstration in Belfast. The occasion was the killing of three children in one family (children of Mairead Corrigan's sister) when the car of a British soldier had gone out of control. The demonstration grew into a movement called the Peace People. The Nobel Peace Prize came to Mairead Corrigan (later Maguire) and Betty Williams.[10]

On the level of the institutional church, forgiveness was forthright. The Catholic and Anglican primates of Ireland stood side by side in England's Canterbury Cathedral, the mother church of the Anglican communion, in a service of forgiveness. They issued a joint press release. Cardinal Daly, for the Catholics, asked Britons' forgiveness "for the wrongs and hurts inflicted by Irish people upon the people of this country on many occasions.... This sense of our need to be forgiven and our willingness to forgive is an essential part of the healing of our memories." Archbishop Eames, Anglican primate, pointed to the grievous sufferings in Northern Ireland: "Both have their regrets and their needs for forgiveness." He appealed for "recognition of the real feelings of vulnerability within the Protestant and Roman Catholic communities." He called for a "united Christian effort" based on a shared recognition that "with God all things are possible."

Bound to Forgive

Following his release from imprisonment as a hostage in Lebanon, Lawrence Martin Jenco, a member of the congregation known as Servites, responded to invitations to talk on his experiences in various parts of the United States. We who knew Father Jenco, or Father Marty as he preferred to be called, were not surprised when his chief message was "Forgive." "The first step in the creation of a new world order," he would say, "may be learning to forgive and to seek forgiveness." He would often begin his talks by reciting the vow of nonviolence of Pax Christi. He told his story in *Bound to Forgive.*[11]

Father Jenco, a staff member in the overseas aid and development programs of Catholic Relief Services, had served in Yemen and Thailand before being sent to Lebanon. As director of the Beirut office of the agency, his task was providing aid for Lebanese civilians, in particular children, trapped in the continuing violence. One morning, his work was interrupted. While he was driving to his office, two armed men halted his car, pulled him into the back seat, and took over the car. Transferred to another car, he was forced to lie on the floor. In a third car he was thrown into the trunk.

A day later, there was another trip in the trunk of the car. The destination was a garage where he was told to stand up with his arms at his side and his feet close together. Packing tape was wound around his whole body to the neck. Obeying the command to open his mouth, he was gagged, and tape was placed so that his mouth remained shut. His captors wound the tape around his head so that the whole body was covered mummy-like. Only his nostrils were left exposed for breathing. He was carried under the bed of a truck in the space where tires are stored. He decided that his body had been wrapped for burial and the truck was the hearse. Breathing was difficult, but with every breath, Jenco recited the prayer of the Russian mystics, "Lord, Jesus, son of God, have mercy on me a sinner."

His body was unwound in his first prison — a small kitchen where he was chained to a radiator. There were half a dozen makeshift prisons after that, all organized by the Hezbollah, the Islamic Party of God. Most of his guards were young extremists. When Jenco showed interest in their beliefs, they would explain them, for example, how *haram* meant foods, like pork, forbidden to them. He found them simple and poorly educated and realized that their knowledge of Islam was very limited. Jenco himself was more acquainted with some aspects of Islam than they were.

There were those who performed great kindnesses. Jenco took pieces of string and fashioned a rosary. Seeing him praying in this way, a young guard took the string of Muslim prayer beads from around his neck and gave it to Jenco. "It contained thirty-three beads," explained Jenco. "It is prayed by holding a bead between thumb and forefinger and giving God a name."

Jenco would go round the beads three times, pronouncing names of God ninety-nine times. He was surprised at how many names he could give to God. Jenco wore the beads around his neck until the moment of his liberation, when they were confiscated. Another man gave him an abridged copy of *David Cop-*

perfield, which he read over and over. For a blessed time, he had another hostage as a cellmate, a minister of the Presbyterian Church, Reverend Ben Weir. Jenco would celebrate the Eucharist with a piece of Lebanese bread, and Weir would celebrate a Communion service based on 1 Cor. 11:17–27, beginning with, "Because the loaf of bread is one, we, though many, are one body, for we all partake of one loaf."[12]

For eighteen months, Jenco endured brutalization, filthy living conditions, and the company of rats, who invaded his cell. Except for the time when Ben Weir and others were brought to his cell, or were jailed nearby, he was in total isolation. Surviving such a period as a hostage left marks on his body, but his spirit bore no vestige of anger or bitterness.

Shortly before his captivity ended, a guard named Sayeed, who had at first brutalized him, sat with him on his sleeping mat. At first he had called his prisoner Jenco, then Lawrence, and finally Abouna, a mark of respect for a priest. Asked Sayeed, "Abouna, do you forgive me?" Jenco replied, "Sayeed, there were times when I hated you. I was filled with revenge for what you did to me and my brothers. But Jesus said on a mountaintop that I was not to hate you. I was to love you. Sayeed, I need to ask God's forgiveness and yours."[13]

In the title of his book, Lawrence Martin Jenco put into three words the Christian teaching of forgiveness: "Bound to forgive." The followers of Jesus have no choice: to remain in the grace of God, or to recover it if lost, we are bound to forgive all injuries, trespasses, and wrongs. We know that only if we forgive the trespasses of others will our trespasses be forgiven by the heavenly Father. "Having forgiven," concluded Jenco, "I am liberated. I need no longer be determined by the past."[14]

The implications of forgiveness for peace should inspire Christians to be apostles of forgiveness, moving, wherever possible, into areas that are crucibles of hate and violence. The Lord's Prayer is a prayer that many people recite daily: it is a prayer recited by the whole congregation after the offering up of the saving cup and the living bread. From the one-person witness of a Will Warren or the message of a Father Marty Jenco, the witness might grow to include two or even three persons (with the Spirit of God in their midst) and then to untold numbers.

In relation to the last petition, "Deliver us from evil" (or the evil one), the evil that we seek to be delivered from varies with each age. In the past, the evil was the plague, a famine, an invading army. Today it includes "the threat of the extinction of our world."[15]

The generations of our time, in praying "Deliver us from evil," are sending up a mighty call to the God of peace to lift from his children the threat of a war that could virtually destroy God's creation.

Part Three

THE DISCIPLESHIP OF PEACE

Chapter 16

Dorothy Day

Christ's Way or the World's Way (1927–59)

Over all the virtues put on love, which binds the rest together and makes them perfect. Christ's peace must reign in your hearts, since as members of the one body you have been called to that peace. —Col. 3:14–15

When the United States declared that it was entering World War II, Dorothy Day announced in the *Catholic Worker*, "Our manifesto is the Sermon on the Mount, which means that we will try to be peacemakers. Speaking for many of our conscientious objectors, we will not participate in armed warfare or by making munitions, or by buying government bonds to further the war effort, or by urging others to these efforts."

In the same January 1942 issue of the *Catholic Worker*, Dorothy Day chronicled her personal response to war. She knelt in the Church of the Transfiguration in New York's Lower East Side and prayed, "Lord God, merciful God, Our Father, shall we keep silent or shall we speak? And if we speak, what shall we say?" These words appeared under the paper's headline, "Our Country Passes from Undeclared to Declared War: We Continue Our Pacifist Stand." Dorothy asserted, "We will print the words of Christ who is with us always — even to the end of the world.... Love your enemies, do good to those who hate you."[1]

The dilemma facing Christians from the fourth century onward, of how to get around the Sermon on the Mount with its hard sayings of loving enemies and praying for persecutors without renouncing Jesus, did not present itself to Dorothy Day and the majority of those in the Catholic Worker movement. These and other hard sayings were at the core of the life of those committed to the movement.

With regard to the Sermon on the Mount, those who justify war relegate it to the realm of private morality and deem it inapplicable to public morality. Christians, however, who embrace the Sermon on the Mount as the heart of their discipleship insist that as believers and citizens they have the right and duty to live by it, and to refuse to accept any orders contrary to it.[2]

The Catholic Worker movement's rejection of participation in World War II, a war just in cause, was shocking to many. Hitlerian totalitarianism and anti-Semitism had been opposed by members of the movement from its founding in 1933. Articles exposing totalitarianism and anti-Semitism became a feature of the group's publication. The Catholic Worker group began to picket the German consulate in New York City in the 1930s to protest the persecution of Jews and dissidents. Dorothy Day helped organize the Committee of Catholics to Fight Anti-Semitism. Gospel nonviolence demanded that the response to massive violence not be in kind. "Our works of mercy may take us into the midst of war," she wrote. "As editor of the *Catholic Worker,* I would urge our friends and associates to care for the sick and wounded, to the growing of food for the hungry, to the continuance of all our works of mercy in our houses and on our farms."[3]

The movement had to be ready for governmental reaction of a negative sort. Here was no Quaker or Mennonite publication, but that of a Catholic group opposing a just war. "Because of our refusal to assist in the prosecution of the war," Dorothy continued in the January 1942 issue, "and our insistence that our collaboration be one for peace, we may find ourselves in difficulties. But we trust in the generosity and the understanding of our government and our friends, to continue to use our paper to 'preach Christ crucified.'"[4]

Dorothy Day saw her peace discipleship not as a special compartment of her life, but as the fruit and expression of a life lived in the light of the incarnation. No aspect of her living in the world was to be outside of the influence of the ministry, death, and resurrection of Jesus, and of his continuing presence at the Table of the Lord.

A Church in Thrall to Just War Teaching

When Dorothy Day entered the Catholic Church as a convert in 1927, she came into a church in thrall to the teaching of justified warfare. Her witness to gospel nonviolence, a term which many, myself included, prefer to the term "pacifism," was prophetic in the extreme. A more unlikely prophet could hardly be imagined. She was a recent convert, a former comrade of the left, who had aborted the child of a failed love affair. A short-lived marriage was followed by a common-law union and the birth of her only child. Her partner, an agnostic with anarchist leanings, would not consent to a wedding ceremony. He made it clear to Dorothy that he could have nothing to do with her or with religion if she embraced it. She thus became an unwed mother.

Could the witness of such a person be accepted as prophetic? Here was a woman invading the preserve of male theologians who for fifteen hundred years had taught the conditions of the just war. Could her peace witness put in question an entrenched position on war and peace that had replaced the earlier tradition of gospel nonviolence?

Separation from her common-law husband, the father of her child, was the first test of her acceptance of Jesus and the church. "Becoming a Catholic," Dorothy related, "would mean facing life alone. I did not want to give up a human love when it was dearest and tenderest. It took me a great many years not to wake up in the morning and reach for human warmth near me. When will I learn to love all men and women with the intense awareness of their beauty, their virtues (strengths) — to see them as Christ sees them?"[5] Her witness grew on the soil of this sacrifice.

Born in Brooklyn, Dorothy Day moved with her family to San Francisco and Chicago, before returning to New York in 1916, when she was eighteen years old. Her father, a journalist, was a literate man whose sports reporting was marked by quotations from Shakespeare as well as literary classics. There were two older brothers, one younger brother, and a younger sister. The family practiced no religion, except for a short period when they attended services in the Episcopal Church. Mrs. Day was of Episcopalian background, and Dorothy was baptized into that church.

During her adolescence in Chicago, Dorothy was the sole member of the family who attended the nearby Episcopalian church. She read the New Testament in an organized way and was devoted to the Psalms. In a letter to a close friend, the fifteen-year-old Dorothy confessed, "I know it seems foolish to try to be so Christlike — but God says we can — why else is his command, 'Be ye therefore perfect'?"[6]

Her attendance at Sunday services soon ceased. She began to read Jack London, whose socialism moved her to think about justice to the poor, as did the work of another socialist, Upton Sinclair. When she read Sinclair's *The Jungle,* with its brutally vivid picture of life in the Chicago stockyards, she decided to wheel the baby carriage of her brother John, fourteen years her junior, into the more miserable areas of the city. She identified with the people who inhabited "the interminable gray streets" rather than with the comfortable people at church. "I used to turn away from the park," she wrote, "and walk over west through the slum districts, and watch the slatternly women and the unkempt children and ponder over the poverty of the homes as contrasted with the wealth along the Shore Drive. I wanted even then to play my part. I wanted to write such books that thousands upon thousands of readers would be convinced of the injustice of things as they were."[7]

Dorothy Day's formal education ended when she was eighteen, after she had spent two years at the University of Illinois at Urbana. While there, she made friends who whetted her appetite for social change, the appetite aroused by her acquaintance with Chicago's mean streets. She attended a few meetings of the University Socialist Club and was elected to the "Scribblers," a club for young people who had shown a gift for writing. One of her closest friends was Rayna Simons, a striking young woman with flaming red hair who eventually went to revolutionary China and journeyed to Moscow to attend the Lenin Institute. When Rayna visited Dorothy in New York City in 1917, they walked the streets of New York with young communists and "sat on the ends of piers singing revolutionary songs."[8]

In New York, Dorothy shared the revolutionary ferment of young workers and artists, finding work on the socialist *Call* and *The Masses.* Also associated with *The Masses* was John Reed, author of the eyewitness account of the Communist revolution, *Ten Days That Shook the World.* Sharing the exultation of those days, she celebrated with thousands of New Yorkers the success of the Soviet revolution, and went about "singing revolutionary songs into the star-lit night." As a comrade of the left, she opposed U.S. participation in World War I, describing her position as "pacifist, in what I considered an imperialist war, though not pacifist as a revolutionist."[9]

The Pain of Separation

The painful separation from the man she loved and father of her child was only part of the pain that accompanied Dorothy's entry into the Catholic Church. There was pain in separating from the friends of the left. Though never a member of the Communist Party, Dorothy had gloried in the struggle to bring a new society to birth, a struggle that was the daily bread of her companions of the left. One of her closest friends was Mike Gold, later to become the editor of the communist *Daily Worker.* It was Mike who introduced her to Eugene O'Neill. In the "Hell Hole," the backroom of a tavern where they often met, O'Neill recited for her Francis Thompson's lengthy religious poem, *The Hound of Heaven.*

When Dorothy recalled to me those days in Greenwich Village, the man she mentioned most often was Eugene O'Neill. Their friendship was undoubtedly close. His recital of "And now my heart is as a broken fount, / wherein tear drippings stagnate" mesmerized her. She recalled O'Neill's repeated question, "Dorothy, are you ready to give up your virginity?" If the questioner had intended marriage, her life might have been vastly different. In fact, he married a friend of hers.

Even at the height of her involvement in the left, surrounded by unbelievers who saw religion, and in particular the Catholic Church, as obscurantist and against human progress, Dorothy felt the need of prayer. "Many a morning, after sitting all night in taverns or coming from balls in Webster Hall, I went to early morning mass at St. Joseph's Church on Sixth Avenue and knelt in the back of the church, not knowing what was going on at the altar, but warmed and comforted by the lights and silence and the kneeling people and the atmosphere of worship. People have so great a need to reverence, to worship, to adore."[10] She quoted one of her frequently read authors, applying the quote to herself: "'All my life I have been haunted by God,' a character in one of Dostoyevsky's books says. And that is the way it was with me."[11]

Leaving the life of Greenwich Village, Dorothy enrolled as a nurse trainee and began a love affair with a patient who was a newspaper man. In the attempt to hold him, she had their child aborted. In any case, he deserted her.

It was not until she met Forster Batterham that she had a stable life, the life of a common-law union. When she discovered that she would have a child, she read the *Imitation of Christ* a great deal and made a firm decision: the child was to be baptized. "I knew that I was not going to have her floundering through many years as I had done, doubting and hesitating, undisciplined and amoral. I felt it was the greatest thing I could do for my child. For myself, I prayed for the gift of faith. I was sure, yet not sure."[12] She began to attend Mass every Sunday morning. In common with other needy women in New York, Dorothy went to the city hospital, Bellevue, for the birth of her child. She named her Tamar, for a leftist friend, and Teresa, for a Catholic neighbor in the next bed.

Tamar Teresa was the name given to her daughter at baptism, and some months later Dorothy presented herself for conditional baptism. Her godmother was a local nun who had instructed her in the catechism. "I loved the Church for Christ made visible," she wrote, "not for itself, because it was so often a scandal to me."[13] She also stated: "Romano Guardini said that the Church is the Cross on which Christ was crucified; one could not separate Christ from His cross, and one must live in a state of permanent dissatisfaction with the Church."[14]

I Lost Faith in Revolution

Five years a Catholic, Dorothy found herself in Washington, D.C., to report for *Commonweal* magazine on a communist-inspired hunger march. She supported herself and her daughter in those days by free-lance writing and managed to keep expenses to a minimum by sharing a tenement apartment in New York's Lower East Side with her younger brother, John, and his wife. It was the height of the Great Depression, with millions of Americans workless, hungry, and without hope.

One-fourth of the nation's workers were unemployed, and destitution stalked their families. Dorothy suffered at being a bystander at the Washington march, while the marchers went by, waving banners calling for unemployment insurance and relief for mothers and children. One banner carried a message with which Dorothy resonated in particular: "Work not wages." She felt that meaningful work was what human beings needed, not simply tasks without meaning, even though they bought bread and shelter.

After the march, Dorothy went to pray in the crypt of the Shrine of the Immaculate Conception on the campus of the Catholic University of America. "I offered up a special prayer," she wrote, "a prayer that came with tears and anguish that some way would open for me to use whatever talents I possessed for my fellow workers and for the poor."[15]

Her passionate desire was to find something to do in the social order besides simply reporting conditions. She longed to help change conditions rather than be a bystander. "I had lost faith in revolution," she announced. "I wanted to love my enemy, whether capitalist or communist."[16] Dorothy had come full circle from her commitment to violence on behalf of revolution.

A Theory of Revolution

Returning to New York, to the crowded four-room apartment, she was ready for a rest when she heard a knock at the door. A man wanted to see her, "a short, stocky man in his early fifties, as ragged and rugged as any of the marchers I had left."[17]

It was Peter Maurin. He had come to the apartment several times while she was in Washington. The editor of *Commonweal* had sent him to see her since he had noticed congruencies between the ideas of Dorothy and Peter. Immediately, Peter began talking, not conversing, but announcing his ideas for the purpose of "indoctrination and clarification of thought."

That night, Peter had to be cut short, but he came back the next day, and Dorothy began to listen to him. For the next four months, he appeared almost every day, bringing, on sheaves of paper, selections from papal encyclicals and gleanings from the early Church Fathers and the saints of the church. They constituted the riches of church teaching gathered by a poor man whose heart and mind were fixed on the need for radical but nonviolent social reform.

Dorothy remarked, "Peter talked you deaf, dumb and blind and believed in repeating, in driving home his point by constant repetition like the dropping of water on the stones that were our hearts."[18]

It is hard to imagine two more bizarrely different persons. A modern American woman in her early thirties, tall, slender, attractive, and a stocky, French, autodidact peasant twenty years older. In the unsearchable designs of providence, their trajectories brought them together to become cofounders of a movement that has influenced the American Catholic Church so profoundly that after the Catholic Worker it was never quite the same church. Peter had been born on a farm in the Languedoc region of southern France — a farm which, it was believed, his ancestors had worked for fifteen centuries. He entered a teaching order of Christian Brothers but left after nine years. Emigrating to Canada, where an attempt at homesteading failed, he crossed over to the United States, probably illegally. He was devoted to St. Francis of Assisi and led the life of a wandering Franciscan. He hoped to see Franciscan ideals of poverty and simplicity reflected in the lives of the followers of Jesus. His wanderings led him from odd job to odd job, finally locating a stable place to live. In point of fact it was above a stable in a summer camp for Catholic children at Mt. Tremper, near Woodstock, New York. In return for year-round work as a handyman, he was given board and room and a small stipend. Whatever funds he had went for Catholic texts and religious books, which he acquired on trips to New York City. He met and talked with anyone who would listen on ways to remake the social order. It was on these trips that he met such people as George Schuster of *Commonweal*, Carlton J. Hayes of Columbia University, and Pitirim Soroken of Harvard University.

Peter came to Dorothy with a theory of revolution, a personalist, communitarian revolution. He called it a "green revolution." It would search out concordances, not hostile confrontation. It would join with the Industrial Workers of the World (IWW), for example, in aiming for "the building of the new within the shell of the old."[19] The new society would heal the breach between gospel teachings and economics and sociology. Social life had become divorced from spiritual values and was driven by the search for profits. Man had become no more than a cog in an immense impersonal machine. The church had the answer to social problems, in its teachings on man as having the infinite dignity of a child of God, and as a co-creator with God through his work. The church, however, was "keeping the lid" on the morally explosive content of its teaching.

Through Peter, Dorothy was introduced to two ground-breaking encyclicals, *The Condition of Labor* by Pope Leo XIII, issued in 1891, and *After Forty Years,* issued by Pius XI in 1931. Leo condemned the misery and wretchedness of large numbers of laboring people while "a small number of very rich men have been able to lay upon the masses of the poor a yoke little better than slavery itself."[20] His encyclical supported the right of workers to associate in unions. The encyclical of Pius, issued during the Great Depression, saw no solution in either individualism or collectivism. "There is a double danger to be avoided," Pius pointed out. "On the one hand, if the social and public aspect of ownership be denied or minimized, the logical consequence is 'individualism,' as it is called; on the other hand, the rejection or diminution of its private and individual character necessarily leads to some form of 'collectivism.' "[21] Pius saw danger in bigness in the economic order. Peter resurrected the traditional teaching of subsidiarity, according to which everything should be accomplished by a smaller, more controllable, and perhaps local entity before consigning it to a larger or more distant entity. The thought of the Catholic Worker movement was always toward smallness, where people were responsible for what was done and the way in which it was done. The New York Catholic Worker never became the board of directors of the movement, laying down rules for other houses. Each house, while it arose out of a response to agreed-upon principles, was free to create its own program. The core of the movement was its recognition of the importance of each person. It prefigured the work of E. F. Schumacher, whose books, including *Small Is Beautiful,* were much discussed in the pages of the *Catholic Worker.*

The Established Disorder

Peter also opened Dorothy's mind to French Catholic thought, in particular, that of Emmanuel Mounier, inspirer of the personalist movement. Mounier described the social system which depended on finance capitalism and on preparation for war as "the established disorder." He included the "bourgeois church," which existed tranquilly within it. Through his journal *Esprit,* he led a struggle to awaken the Catholic Church from its "bourgeois lethargy" and to its obligation for heroism in creating a new order of justice. "Injustice!" he wrote in *Esprit,* "thousands of men ignore it in complete tranquility.... We will haunt their nights, our nights, with its hoarse voice." Peter took from Mounier a concept of revolution that was personalist and communitarian.[22] Peter also synthesized the work of Jacques Maritain, philosopher and animator of an updated concern for Thomas Aquinas.

Peter talked of the three pillars for the remaking of society: (1) cult, englobing the teachings of Jesus, the Mass, and sacraments; (2) culture, the wisdom of the ages, of the saints, of the Fathers of the Church, martyrs, theologians, and philosophers; and (3) cultivation, meaning a return to and respect for the land and for all work, especially crafts and manual work.

The three pillars would be advanced, first, by round-table discussions for "clarification of thought," one of Peter's favorite expressions; second, by houses of hospitality, where the performance of the works of mercy would bring together those who have and those who have not for mutual aid; and, third, by agronomic universities, where "scholars would become workers and workers would become scholars." In a time of mass unemployment, Peter was fond of saying, "There is no unemployment on the land." Peter envisaged Dorothy as a St. Catherine of Siena who would convince the American bishops of the need to implement his program and, at least, open a house of hospitality in each diocese.

The round-table discussions began immediately in the kitchen of the apartment on East Fifteenth Street. People visited and stayed to eat whatever was available. One idea of Peter's for spreading the ideas of the green revolution was to publish a paper. This idea appealed to Dorothy, a writer herself and a member of a journalistic family. Two of her brothers had joined her father in newspaper work. She discussed the practical aspects of it with Peter, and found that he would not involve himself with such matters. "I enunciate the principles," he said, and explained the need for the dependence upon the providence of God for the details and the wherewithal of putting flesh on the principles.

A Tabloid at Union Square

As soon as she had fifty-seven dollars, the sum of payment for an article in *America* magazine and a few gifts from friends, Dorothy arranged for the printing of an eight-page tabloid. It contained seven contributions from Peter, the phrased paragraphs that came to be known as "Easy Essays." One essay dealt with money-lending:

> Christ drove the money-changers
> out of the Temple
> But today nobody dares
> to drive the money-lenders
> out of the Temple.
> And nobody dares
> to drive the money-lenders
> out of the Temple
> because the money-lenders
> have taken a mortgage
> on the Temple.[23]

Dorothy's burning love for the poor and voiceless lit up the pages. With a searing pen she described the exploitation of black labor in Mississippi by the U.S. War Department and the oppression of women in sweatshops and factories. Twenty-five hundred copies of the tabloid were off the press by May Day 1933, when Union Square, between Fourteenth and Eighteenth Streets on the East Side of New York City, was thronged with fifty thousand people at the annual communist rally. Dorothy Day, with three young men, moved into the surging crowd, selling the papers for a penny a copy or just handing out the tabloid, which was similar to the communist *Daily Worker.* It was called the *Catholic Worker.*

The word "Catholic" evoked hostility from some of the crowd, and two young men took off, leaving Dorothy with a single co-worker.

The short editorial spoke of compassion for the unrelieved misery, and even despair, of the men, for they were almost all men, congregated in the square. In those days, the Catholic Church, whose members were, in the main, workers, was hard put, through its Catholic Charities organizations, to provide funds, as well as food, clothing, and shelter, to countless among the near-destitute. Home relief was no more than a pittance, and the Catholic community — immigrants and the children of the immigrants — was savagely hurt by the Depression. Dorothy's message was that hope came not only from revolutionary sources (though for countless Americans the Soviet Union was their beacon, their dream of material salvation) but from the age-old teachings of the church.

The editorial began:

> For those who are sitting on park benches in the warm spring sunlight. For those who are huddling in shelters trying to escape the rain. For those who think that there is no hope for the future, no recognition of their plight — this little paper is addressed. It is printed to call their attention to the fact that the Catholic Church has a social program — to let them know that there are men of God who are working not only for their spiritual, but for their material salvation.[24]

May 1, 1933, is considered the birthday of the Catholic Worker movement. Dorothy had decided that Peter was to be her mentor. The bond between the two was a search for a synthesis between the teachings of Jesus and the way his followers should live. Jesus' commandment of love should undergird all aspects of life — and love finds its expression in the works of mercy.

Hospitality

The fleshing out of one aspect of Peter's program came about as a response by Dorothy to a tragedy. She recounted to me that a sort of communal kitchen grew as people came to the apartment to discuss issues and volunteer with the putting out and distribution of the *Catholic Worker.* Hungry people came for coffee, bread, and soup. Some of them asked for shelter. She remembered two women who had been sharing meals. One evening they asked if they could stay the night. There was simply no space. The homeless women went out into the street. When one of the women came to the apartment some days later, Dorothy asked her about her companion. The woman revealed that her friend in despair had thrown herself under a subway train. With only five dollars in her hand, Dorothy walked down Fifteenth Street, from number 436 where they were, until she found a vacant apartment. She rented it with a five-dollar deposit, and it provided shelter for six homeless women. This was the first move toward a house of hospitality.[25]

Dorothy and Peter were at one in viewing "the daily practice of the works of mercy" as the very ground of the Christian life. Peter believed in direct action for social change, and for him, the performance of the works of mercy across all barriers was the most direct action of all. The works of mercy were to be carried out even at the cost of personal sacrifice and rested on the bedrock of voluntary poverty. Dorothy, however, was the one who emphasized the obvious: the daily practice of the works of mercy should not go into limbo or be interrupted when different groups within the human community decided to wage war on each other. While Peter was certainly against killing and war, it was Dorothy who proclaimed the message of gospel nonviolence, or pacifism, as she boldly announced it, in the paper and in the acts of witness that came to be characteristic of the movement. Dorothy and Peter were at one on these points, as they were in appreciating the anarchism of Peter Kropotkin, which downplayed the nation-state and promoted mutual aid instead of competition, and personal responsibility instead of dependence on government.

The Gospel Ideal of Pacifism

In *Breaking Bread,* Mel Piehl expressed the view that "after 1945, it was the issue of pacifism that most effectively represented the Catholic Workers' gospel nonviolence.[26]

Many would question Piehl's thesis that it was "after 1945" that pacifism stood for the movement's gospel idealism. The other aspects of gospel idealism in the movement were closer to traditional Catholic expressions of piety. These included the ideal of poverty and simplicity of life, recommended not only to those under vows in religious life, but to all the followers of Jesus in accordance with their state in life; the overall command to love God and our neighbor as ourselves, though honored in the breach, was not new to the ears and hearts of churchgoers.

The communitarian ideal, as lived in communities of Benedictines, Franciscans, and hundreds of other vowed congregations and societies, was also shared with lay "Third Orders" attached to the Benedictines and Franciscans. While Dorothy led those in the Catholic Worker movement to carry the commandment of love to what some considered the extreme in such matters as hospitality, there was no clash with commonly agreed upon teachings.

Pacifism was another matter altogether. The word had a sort of heretical ring to Catholics; it was not part of the vocabulary of faith. Ron Musto points out that as late as 1967, the *New Catholic Encyclopedia* concluded, "It is clear from what has been said that absolute pacifism is irreconcilable with traditional Catholic doctrine."[27] Pacifism was confused with passivity in the face of aggression, oppression, and evil. It might be accepted from small, tightly knit communities of Mennonites or Brethren whose distinctive garb and way of life had become part of the American historical process. Pacifism might be honored in the case of the Quakers, whose peaceful overtures to the Indians had been so admirable and whose opposition to war was respectable.

For Catholic citizens in the 1930s, most of them of recent immigrant stock and grateful to the United States for a new life of freedom (even though the Depression years were bitter), the thought of saying no to their government's call for military service was close to heresy, if not actual heresy. With the exception of a few bishops, the entire American Catholic community could be seen as in full and unquestioning obedience to just war teaching. They might not have known the conditions that permitted a given war to be called "just," but they believed that such conditions were in the safekeeping of their church leaders and were formulated somewhere in the theology books in seminaries.

In May 1936, the *Catholic Worker* announced that it was a pacifist paper, opposing class war, class hatred, and imperialist war. Replying in advance to those Catholics who equated pacifism with cowardice, Dorothy wrote:

> A pacifist who is willing to endure the scorn of the unthinking mob, the ignominy of jail, the pain of stripes and the threat of death cannot be lightly dismissed as a coward afraid of physical pain....
>
> A pacifist even now must be prepared for the opposition of the mob who think violence is bravery. A pacifist in the next war must be ready for martyrdom.[28]

The implications of the pacifist stance became starkly clear after the outbreak of the savage civil war in Spain in July 1936. It was to last thirty-three months and cost as many as a million lives. Catholic publications were filled with accounts of outrages against priests and nuns, including tortures and executions, and burnings of churches. Except for some Basque Catholics who sided with the Loyalists, the church in Spain put its hopes in a victory of the Insurgents, led by General Francisco Franco. The U.S. church followed the lead of the church in Spain. Support of the Insurgents took on the overheated rhetoric of a crusade. The *Catholic Worker* profoundly shocked most of its readers by announcing its neutrality as between the warring sides.

The Use of Force

In the November 1936 issue of the paper, Dorothy discussed "the use of force" with relation to the Spanish Civil War. She reminded the readers how difficult it had been for the disciples to accept the "hard saying" that they must follow Jesus in laying down their lives for their friends. She used a powerful example:

> Christians, when they are seeking to defend their faith by arms, by force and violence, are like those who said to Our Lord, "Come down from the Cross. If you are the son of God, save yourself."
>
> But Christ did not come down from the Cross. He drank to the last drop the agony of His suffering, and was not part of the agony, hopelessness, the unbelief, of his disciples.... If 2,000 have suffered martyrdom in Spain, is

> that suffering atoned for by the death of 90,000 in the Civil War? Would not the martyrs themselves have cried out against more shedding of blood?[29]

Yet she did not condemn those in Spain who took up arms and engaged in war. She wondered what each person would do if confronted with the dilemma of taking part or not taking part in Spain's travail.

Her position, however, was clear. At the height of the carnage in Spain, she repeated her opposition to the use of force. "As long as men trust in the use of force, only a more savage and brutal force will overcome the enemy.... We must stand opposed to the use of force." She admitted that it was inconceivably difficult to say what she was saying. "It is folly — it seems madness — to say as we do, 'We are opposed to force as a means of settling personal, national or international disputes.'"[30]

Dorothy gathered some cohorts for her position on the Spanish Civil War, chiefly from European Catholicism. They included the philosopher Jacques Maritain and the novelists Georges Bernanos and François Mauriac. "Spanish Catholic Flays Both Sides" was the title of the front page article in the December 1938 issue of the paper. It appeared originally in *Esprit,* the publication of Emmanuel Mounier, who sent it to the *Catholic Worker.* The article, which condemned the bloodletting of each side, expressed Dorothy's point of view.

Many Catholics saw Dorothy's refusal to support the Spanish Insurgent cause as coming from her leftist past and attacked her as a communist. She was saddened by the fact that one of her oldest and dearest friends, Mike Gold, a convinced communist, considered her pro-Franco for not declaring for the Loyalists. Neither side glimpsed her vision, a vision that rose above the issues that divide humankind. She attempted to clarify that vision in a restatement of the "aims and purposes" of the *Catholic Worker.* She wrote:

> We believe that all men are members or potential members of the Mystical Body of Christ. This means Jews, Gentiles, Black and White. This means our enemies as well as our friends. Since there is no time with God, and since we are told that all men are members or potential members of Christ's Mystical Body, that means that now at the present time we must look upon all men with love. We must overcome all evil with good, hatred with love.
>
> Hence our work to combat anti-Semitism; to combat the use of force as a means of settling disputes between men and nations.[31]

When the Catholic Workers picketed the German consulate in New York City, they carried the sign, "Spiritually we are Semites."

Laypeople Fight the Wars

Dorothy Day continued her struggle to lead Catholics toward a difficult step, that of unlearning a traditional teaching, justified warfare. Despite the fact that she was making little impact, she took every path possible to break the tie of the individual Catholic to a government's order to kill in war.

When she heard that hearings were to be held in Washington before the

Military Affairs Committee of the Congress to discuss the adoption of the Compulsory Military Training Law, she decided that the Catholic position should be heard. Along with a community member, Joseph Zarrella, she journeyed to Washington, ready to present to the committee the anticonscription stance of the movement. Having studied the draft of the proposed law, she pointed out at the congressional hearing that it did not make provision for Catholic conscientious objectors. A senator informed her that the right of conscientious objection was adequately covered in the draft, to which she replied, "It does not protect Catholics. It may protect the Quakers, the Mennonites, the Dunkards, but not Catholics."[32] She told the committee that if an unjust law were passed, she and her movement would consider it a duty not to obey. They would run the risk of having the movement wiped out by imprisonment and fines.

There was one clerical spokesperson, Monsignor George Barry O'Toole, for the right of Catholics to be recognized as conscientious objectors. O'Toole, a professor of philosophy at the Catholic University of America, had been contributing articles on conscription to the *Catholic Worker* between October 1939 and November 1940. In January 1941, the *Catholic Worker* reprinted his articles in a booklet entitled *War and Conscription at the Bar of Christian Morals.* Several clerics testified on behalf of the church for exemption from the draft for seminarians and members of congregations of Christian Brothers. A cleric from National Catholic Welfare present at the hearings questioned Dorothy Day's right to speak for Catholics on this issue. "We are speaking for lay people," she responded, "and they are the ones who fight the wars."[33]

Dorothy's intervention on behalf of Catholic conscientious objectors was of no avail. The Selective Service Act passed in August 1940, ignoring the recommendation from Dorothy and representatives of the mainline churches that the right of conscientious objection should apply to those whose beliefs encompassed opposition to even a possibly just war.

Though there were only two, possibly four, Catholic conscientious objectors listed or recognized in World War I, Dorothy expected more objectors in World War II.[34] Draft boards, four thousand of them, routinely denied the status of conscientious objection to those who held to just war teaching. Many considered the war in question, World War II, unjust according to the conditions of the just war, in particular the matter of the means used to prosecute the war. This position, termed selective objection, was not recognized for conscientious objector status. Even Catholics who no longer followed just war teaching, but were against all wars, rarely qualified for conscientious objector status. Their draft boards told them that their church taught just war, not pacifism.

World War II

The World's Way and Christ's Way

After World War II broke out in Europe, the Catholic Worker movement was riven by the issue of pacifism, with some of the key members of the community dissenting from a position of absolute Christian nonviolence. In September 1939, Dorothy wrote, "We must choose sides now; not between nations at war but between the world's way and Christ's way." By 1940, there were thirty Catholic Worker houses throughout the nation. Not all the community members saw the issue in the same starkly clear way as Dorothy. Some Catholic Worker houses refused to distribute the New York *Catholic Worker,* and one group actually put it to the torch. Dorothy wrote on August 10, 1941, to all the houses, explaining that she could accept that not everyone in the Catholic Worker movement stood with her on the issue of pacifism. She pointed out, however, that if any house refused to distribute the *Catholic Worker,* it should separate itself from the movement. Tom Sullivan referred to this as Dorothy's "encyclical letter."

A week-long retreat held later that month at Maryfarm, the Catholic Worker farm in Easton, Pennsylvania, brought together many members of the various Catholic Worker houses. They prayed, worshiped, and discussed the future of the movement. Some returned to their houses still opposed to pacifism. Much correspondence followed as Catholic Workers separated themselves from the movement and responded to the draft for combat in World War II. Among these were John Cogley and Tom Sullivan of the Chicago house, which closed in April 1941. In a letter written to Dorothy in November 1941, Cogley revealed how pacifism was viewed in terms of heresy. "Profound disagreement is a wall between people," he wrote, "and it rears higher every day. How I wish you weren't a heretic. And sometimes how I wish that I were one too."[35]

Weapons of the Spirit

As World War II tightened its grip of horror around more peoples of the globe, including Americans, the life of the Catholic Worker movement waned. Dorothy, however, never wavered in her faith in the movement.

With Arthur Sheehan, an editor of the *Catholic Worker,* she sponsored the Association of Catholic Conscientious Objectors. Arthur was exempt from the draft as a recovered tuberculosis patient. Such Catholic objectors as there were made their way to the Catholic Worker house in New York City. The Friends, Mennonites, and Brethren operated camps for conscientious objectors in an agreement with the National Service Board for Religious Objectors. Why not a camp for Catholic objectors? With the wild courage that characterized her, Dorothy located an unused Forest Service Camp near Stoddard, New Hampshire. With minimal refurbishing, a camp for those Catholics who had achieved conscientious objector status opened its doors — rickety in all likelihood — in October 1942. The whole camp was described as a ramshackle place. Clothing and food were collected through the Catholic Worker and shipped to the camp. The fierce winter snows of New Hampshire necessitated a more habitable center. One was found at Warner, New Hampshire. Arthur Sheehan, with the sponsorship of Dorothy Day, edited a newspaper published on an irregular schedule, the *Catholic CO,* for the Association of Catholic Conscientious Objectors.

Nothing, not even the heroic efforts of Dorothy Day, could save the camp experiment. The Warner camp closed in March 1943, with most of the conscientious objectors being transferred to a camp operated by the Friends. A monthly stipend of fifteen dollars was supposed to be paid to each objector in civilian service. Since there were no Catholic funds available for the CO programs, these funds were paid by the Mennonites and Friends. It is estimated that the total of Catholic conscientious objectors in World War II was 135, of whom sixty-one were sentenced to prison terms served at twenty-eight federal prisons. Eighteen served their prison sentences in Danbury Federal Prison.[36]

Among the seventy-five Catholic conscientious objectors who passed through Stoddard and Warner was Gordon Zahn, who recalled his experience in *Another Part of the War: The Camp Simon Story*.[37] Though Catholic conscientious objectors were hardly heroes in their own community, some Catholic institutions accepted them for alternative civilian service, in particular, the Alexian Brothers Hospital in Chicago. Gordon Zahn, among others, joined the staff at Rosewood Training School for the Mentally Handicapped in Maryland.

The masthead of the *Catholic CO* carried a legend that expressed the spirit of Dorothy Day and Arthur Sheehan:

> We hope that war will be overcome through the church, and even if, after two thousand years, this hope is still unfulfilled, we still hope and go on knocking at the door like the unfortunate man in the gospel.[38]

Dorothy contributed an article entitled "Women and War" to the newspaper. She addressed the charge that she, as a woman, was "placing burdens on the backs of others that I did not have to bear myself." Her response was that women should intensify their antiwar, anticonscription efforts. It was still possible to avoid the draft. Many men who had registered their views with the Catholic Worker were never drafted. These should be encouraged, as well as any young men who had doubts about fighting or killing. In any other war, she hoped to see "a mighty army of Catholic CO's using the weapons of the spirit."[39]

Dorothy Day was often asked if those at the Catholic Worker thought that "a man is in mortal sin for going to war." She replied:

> To my mind the answer lies in the realm of motive, the intention. If a man truly thinks he is combating evil and striving for the good, if he truly thinks he is striving for the common good, he must follow his conscience regardless of others but he always has the duty of forming his conscience by studying, listening, being ready to hear his opponents' point of view.[40]

A Time of Solitude

Of a sudden, in 1943, with World War II bringing death and desolation to even larger areas of the globe, Dorothy decided to absent herself from the Catholic Worker. She announced that she would give a year to solitude and prayer, removing herself from the responsibilities of the movement. She turned over the editorship of the paper to Arthur Sheehan and took her name from the masthead.

Dorothy explained that the decision had arisen from her participation in a retreat directed by the Reverend John J. Hugo at Maryfarm, in Easton, Pennsylvania. Those who made the retreat, given repeatedly by several priests, knew that in the eight days of silence, the focus was on the ascetic life. Dorothy had taken what she called "the famous retreat" many times and called it "the bread of the strong" on the spiritual level. The original retreat was preached by a Canadian priest and was originally intended for priests who left their parish concerns for eight days of silence. Because of its rigor, the priest who preached the retreat was silenced. The retreat, emphasizing the return to "radical Christianity," was introduced into the United States through the Catholic Worker. Father Hugo asserted that "a true disciple cannot escape the cruciform pattern of Christian living." To Dorothy, the retreat was "stimulating, glowing, alive, challenging," a source of spiritual direction. "We were learning how to die to ourselves, to live in Christ, and all the turmoil of the movement, all the pruning of natural love, all the disappointment were explained by the doctrine of the cross in 'the folly of the cross.'"[41]

Those who made the retreat, as I did on one occasion, know that it dwelt on pruning, on the need to "give up" what might have been dearest in our lives, but what would hinder us from union with Jesus. It was undoubtedly unsettling for some, who came to it with the sense of guilt and scrupulosity often associated with Catholic education. For Dorothy it meant strength and a greater than ever dependence on the providence of God.

The year of solitude began in September 1943.

Dorothy's hermitage was in Farmingdale, Long Island. She was given hospitality by the Sisters of St. Dominic in a building formerly used as a school. She attended early Mass with the Sisters and spent time during the day in their chapel. News of destruction in far parts of the world fed headlines, and grief over sacrificed lives, of combatants and noncombatants, filled hearts. The flow of invitations that had been coming to Dorothy to speak on peace to parishes, schools, and seminaries had petered out. The hours of solitude, in which she took her meals alone and sat at her typewriter without interruption, must have been a severe and even painful change from eventful days on the road or at the Catholic Worker.

The June issue of the *Catholic Worker* brought its readers during a summer of war a center-spread with voices of the early church that were muffled when Christians were allowed to join the world's war-makers. Under the title "Patristics and Peace" were printed the famous questions of Pope St. Clement (d. 99):

> Why is there strife and anger and disunion and war among you? Have we not one God, one Christ? Is not one Holy Spirit poured out on us? Have we not one calling in Christ? Why do members of Christ tear one another? Why do we rise up against our own body in such madness? Have we forgotten that we are members of one another?

Other telling peace voices were cited from the Fathers who were clearly disregarded by the Christians engaged in World War II: St. Justin (d. 167), Origen (d. 240), St. Gregory of Nazianzen (d. 390), St. Cyprian (d. 304), St. Clement of

Alexandria (d. 217), St. John Chrysostom (d. 400), St. Basil the Great (d. 379), and Pope St. Leo the Great (d. 461).

In calling up ancestral voices prophesying peace, Dorothy and Arthur Sheehan wanted, through the paper, to remind its readers that the war then crucifying humanity would not have been blessed by the early church, nor would the church have made allowances for its members to take part in it.

Dorothy was able to visit her mother farther out on Long Island, and had time to spend with her daughter, Tamar, a student at the Farmingdale School of Applied Agriculture. As my sister, Mabel Egan, was Tamar's roommate, I visited them at Christmas that year, and spent some time with Dorothy. I found that she was the center of a little community of three, and was a mother to Tamar and Mabel. She relished the evening when they came to her room still in their barn clothes and smelling of the cow barn and pig pen. She would have hot cocoa for them and often fresh baked bread. The young women, Tamar eighteen and Mabel a few years older, seldom cleaned their room adequately. They were up at dawn to do the milking and then went to other duties at the farm and school. They would return from chores, they told me, to find Dorothy had slipped in to set their room to rights. Dorothy was engaged in writing some reflections on Peter Maurin.

Dorothy seemed glad to have company, and we sat a few evenings in her "cell," a long, narrow room with a row of sinks and a stove. It had been used for cooking classes. In the corner was a sleeping cot. On Christmas Eve, we talked about the countless men, women, and children forced to bear the cross of suffering in wars whose causes they did not understand. Dorothy was concerned not only with the anguish of the day but with the unimaginable aftermath, the hopelessness and hunger after the killing had stopped. I reminded Dorothy that many people owed their pacifist convictions to her writings and example. In my case, my leanings toward gospel nonviolence, nurtured by my mother, had been solidified by Dorothy.

It was to the Catholic Worker movement I had come when freed from family obligations. I wanted to throw myself into work that ran counter to the spirit of the time — a time when humanity was hurling itself into the abyss of violence. I did not remind Dorothy how it happened that I did not join the Catholic Worker. The one time I met her was at Mott Street, when I informed her that I was ready to join the Catholic Worker. She gave me a welcoming smile and said, "Good. Come every Wednesday and help with the mail." I did not explain that I had burned all my bridges behind me and was free as air. My obligation to my brothers and sisters, orphaned by the early deaths of our parents, had been taken over by an uncle. It was then, after being turned down, as I saw it, by Dorothy, that a new Catholic agency opened the door to my future life.

The work allowed me to follow Dorothy's advice to take the works of mercy as far as possible "into the midst of war." The agency was the newly founded overseas aid agency of the American Catholic community, called at that time, War Relief Services.

I had just come from my first assignment in a remote camp in Mexico, near the town of Guanajuato. This was referred to in chapter 1. I told Dorothy about the fifteen hundred Polish refugees from Siberia, deported there during the Hitler-Stalin Pact of September 1939 to June 1941. I expected a tidal wave of refugees after the end of the war. We both agreed that only an immense explosion of compassion and practical works of mercy could meet the needs of those victimized and displaced by the works of war.

As twilight fell, we did not turn on the light, but went on talking about war and peace and human anguish until it was time for midnight Mass in the parish church of Farmingdale. I counted it a blessing that I was able to attend Mass on the eve of a wartime Christmas with Dorothy Day. I do not think that either of us were aware that during the year 1943 when humanity was rending itself apart, the Holy See issued an encyclical reemphasizing the teaching of the oneness of humankind in the Mystical Body of Christ.

The message of peace — and as we saw it, Christian pacifism, so small, so unheard, while millions were engaged in the works of war — was still being announced at altars throughout the world. The message of the angels announcing peace rang out in the old Latin Mass in that little parish church:

> Gloria in excelsis Deo,
> et en terra pax hominibus bonae voluntatis.
> Laudamus te.
> Benedicamus te.
> Adoramus te.
> Glorificamus te.
> Gratias agimus tibi propter magnam gloriam tuam.
>
> [Glory be to the Lord on high,
> and on earth peace to men of good will.
> We praise thee.
> We bless thee.
> We adore thee.
> We glorify thee.
> We give thee thanks for thy great glory.]

Dorothy prayed for all human beings, especially those who were suffering and dying. She prayed for all associated with the Catholic Worker, and mentioned by name Joseph Zarrella and Gerry Griffin, then driving ambulances in the "theater" (strange word) of war in the Middle East. Those combatants whose lives contained memories of Christmas and the "Gloria" might have been trying to relive those memories that very hour on the battlefields. For Christians who were able, as we had been, to approach the Table of the Lord on that holy night, would they not wonder how those who shared the one Bread to form one Body could face each other as enemies by the orders of governments?

Within a matter of weeks after taking leave of Dorothy, I was at the border between Spain and France, looking northward toward a continent locked in violence. I was ready to join other voluntary agencies in receiving those fortunate ones who crossed the border — either on foot or, as in the case of the hunted children of the Jews, carried over the Pyrenees by trusted couriers.

Dorothy Day's year of solitude turned out, not surprisingly, to be less than a time of complete solitude and less than a year in length. In July 1944, after six months away from the Catholic Worker, she was back on Mott Street in New York's teeming Lower East Side. A life of even partial solitude was not for her. As she was to say many times, "We hear about desert fathers, but we never hear about desert mothers. Women are meant for community."

On April 19, 1944, Tamar, Dorothy's daughter, was married to David Hennessey in St. Bernard's parish church near Maryfarm in Easton. Mabel Egan was her maid of honor. The wedding breakfast was at the farm, and Peter Maurin was present. Tamar's fear was that having an audience would move Peter to rise to give one of his talks to make his "points." He did indeed rise, but was discouraged by Dorothy from any lengthy indoctrination.

World War II had another year to run, and the paper carried the message of love, of overcoming evil with the weapons of the spirit, to the last day of bloodletting. When peace was declared in Europe in May 1945, the *Catholic Worker* carried in the center column of the front page the Beatitudes of the Sermon on the Mount: "Happy the merciful, Happy the meek, Happy the Peacemakers, children of God." Also on the front page was a headline "PEACE NOW WITH JAPAN." During May, June, and July of 1945, it might have been possible to negotiate a peace with Japan, already on its knees from conventional bombings. Such a peace would have forestalled the unleashing of nuclear terror on humankind in Hiroshima and Nagasaki.

The Catholic Worker — Wounded Movement

The war that inflicted such deep wounds on the human community and on its planetary home left a wound on the Catholic Worker movement. The circulation of the paper, including bulk mailings, was reduced to about fifty thousand, and only ten Catholic Worker houses managed to keep open, some of them feebly.

Yet the movement had not died, and the paper, the only pacifist Catholic paper published anywhere in the world, was still being published from Mott Street in New York's Lower East Side. The fear that the U.S. government would take some action against the movement did not materialize. Dorothy herself, however, was considered so dangerous that in 1941 she was put on a U.S. government list to "consider for custodial detention in the event of a national emergency."

With the end of war's indiscriminate slaughter in Europe came heart-stopping proof of the unspeakably discriminate slaughter of Europe's Jews. Photographs of piled-high corpses, the remains of those not consumed in the ovens of Auschwitz or other annihilation camps, were spread across newspapers. Dorothy's closeness to Jews was part of her life from her college days when her treasured friends were Jewish. The revelation in cold print of what had been done to the Jews shook Dorothy to the core of her being. One day in 1945 she turned to me, her face a mask of suffering, and pondered aloud, "If I had known all this, known it while it was happening, would I have been able to maintain my pacifism?" A while later she added, "But all the violence didn't save the Jews."[42]

Dorothy's tragic awareness of the fragility of life after World War II was expressed many times. She wrote in the September 1945 issue of the paper:

> Wherever we go there is talk of the atomic bomb. All are impressed with the immense death, not only for themselves but for their dear ones.... We can only suggest one thing — destroy the two billion dollars worth of equipment that was built to make the atomic bomb, destroy all the formulas, put on sackcloth and ashes, weep and repent.

New Life

The *Catholic Worker* took on new life after 1945. Dorothy was joined by young men released from the service — Tom Sullivan, formerly of the Chicago house, and Jack English, of the Cleveland house, who had been a prisoner of war in Romania, came to the New York house. The conscientious objectors Gerry Griffin and Robert Ludlow came to take part in the work. Arthur Sheehan and Stanley Vishnewski, both exempt from military service, were part of the New York house. The circulation of the *Catholic Worker* was gaining, and new houses and farms began to sprout across the country.

Among the women, Jane O'Donnell, Helen Adler, Irene Mary Naughton, and Marge Hughes worked with the homeless on the Bowery and took turns in meeting the needs of the Catholic Worker farm, now in Newburgh, New York. The *Catholic Worker* printed appeals that reached it from hospitals, schools, and orphanages in war-afflicted areas. Dorothy herself inserted calls to "Feed the hungry children! We are Herods, we kill children. There is no peace while the world starves."

It soon became clear that the peace following World War II was only a continuation of war by others means, chief of which was the building up of fear against an ally now seen as an enemy.

It did not take long to see the outlines of the Cold War, and Dorothy Day expressed her opposition as she had to the "hot" war. As early as 1947, the *Catholic Worker* asked, "What is our stand on Russia?" and quite predictably the answer was, "The Russians are our neighbors, our brothers in Christ." Dorothy wrote, "To those who call us isolationist, we must remind them that the Good Samaritan did not leave the poor traveler by the road and run after the robbers. He ministered to the wounded, fed and sheltered him."

When the Cold War exploded into a hot war at the other side of the world, Korea, Dorothy expressed her heartbreak at seeing once more the lists of casualties in the *New York Times*. She sensed the fear in the people's hearts that the conflict would become more widespread, but she repeated unequivocally her peacemaking vocation. "Once again, we must reiterate our absolute pacifist position. We believe that not only atomic weapons must be outlawed, but all war, and that the social order must be restored in Christ."[43]

We Live in the Shelter of Each Other

During the 1950s, peace movements were challenged by the growth of nuclear weapons. How to concretize the reality that the two superpowers were putting the human race in jeopardy by their nuclear arsenals and all other preparedness for war? The U.S. government embarked on a policy which gave the peace movements an opportunity to make public protests and to nourish dissent. The policy was enshrined in the Civil Defense Act and consisted of trying to prove to American citizens that they could be protected against nuclear attack through Civil Defense drills. In New York City, specific centers were designated as "shelter areas." Citizens were to rush to them at the sound of the siren, and remain there until the signal ended.

A full-scale dress rehearsal of the Civil Defense organization took place in 1955. In Washington, D.C., "Operation Alert" began at noon on June 15 and ended at 2 P.M. the following day. President Eisenhower and some fifteen thousand key members of the executive branch rushed to secret shelters three hundred miles outside of Washington. In New York City, according to Civil Defense authorities, a hypothetical hydrogen bomb (equivalent to five million tons of TNT) would explode over the city. In the mock thermonuclear attack, from 2:05 to 2:15 P.M., 2,991,280 New Yorkers would be killed, and 1,776,899 injured.[44] Subway stations were considered shelter areas. Children in schools were ordered to remain in their seats while putting their heads between their knees.

Dorothy was a leading figure in protests that revealed the hideous nightmare of the threatened use of nuclear weaponry. Together with a group of pacifists she decided not to take shelter. This would demonstrate the sham of Civil Defense and make a public witness to nonviolence. The protest was conducted according to the concept of satyagraha as described by Gandhi.

The office where I worked was then on the seventy-fifth floor of the Empire State Building. We knew what to do when the siren sounded for the drill: we were to leave our desks and repair to the hallway to an area marked "Shelter Area." We were to remain there until the siren stopped. The office staff obeyed the law, though they realized that the Empire State Building would be a prime target in the event of an attack. In fact, my office, Catholic Relief Services, had known the impact of a plane on July 28, 1945. A B25 bomber, off course in a fog, crashed into the office then located on the seventy-ninth floor. Ten staff members died.

Those who choose acts of satyagraha must announce it in advance, perform it without hatred of the opponent, and be ready to suffer for it. For Gandhi, satyagraha was the argument for willingly accepted suffering. Dorothy was one of the signers of a letter to the office of the mayor of New York City stating, "Such publicized civil defense tests help to create the illusion that the nation can shield people from war's effects. We can have no part in creating such an illusion."[45] Dorothy and the other three signers, Ralph Di Gia for War Resisters League, Kent Larrabee for the Fellowship of Reconciliation, and A. J. Muste for Peacemakers, were to walk across City Hall Park, fronting the mayor's office, and then remain in the open until the siren was stilled. Such disobedience entailed the penalty of a possible year in prison or a fifteen-hundred-dollar fine, or both.

Civil Defense officials were ready for the resisters, who came to number twenty-eight in all. They were crowded into police wagons and taken to jail. Nine hours after their arrest, the twenty-eight pacifists were arraigned in night court. Dorothy was one of the protesters accused by the judge as "murderers" for their part in contributing to the "utter destruction of three million people theoretically killed in our city." All the defendants were released on bail after thirty-five thousand dollars was raised on their behalf by sympathizers.

The trial was delayed until November of that year. This time, the judge, singling out Dorothy, had good words to say about the Catholic Worker's programs for the poor and homeless. "I hoped she hadn't pleaded guilty," he stated. "I would have found a way of acquitting her. . . . I wish you didn't plead guilty. The next time you come before me, if you do — you plead not guilty."

The judge did not understand Dorothy, her reasons for bearing witness, nor her willingness to pay the penalty. She participated in three more protests against Civil Defense drills, calling for jail sentences of five days each and one sentence for thirty days.

In the protest of June 1957, the Civil Defense office was duly notified, in advance, and a letter was delivered to the mayor's office in city hall. The number of protesters had grown, so that only those who sat down in a prominent place in City Hall Park and courted arrest would be taken away in the police wagons.

By the act of sitting immovable in City Hall Park, Dorothy Day was asserting: "We were setting our faces against things as they are. But especially we wanted to act against war, nerve gas, guided missiles, testing and stockpiling of nuclear bombs, conscription, the collection of income tax — against the militarization of the state."[46]

The judge handed down a sentence of thirty days, and Dorothy found herself in the Women's House of Detention in Lower Manhattan. Along with Deane Mary Mowrer, a member of the Catholic Worker community, and Judith Malina, Dorothy plunged into the life of a city prison crowded with over five hundred women. Many, she found, were prostitutes, drug users, and even murderers. Their cells were like cages, with gates composed of five heavy steel bars.

Dorothy was assigned to the prison laundry, a stifling place in the torrid heat of the New York summer. The noise assaulted her whole being: the clamor of steel gates, seventy to a floor, the blaring of the television in the recreation room, the banging of pots, pans, and dishes, and the incessant shouting, even screeching, of the women began to unsettle her nerves. One of the prisoners told Dorothy that as they were treated like animals, they felt they could act like animals. "Animals, however," Dorothy commented sadly, "are not capable of the unmentionable filth that punctuates the conversation of prisoners, so that these prisoners are, in a way, pushed below the animal level. I can only hint at the daily, hourly obscenity that pervades a prison."

Dorothy attempted to transcend the filth without denying its existence. She wrote in her notebook:

> In trying to see Christ in our sisters, and loving them in their suffering, we are not oblivious to their faults. This should not be sentimentalism but true love because primarily, we love them because Jesus loved them. I recognize the fact that outside, stupefied with drugs or ugly with drink, they would be hard to love. They showed us pictures of their children and their faces were alive with love and longing. Afterwards, they lay sorrowing on their beds. But many times, too, they were triggered by some affront or injustice, screaming or flaring into temper or foul language, and their rage was such that others kept silent until their dark mutterings died down like the thunder of a summer storm.... On our detention floor were six women waiting trial for homicide. They had stabbed or shot drunken husbands. One somber woman had hired another man to kill her husband for $100. But there, mingling with them, all the day with gates open and corridors free, we were sisters. We saw in ourselves our own capacity for sin, violence, and hatred.[47]

One of the prisoners noticed that Dorothy read from the same book in the morning and at night. Dorothy explained that she was reading from the Psalms. In the morning, they gave her strength for the day. The woman explained that all she wanted in the morning was her "fix" (of drugs) and asked Dorothy if the Psalms were like her "fix." Dorothy said that one might look at it that way.

Dorothy's cellmate was Judith Malina, a small, attractive actress and founder with her husband, Julian Beck, of the avant-garde Living Theatre. Her presence with Dorothy was a protection against women who preyed on other women. Commenting on those days long afterward, Malina remarked, "Not a month has passed since those days with Dorothy that I have not reflected on my experience with her. I live with her example of compassion for all. I remember her vision that each person is a gift of God. Her extreme optimism encourages me. It reminds me that whatever we do counts, and that sometimes we can move mountains. When my compassion falters, I feel that I have failed Dorothy."[48]

Only Dorothy Day would have described her jail time in terms of the works of mercy. "For years we at the Catholic Worker performed all the works of mercy except visiting the prisoners.... We visited prisoners by becoming prisoners ourselves for five years running, until Civil Defense authorities dropped the compulsory drills."[49]

Each year the number of protesters increased, until in 1959, more than two thousand persons gathered in City Hall Park and refused to take shelter at the sound of the siren. Only a token number were arrested, and the rest of us (I had rushed to the park during my lunch period) protested by our presence at the fringes of the crowd.

By 1961, the Civil Defense drills were a thing of the past.

Somehow it was fitting that Dorothy should take the lead in protesting the false shelter against the works of war by huddling in spots marked "Shelter Areas." She had pioneered in opening real shelters, physical and spiritual, to those almost inconsolably lonely in one of the world's great cities. Her example

Chapter 17

Dorothy Day

Love Is the Measure (1960–80)

It was a dear delight to Dorothy that in the 1960s she was able to travel about the world for peace. The decade was climaxed by an invitation in 1970 to speak in Australia, which then had conscript soldiers fighting in Vietnam.

In 1962, her destination was Cuba, led by a man who had achieved power in 1959 and in 1961 declared himself a Marxist-Leninist. Dorothy went as a journalist and was one of the last journalists allowed into Cuba in that period. Cuba was being presented in the American press as an enemy of the United States. The *Catholic Worker* had aroused controversy by publishing letters from a young man in Cuba praising the Castro revolution and the regime that followed. It was his view that Dorothy, were she in Cuba, would be teaching "socialist morality of generosity and sacrifice." She replied that she would not be teaching in socialist terms but would "certainly try to speak always in terms of the generosity and sacrifice of Jesus Christ, our brother and our God."[1] She went further and said that for the Christian, the issue was not revolution, but

> to grow in love, to rejoice and be happy and thankful even, that we are living in such perilous times and not just benefiting unwittingly by the toil and suffering of others — rejoicing even that there is every sign that we are going to be given a chance to expiate here and now for our sins — and so to help the revolution and convert the revolutionaries. This is the dream worth dreaming, and the only kind of vision powerful enough to stand side-by-side with Marxist-Leninism.[2]

When her visa arrived, Dorothy traveled to Cuba as a sort of peace pilgrim. She described the Cuban people in human terms — their extreme poverty and their hope for change. One Cuban mother spoke to Dorothy about being troubled over sending her children to schools where Marxist-Leninism was being taught. Dorothy replied in part by pointing to cases in the United States where black children were jeered for entering schools that were formerly reserved for white children. She said that Christians must be ready for jeering and contempt and be ready to give a reason for the faith that is in them. With regard to the singing of the "International," as translated for her by a Cuban mother, she said that most of the verses could be joined in with enthusiasm: "Arise poor of the world, on your feet, / slaves without bread — Let us change the face of the earth." However, when it came to the third verse, "No more Supreme Saviors, no Caesar, no bourgeois, no God," Christians could decide to sit down in protest.

Her advice, as contained in her reporting on the month-long visit, was to just "find concordances, as our Holy Father has urged, rather than to seek out heresies, to work as far as one can with the revolution."[3] She commented that what she saw as a regrettable concordance already existed between Marxism and the majority of Catholics, that "both believe that there is nothing nobler for a young man than to bear arms for his country."[4]

Spark of Light, Center of Love, Vivifying Leaven

When Pope John XXIII issued *Pacem in Terris* (*Peace on Earth*) on April 11, 1963, Catholic peacemakers took heart. The encyclical letter, addressed to all humanity, appealed to peace-minded people of diverse traditions. The encyclical surprised even members of the church hierarchy. An archbishop, Lorenz Jaeger, of Paderborn, Germany, asserted that with it the church had reached "the end of the Constantine era," an era marked by alliances between the spiritual and temporal powers. It broke through all narrower goods to reach the universal common good, the good of the whole human family, yet preserving the rights and dignity of the individual human person.

A group of fifty women representing different strands of opinion and different religions, or none, decided to journey to Rome as "mothers for peace" to thank the aged pope in person for his peace message. Dorothy agreed to join them, her fare paid by two fellow pilgrims, Mrs. Hermene Evans and Mrs. Marguerite Tjader-Harris.

For Dorothy, the fact that the letter was issued on April 11, during Easter week, the feast of the commemoration of the Last Supper, was a great joy. Her heartfelt thanks included the realization that the Holy Father had not mentioned the just war, and had opened the door to a concordance with those living under communism. She had hoped that the group would be able to express their gratitude in person in a private audience. Along with Hildegard Goss-Mayr, the secretary of the International Fellowship of Reconciliation, she drafted a letter to Pope John and deposited it with one of his secretaries. There was no reply. The group, among them Yoko Moriki, a young woman whose pitted face was an indelible reminder that she was a survivor of Hiroshima, had prayed for a face-to-face audience. Instead, they were asked to present themselves at a public audience in the Basilica of St. Peter's. The Holy Father, carried around the columns of the basilica on his *seda gestatoria,* was conducted to his throne. He gave his blessing to all present. Then the names of the pilgrim groups from various countries were read out.

"And then the Pope began to speak," Dorothy wrote, "and the words that fell from his lips seemed to be directed to us, to one group, speaking as he did about 'Pilgrims of Peace' who came to him and his gratitude for their encouragement." A young woman translated his words as he spoke, and kept beaming at the group, as did other members of the audience. They were wearing buttons, almost as large as saucers, bearing the legend "Madre per la Pace." The young woman, indicating first the pope, and then the group, wanted to let them know that he was speaking to them especially. "It seemed too good to be true, and if all those around us had not kept assuring us that he was speaking to us, I would have considered it but a coincidence. Our messages had reached him, we felt, impossible though it had seemed they would."[5]

During the return voyage on the Italian liner *Vulcania,* prayers were led at daily Mass for the health of Pope John XXIII. On the day of Dorothy's return, June 3, 1963, she received the news of the death of the beloved pope. His message in *Peace on Earth* was that every believer "must be a spark of life, a center of love, a vivifying leaven amid his fellow men."[6]

With Peacemakers in England

In the fall of 1963, Dorothy was in Europe again, this time invited to speak at a Pax conference in England. The site of the Pax conference was Spode House, the former seat of the Spode China family, near Birmingham. As secretary of the American branch of Pax, I was part of the proceedings.

Besides being the heart of the Catholic Worker, Dorothy became the godmother of Catholic groups that carried peace messages into the wider Catholic community. Pax was one of the groups. The American branch came to birth in 1962 as an affiliate of Pax, England. Almost at the same time, the Catholic Peace Fellowship was founded as a member of the Fellowship of Reconciliation. Pax, though led by those faithful to gospel nonviolence, accepted as members those who believed in the possibility of the just war. The Catholic Peace Fellowship was expressly pacifist. For seven years, the summer conferences of Pax were held at the Catholic Worker farm in Tivoli, New York. Dorothy was the luminous center at these conferences and spoke at each one. As many as two hundred people converged at the farm for the Pax Tivoli conferences on such topics as: "Mercy and Revolution," "Peace beyond Vatican II," "The Duty of Resistance," and "The Wider Identity." Speakers included Archbishop Thomas D. Roberts on "War and Christian Conscience Today," Dr. Seymour Melman on "Disarmament and the War on Poverty," and Dr. Tom Stonier, nuclear physicist, on "Nuclear Disaster." Erik Erikson addressed the topic of "The Wider Identity." Since he could not be present, he sent us his paper, in which he laid out that humankind is being given a chance, possibly its last chance, to become what it is, one species, above all the divisions of nations, tribes, castes, or classes.

When Cesar Chavez of the United Farm Workers was invited to speak at the 1969 Pax Tivoli conference commemorating the centenary of Gandhi's birth, he sent one of his co-workers. Marion Moses spoke for Chavez and the farmworker movement. Moses maintained close contact with Dorothy and the Catholic Worker over the years and addressed the movement on her work as a leader in occupational safety measures for agricultural and industrial workers.

Outdoor Masses marked the conferences as well as the showing of peace films and music supplied by such pianists and composers as Mary Lou Williams. When Pax became the matrix of Pax Christi–USA, Dorothy was both mentor and sponsor.

Dorothy Day was well known in the British Isles, since many young men and women had come from Britain to volunteer in summers or shorter periods at the New York Catholic Worker.

Eager to meet Dorothy, participants came to the Spode House gathering from all parts of England, Wales, Scotland, and Ireland. Dorothy addressed the topic "Fear in Our Time." Her talk was an extensive one, covering the importance of community, the necessity of a certain degree of poverty for peacemakers, and the refusal to participate in Civil Defense drills as part of a policy of psychologizing for war. She began to talk about death in a way I had not heard before. She talked of dispelling the fear of death through faith in the resurrection of Jesus:

> This is what converts expect when they come into the Church. They find it in the lives of the saints who accept the idea of death in whatever form it takes. We say all these things in our prayers and do not mean them. But God takes us at our word, fortunately, and so we are saved in spite of ourselves. We are just dragged in by the hair of our head. This is the message we try to give at the Catholic Worker. It is painful to speak of and that is one of the reasons we rejoice in tribulation. We rejoice in suffering and so we can speak in these terms. It is not morbid. We have been called necrophiliacs. We have been accused of having a sordid and morbid delight in the gutter and so on, the worship of ash cans. The fact of the matter is that God transforms it all, so out of this junk heap comes beauty.[7]

The Anarchist Society of London invited her to talk to the society, moved undoubtedly by the reference to Kropotkin's brand of anarchism in the *Catholic Worker.* An ancient hall, the Dryden Room, was filled with persons who seemed to relish her emphasis on voluntary associations as a way of turning away from dependence on the state. She talked against the intrusive power of the state, in particular, its war-making power. One of the questions addressed to her expressed a wonder that she could harmonize her membership in the Catholic Church with its role in history and her view on anarchism. It was a confrontational query, as though the questioner wanted an argument or a defense of the Catholic Church by the speaker. Dorothy mounted no defense of the church but disarmed the questioner by repeating one of her favorite descriptions of the Catholic Church. It was Romano Guardini's statement that the church was the cross on which Christ was crucified. One needed, she replied, to be in permanent dissatisfaction with the church.

When we passed the American embassy in London, we found that it was surrounded by a ring of protesters who signaled their opposition to American nuclear stockpiling.

We met with Hugh Brock, the editor of the pacifist weekly *Peace News* and a leader in promoting the Easter peace marches, which brought throngs of people to Aldermaston to protest Britain's nuclear arsenal. After their presence at the famous military center at Aldermaston, they returned to gather at Trafalgar Square. In 1961, Brock told us, there were over fifty thousand people in Trafalgar Square. Brock put himself at Dorothy's service for anything she wanted to do in Britain.

"The one thing I want to do in London," she replied, "is to see Muriel Lester again." Lester, the traveling secretary of the Fellowship of Reconciliation, had met with Dorothy several times in New York City. They had made a retreat together at the retreat house of the Anglican Companions of the Holy Cross.

We traveled to Lester's cottage at the edge of the New Forest outside London. Lester, thirteen years Dorothy's senior, had followed a career dramatically paralleling that of Dorothy. Deserting her bourgeois upbringing, she had chosen voluntary poverty and opened a center for the impoverished in London's swarming East End. A devout Anglican, she became, like Dorothy, a "Sermon on the Mount pacifist." This led her to oppose militarism and British imperialism. From this arose her close friendship with Gandhi. When Gandhi came to London in 1931 for the Round Table Conference, he refused to be a guest of King George and decided to stay at Lester's center, Kingsley Hall, for ten weeks. It was a community center similar to a Catholic Worker house of hospitality. Lester questioned Dorothy about her recent jail experience and reminisced about her own arrests and internment. Lester had journeyed around the world nine times in her work of witnessing to gospel nonviolence and works of justice.

On the mantelpiece of the simple living room was a letter framed from Gandhi: "I had written to him telling him of my worries about his health," Lester explained. "This letter was written just before Gandhi undertook one of his longest fasts. He asked me not to be concerned about his health and not even about his survival."

The two women talked of campaigns lost and won, antiwar, anticonscription, antipoverty. At the outbreak of World War I, Muriel Lester stated, "we refused at Kingsley Hall to pronounce a moratorium on the Sermon on the Mount." I found myself marveling at their strength and vision in challenging the tradition of the just war, a respected tradition of both Catholic and Anglican communions.

The two women were uncannily alike in facial contours, each with chiseled features and high cheekbones. Muriel Lester's eyes were a sapphire blue while Dorothy's were a clear gray-blue and slightly slanted. Both were intensely alive and involved in the history of their time, Muriel at seventy-nine years of age and Dorothy close to her sixty-sixth birthday.

Both saw the discipleship of peace for the followers of Jesus as grounded in a love that demanded simplicity of life, closeness to the poor, and willingness to protest against the nation-state even at the cost of imprisonment.

What Might Have Been

On our trips, we talked about what might have been. I asked Dorothy what course her life might have taken if Forster Batterham had agreed to marriage. She had once said that she wanted to be a joyful mother of many children. I was sure that after the ferment of her youth, and her teenage dream of writing books that would wake up the conscience of the world on behalf of the poor, she would not have settled into placid domesticity.

"I would have written many more books," she replied, "many books." She went on to say that they would probably be novels about people caught in the trap of poverty and beaten down by wars and injustice. She would describe what they suffered and what they achieved. "With the Catholic Worker, that was out of the question," she said. Her one regret was that the community life at the Catholic Worker prevented her from doing more writing — yet she did succeed in publishing several books and hundreds of articles.

An omnivorous reader, Dorothy began in her late teenage years an involvement with the Russian novelists that colored her life. Tolstoy, Turgenev, and, above all, Dostoyevsky reflected some of her attitudes toward living, and she would refer to them frequently, quote them, and retell stories from their works in her talks. In her talks she often mentioned the advice of a monk, Father Zossima, in *The Brothers Karamazov* to a woman who could not love unless she was awarded praise and gratitude. "Love in action is a harsh and dreadful thing compared to love in dreams," the monk tells her. "Love in dreams is greedy for immediate action, rapidly performed and in the sight of all." He points out that people give their lives in an ordeal that draws applause as long as it is soon over. "Active love is labor and fortitude, . . . and just when you see with horror that in spite of all your efforts you are getting further from your goal instead of nearer to it — at that very moment you will reach and behold clearly the miraculous power of the Lord who has been all the time loving and mysteriously greeting you."

The faith in the ever-present power of God was what fed Dorothy's constancy in serving the poor and struggling for peace.

The Point Is to Change It

One of Dorothy's books, *Loaves and Fishes,* which had been published earlier in the United States, was brought out in London while we were there. At a gathering organized by Victor Gollancz, the publisher, Vera Brittain came to greet Dorothy. Her account of the blighted lives in Britain in World War I, *Testament of Youth*, spoke to a whole generation and sold millions of copies. The antiwar message of Vera Brittain was dramatized in a series of films produced by England's *Masterpiece Theatre* and shown worldwide. One of the guests at the gathering invited us to lunch with him at the House of Lords. He was Lord Pakenham, who explained to us that he was the first Catholic speaker in the House of Lords since the time of Elizabeth I. While we were waiting for Lord Pakenham to meet us in an anteroom of the House of Lords, we picked up some crested note paper. Dorothy used it to write to a member of the New York community who, Dorothy said, had little faith in himself. She reminded him that he was a brother to Jesus and therefore to the son of the king of heaven.

When we were invited to the wedding of two Pax activists, Angela and Adrian Cunningham, we found ourselves in a simple, rented facility in Highgate Park in company with a large contingent of the London peace community. I remembered from my studies that Karl Marx was buried in Highgate Cemetery. We were told it was within walking distance, and a guest offered to walk with us. We made our way through a gentle mist and soon stood before the enormous carved head of Marx on its ponderous marble base. On the marble slab were fresh red roses. Dorothy was silent and seemed to be in prayer. She told me afterward that she prayed for all those named on the grave plot, particularly for Jenny, Marx's wife, who lacked money to pay for a coffin for a dead child, and for his daughter Eleanor, who committed suicide.

We meditated on the powerful words from the *Theses on Feuerbach* carved into the marble base:

> The philosophers have interpreted the world in various ways.
> The point, however, is to change it.

A thought stimulated by this assertion crossed my mind later. The theologians have interpreted the just war in various ways. The point, however, is to change it to gospel nonviolence.

Fasting in Rome

Dorothy went by ship to Rome in September 1965 for the last session of Vatican II. On board the Italian liner *Raffaello* were thirty-five council-bound American bishops, along with a great many priests. Her custom of daily Mass was never more easily satisfied. She also had time to talk of peace, war, and conscientious objection with bishops, notably Bishop Mark McGrath of Panama. She brought with her three hundred copies of the *Catholic Worker,* the issue on Vatican II that had been already airmailed to every Catholic bishop in the world (as described in chapter 10). In Rome, the Notre Dame Hospitality Center had found her a room. Barbara Wall of English Pax, who with her husband, Bernard, had started a Catholic Worker center years earlier, and Richard Carbray, volunteer secretary of Archbishop Roberts, were ready to shepherd her about Rome.

The penitential procession that opened Vatican II was impressive to Dorothy. The date marked the commemoration of the Feast of the Holy Cross of Jesus. The bishops made their way in solemn procession from the Church of the Holy Cross to St. John Lateran, considered the mother of all Christian churches. One of the litanies she remembered was the "Parce Domine" (Spare us Lord). Among the people she met with were Dr. Joseph Evans and his wife, Hermene, of Chicago, both supporters of the Catholic Worker and other peace causes. They had just come from a visit to the grave of Franz Jagerstatter, in Austria. With them was Gordon Zahn, whose account of the witness and execution of Franz for refusing participation in Hitler's war served as a beacon to war resisters around the world. Archbishop Thomas D. Roberts, to whose jubilee celebration as a Jesuit Dorothy was invited, had placed some of the ashes of Franz's cremation in his bishop's cross.

A Paulist priest, Ellwood Kieser, who later produced a dramatic film on Dorothy's life, tape-recorded an interview with her. She gave a talk at a Monday afternoon gathering at the headquarters of Cardinal J. L. Suenens of Belgium.

Also speaking was Frank Duff, founder of the Legion of Mary, a lay group which had spread from Ireland to many countries of the world. Dorothy told of her encounters with volunteers from the legion at the New York Women's House of Detention.

When I arrived in Rome by plane a week after Dorothy, I found her in a simple room at Via Napoleone III. Her lodgings were with an ordinary Italian family who charged a dollar a night. She felt comfortable there, as it reminded her of the Lower East Side.

The fast of the nineteen women for the peace deliberations of the world's bishops was to begin on October 1. We went over our lobbying aims while Dorothy readied herself for ten days of prayer and fasting. On the eve of the fast, we were the dinner guests of a priest whom I knew from my work with Catholic Relief Services. Dorothy said that she felt guilty eating in a deluxe restaurant. I mentioned that before the fast of Lent, Catholics enjoy *carnivale,* the farewell to meat and celebrations. She settled back for a pleasant evening and later commented that she "certainly felt better for it during those first days of my fast, which were supposed to be the hardest."[8]

The next morning we met at St. Peter's Basilica with Barbara Wall of Pax, England. Barbara's *tessera* (pass) gave us entry to the council Mass with the bishops in the great basilica. Dorothy used the time of waiting to go to confession. Lines of priests, nuns, bishops, and laypeople were waiting on either side of the open confessionals. "How the Council has broken down barriers between clergy and laity," she remarked, "and how close the bishops seem when they are together from all parts of the world, at home in Rome, and not set apart alone and distant on episcopal thrones and in episcopal palaces."[9]

The Mass that morning was in the ancient Syriac rite, the chanting and music filling the basilica with a sound that was to us mysteriously moving. Barbara and I joined Dorothy for the last cup of coffee before the fast. Giving up coffee to start the day would be one of Dorothy's most difficult sacrifices since she never traveled without powdered coffee for her morning cup. Lanza del Vasto, called Shantidas (Servant of Peace) by Gandhi, met us in the garden of the Cenacle convent in a suburb of Rome. Towering and patriarchal, his flowing white beard and hand-spun woolen overgarment setting him apart from the generality of folk, Lanza del Vasto led the group in prayer. Following the Lord's Prayer, the group recited the Peace Prayer attributed to St. Francis. The reading of the Beatitudes by the whole group was the occasion for our parting with the nineteen women who went to their rooms to embark on the ten-day fast.

The fasters followed a schedule, beginning with daily Mass at 7:15 in the morning, followed by short prayer in common as on the first morning. From nine in the morning until noon, they remained in their rooms in silence, praying, reading, and writing. Throughout the twenty-four hours, one of the fasters maintained a vigil of prayer in the chapel. At noon, the women gathered in the garden and read together. One of the books was a work of Martin Luther King. At four in the afternoon, there were lectures, generally by priests, and at six came the doctor to check on the physical condition of the fasters. When I visited the Cenacle, I found that Dorothy was bearing up well, but that Chanterelle, the wife of Lanza del Vasto, became so weak that she was bedridden. Dorothy mentioned in passing that what tempted her sorely was the odor of coffee wafting through the morning air. It was something she could not escape. Dorothy has written of the special pains that she experienced for the first time, pains that seemed to pierce to the marrow of her bones as she lay down at night. These pains, reaching into her very bones, were given to her, she felt, so that she could have a hint of the suffering of the world's hungry. "It was a small offering of sacrifice," she wrote, "a widow's mite."[10]

Visiting Dorothy from time to time were several bishops, among them Archbishop Thomas D. Roberts of Bombay and Abbot Christopher Butler of Downside Abbey in England. They brought her word of developments in the peace discussions in the basilica, as did Barbara Wall and the Companions of the Ark. We regretted that Archbishop Roberts was not called upon to give his intervention orally with its moving example of Franz Jagerstatter. A serious slip-up had occurred, because Roberts had submitted his request with far more than the seventy signatures of fellow bishops as required for a bishop to be given the floor. Only cardinals could take the floor at will.

Burn Yourselves, Not Your Draft Cards

The fast over, Dorothy returned to New York in mid-October 1965 to find that the turmoil of the Vietnam conflict was putting its mark on the Catholic Worker movement. Young volunteers searched for new ways to protest against what Emmanuel Mounier called "the established disorder." A theme of the day was not to trust anyone over thirty, and a few suggested that Dorothy should cede her authority in the movement to young militants. The leadership of the Catholic Worker was ultimately in Dorothy's hands, though volunteers took over after decisions were arrived at by common consent. It was the young volunteers who wanted to "hit the bricks" with protests that would be blazoned in the headlines or would light up television screens. To the young people, a quiet witness like that in Rome, though aimed at moving a worldwide church closer to being a peace church, seemed of little account. The church had less relevance for many, and the tenor of their lives indicated a growing freedom from its restraints. In the world of the 1960s, promiscuity began to be taken lightly even among church peace activists.

When I returned from Rome in November, I discovered that Dorothy was alarmingly drained of energy. She had not been prepared for the hostility against her by some of the younger people in the house. Even more disturbing to her was the hostility to the institutional Catholic Church and its rules of conduct. Her spirits, buoyed up in Rome by the days of prayer and sense of community, simply plummeted.

On November 6, 1965, she stood on a platform in Union Square to express her support for young men protesting the Vietnam War by publicly burning their draft cards. Among the burners was Tom Cornell, a former managing editor of the *Catholic Worker* and co-secretary of the Catholic Peace Fellowship. A law had been passed threatening prison for this act.

"I speak today as one who is old," she told the group, "and who must endorse the courage of the young men who themselves are willing to give up their freedom." Cries of "Moscow Mary" came from one part of the crowd. Counterdemonstrators saw the Vietnam War as a crusade against the spread of communism and opposition to it as somehow traitorous. "Burn yourselves, not your draft cards," they shouted. Sharing the platform with her was Abraham J. Muste, the eighty-year-old patriarch of the peace movement and a leader of the Fellowship of Reconciliation. Dorothy reminded the group of protesters that by their presence and their affirming of the draft card burning, they were also breaking the law and were liable to a penalty.

"I speak today as one who is old" was surprising, coming as it did from a woman usually filled with energy and youthful hope. It also had a special poignancy. Union Square on the Lower East Side was the same location where, on a spring day over three decades earlier, the young Dorothy Day had launched the Catholic Worker movement with the first issue of the *Catholic Worker.* In two days, November 8, 1965, she would enter her sixty-eighth year.

On November 9, just before dawn, Roger La Porte, twenty-two years old, sat in front of the UN headquarters, drenched himself in gasoline, and made a flaming torch of his body. To people who reached him, he said, "I am a Catholic Worker. I am antiwar, all wars." Then he lapsed into a coma. Attempts to save the mortally burned young man were fruitless. The Catholic Worker and Dorothy Day herself were besieged by the press and television newsmen. Dorothy had never met Roger La Porte, who was not a member of the Catholic Worker community but rather a volunteer who came to help at mealtime. She herself came in for bitter attack and questioning, even from friends like Thomas Merton. Exactly twelve hours after La Porte had pierced the dawn with the flames of his body, a total blackout due to a power failure descended on New York City. During the darkness, Dorothy had time to reflect on La Porte's immolation and the furor it aroused.

When I saw Dorothy, I tried to console her, to raise her flagging spirits. She turned to me with a leaden look. "These times are hard," she murmured, as though with an effort. "I have never felt so low, so rejected in all my life."

The question of suicide clouded the witness of La Porte. He lived thirty hours after his immolation. Dorothy was immensely relieved to learn that he had gained consciousness before his death, and had made his confession to a Carmelite priest. The priest related that La Porte said, "I want to live," and had made the most devout act of contrition he ever heard.

Discussions at Vatican II continued through November on the document containing the peace statement. We were informed that the right of conscientious objection to military service was secure. We heard of some problems with the peace statement arising from the objections of American bishops (as recounted in chapter 10), but then came the word that *The Church in the Modern World* had been voted on affirmatively by the bishops of the world. News of the final vote came on the radio early on December 7, 1965. Dorothy's usual resilience reasserted itself. She commented, "The happy news on the radio this morning is that the Vatican Council has passed with an overwhelming majority vote the schema on 'The Church in the Modern World,' included in which is the unequivocal condemnation of nuclear warfare. It is a statement for which we have been working and praying."[11]

Whether the witness of prayer and fasting or the mailing of the *Catholic Worker* to the world's bishops had any effect on the final outcome can never be assessed or known. What is important is that it occurred and took place as a lay action, much of whose inspiration came from Dorothy Day.

Dorothy Day was in Rome again in 1967 to take part in the Congress of the Laity. As usual, her way was paid by a generous friend. She took part in workshops on poverty and peace, and was chosen as one of two Americans to receive the Eucharist from Pope Paul VI in St. Peter's Basilica. Perhaps to indicate that the Vatican was not leaning too much to the pacifist side, the other recipient was a member of the U.S. military, a well-known astronaut.

During our stay in Rome, I took off by air to spend a week in Israel, but could not persuade Dorothy to take a plane. While I was gone, she pondered with regret the fact that her fear had prevented her from visiting the holy places. She had always longed to visit Israel's kibbutzim and moshavim settlements. When it came time to return to New York, with a stopover in England for a Pax meeting, she took to the air for the first time in her life, and never hesitated again to take planes.

Around the World for Peace

Australia had drafted troops battling in Vietnam, and opposition to military involvement had been building up there. Dorothy was invited to be the main speaker at an anti-Vietnam moratorium in Sydney in August 1970. I was invited as a representative of Pax. With the invitation came round-the-world airline tickets. If Dorothy harbored any lingering distaste for flying, she did not express it.

Over the Pacific Ocean, we crossed the international dateline, thus missing the date of August 6, 1970. I rejoiced that we would miss the anniversary of Hiroshima. Dorothy remonstrated, "Woman [she always addressed me as 'Woman' when we disagreed], don't rejoice. We are missing the feast of the Transfiguration of Our Lord."

The Vietnam Moratorium Committee was jubilant at the turnout in Sydney's town hall; every seat was taken, and the balconies overflowed with people, mostly young and vociferous. The supporters of the moratorium filled the streets, carrying banners and lighted candles. They saw the general support as a serious blow to the government's war policy and as a step toward winning Australia's disengagement from Vietnam.

When Dorothy stood up to speak, she was cheered long and loudly. The witness of the seventy-three-year-old woman from half a world away seemed a powerful symbol of war resistance and civil disobedience. She did not confine her speech to war, but told of the ideas of the Catholic Worker movement, of the need for community, and of love expressed in the works of mercy. Feeding and sheltering the poor, as voluntary efforts, were direct, revolutionary actions when they challenged the dehumanized, competitive system that people took for granted. The Catholic Worker, she explained, was against class war and race war and had actually opposed the major wars since its founding — World War II, the Cold War, the Korean War, and the Vietnam War. She talked of the need to support conscientious objectors and absolutist resisters. She was voicing what peace and resistance groups wanted to hear.

When I spoke of my experiences as part of a relief mission to Vietnam in 1955, I mentioned that whenever that war was over, there was going to be an exodus of Vietnamese. I was aware of the deep rift in Vietnamese society and of the diehards on both sides. I asked all present to consider programs that would lead to the maximization of the security of the innocent, in particular help to asylum-seekers. I asked if Australia would be willing to help those who would flee Vietnam. From the balcony arose a shout, "Abolish white Australia." It was taken up by many voices and led to general applause for the concept.

A peace priest, Roger Pryke, had arranged a series of seminars on peace and justice with Dorothy as leader. On that trip, I realized that I had to protect Dorothy from overexertion. She had what doctors called an enlarged heart. The daily schedule of seminars, speeches, and interviews was heavy. People from various religious affiliations and social justice groups had signed up to participate. We arranged it so that all seminars took place in the morning hours. After a rest in the afternoon, we would follow a lighter schedule, conferring with conscientious objectors, with aboriginals, and with groups supporting them. Evening speeches and meetings were on my schedule.

We met with the founders of the Australian Catholic Worker movement, which had been founded in 1935. They had edited a lively *Catholic Worker* magazine for many years. A community, headed by the Reverend John Heffey, expressed the Catholic Worker communitarian ideas in Gaddysdale, near Sydney. Faithful followers of Peter Maurin and Eric Gill, they had designed and hand lettered the posters announcing Dorothy's appearance. The community members took Dorothy to their center while I went to Melbourne to give talks and take part in seminars.

Dorothy's presence encouraged the mounting antiwar sentiment in Australia. Little more than two years later, in December 1972, Australia responded to public sentiment by disengaging itself from the Vietnam conflict and bringing its troops home.[12]

India

The Australian trip brought the special gift of allowing Dorothy her only visit to India. I sent word ahead to Mother Teresa, with whom I had worked through Catholic Relief Services, and there she was, at the Dum Dum Airport of Calcutta, waiting to garland Dorothy with fresh flowers. How alike these two visionaries were. Mother Teresa served the dying of the scourged city, seeing each one "as Jesus in a distressing disguise." Dorothy reminded Christians that Jesus likened salvation to "how we act toward him in his disguise of frail, ordinary humanity." The young novices of the Missionaries of Charity were gathered together to hear Dorothy talk of work among the poor of another metropolis, and of her jailing when she refused to take cover in a mock air bombardment of New York. The Gandhian struggle, in which many women took part, including Mother Teresa's friend Padmaji Naidu, had opened their minds to the idea of imprisonment for conscience. At the end of her talk, Dorothy was the recipient of a gift which was, as far as I know, not given to any other lay woman — a crucifix attached to a safety pin. Mother Teresa affixed it to the blouse over Dorothy's left shoulder, exactly as it was attached to the habit of Mother Teresa herself and of every member of the Missionaries of Charity.

I had also sent word to Delhi to a friend and fellow-Gandhian, Devendra Kumar Gupta, who had headed the Gandhian Centenary Committee. He welcomed Dorothy and told her he had been a faithful reader of the *Catholic Worker* for many years. Devendra saw the many concordances between the ideas of Peter Maurin and of Gandhi. As a young man, he had been a member of Gandhi's ashram at Wardha, but at Vinobha Bhave's suggestion, he had taken his family to live and work in a poor village. Devendra took us to the Gandhi Museum and to the Rajghat, the cremation place of Gandhi. We stood in bright sunlight before a slab of black marble and read the Gandhi talisman:

> Recall the face of the poorest and most helpless man you may have seen and ask yourself if the step you contemplate is going to be of any assistance to him. Will he be able to gain anything by it?

The parting gift made by Devendra to Dorothy was a significant one. Gandhi had seen the spinning wheel as a symbol of India's self-reliance and of the nonviolent movement for *swaraj* (self-rule). The Gandhian movement wanted Dorothy and the Catholic Worker to have the newest, smallest, and most compact spinning wheel yet invented.

La Causa: *The Last Prison Witness*

Cesar Chavez and the United Farm Workers Union had been close to the Catholic Worker movement since the union's beginnings. When the first group of farmworkers came to New York in 1968 to promote the grape boycott, they found warm hospitality at the New York Catholic Worker. It was for *la causa,* the cause of the United Farm Workers, that Dorothy made her last prison witness in 1973. Dorothy and I had taken a plane to San Francisco on July 30 with several purposes in mind, one of which was attendance at the fiftieth anniversary gathering of the War Resisters League, the group which promoted social and international peace from a humanist and Gandhian standpoint. Joan Baez, the folk singer and peace activist of note, had invited Dorothy to spend a week with leaders of the farmworkers at the Institute for the Study of Nonviolence. Joan had founded the institute and had shown her dedication to the cause of peace by refusing to pay taxes and by jail terms for civil disobedience.

Plans changed when a message came from Cesar Chavez that the future of the union was at stake. A strike had been called at the vineyards, and the growers had obtained injunctions against picketing in some areas. We met with Joan at the institute, where she gave a short solo performance in Dorothy's honor, and then we all threw ourselves into the program of *la causa.*

Dorothy saw the United Farm Workers movement as one that pointed to a new social order and was at the same time part of the peace movement. The grape boycott, in which we took part in order to improve the conditions of farm labor, was a method of nonviolent resistance. It took its example from the actions of denuded Irish peasants who had once had nothing with which to bargain (against their landlord, Captain Boycott) but their labor.

"Nonviolence," Cesar Chavez told the workers, "tests our patience, our power to love." He modeled the tactics of the movement on those of the Mahatma Gandhi. His inspiration came from the example of Jesus.

We were ready to picket and were given sleeping space on mattresses at the farmworkers' clinics. We got up at farmworker time, before 3 A.M. and made our way before the light of dawn to places along the road at the edge of the vineyards. Before us, as far as the eye could see, were straight rows of carefully tended vines. Soon, the men and women appeared to perform "stoop labor," the task of grape picking.

At one vineyard, we joined in the call to the workers to come out of the field. "Come out to your brothers and sisters. Come out for *la causa.* Come out for your children," challenged the union organizers through bullhorns. Dorothy and I tried to reach those nearest to us. "Have courage. Stand with the union." I was able to talk to them in Spanish. To our immense joy, one morning fourteen farmworkers, all men, walked out of the vineyard right then and there and stood with us.

The Teamsters Union, contesting with the United Farm Workers for contracts in the fields, was very much in evidence. A powerfully built teamster carrying a long stave rushed up to us. He fixed his eyes on Dorothy and shouted in emotional tones, "Jesus Christ was not nonviolent, Dorothy Day."

Joan Baez and Daniel Ellsberg, who leaked the Pentagon Papers, joined the picket line. Joan often took out her guitar to sing "De Colores," which became almost an anthem of the farmworkers. Daniel Ellsberg told us that one of the persons from whom he drew courage during his two-year struggle with the law was Cesar Chavez. He had also found strength in the witness of Franz Jagerstatter. Ellsberg had courted imprisonment by his act of conscience in making public the Pentagon Papers with secrets about the Vietnam War. Dorothy had brought with her a gift she had received for her birthday on one of our Pax trips to England. It was a folding chair-cane such as English people take to the races. At times, she would sit in the picket line and chat with the police, who stood in imposing lines at the struck vineyards. One day she told the police that on the following day she would read the Sermon on the Mount to them. A photograph of Dorothy Day on her chair-cane, calmly facing two policemen whose burly backs framed her right and left, was widely published. It became a symbol of revolutionary patience.

Dorothy did not return to those policemen or that picket line. Over the loud-speaker came an announcement that Dorothy Day was to meet Cesar Chavez at a certain vineyard. She was whisked away in a pickup truck. Cesar asked her if she would picket alongside him and other farmworkers at a vineyard where the injunction was in force. She was soon arrested with Cesar and was carrying her chair-cane when she was led off to the police van. Thousands were now in jail for resisting what they considered an unconstitutional ban on picketing.

The jail in Fresno was full, so Dorothy, with about a hundred others, was taken to a prison farm located on agricultural land not far from Fresno. There were priests and religious Sisters as well as women farmworkers in the group. Dorothy rejoiced in daily Mass and recitation of the rosary. There were prayer vigils at night for the outcome of the meetings Cesar was holding with the growers and the Teamsters Union. During a weekend while Dorothy was in jail, I flew to Asilomar to fulfill our promise to attend the fiftieth anniversary gathering of the War Resisters League.

When Joan Baez visited the prison farm, she was allowed to give impromptu concerts. Her pure soprano voice reached all the prisoners from an open-air courtyard. One of Dorothy's prized possessions was her green prison uniform. She was allowed to keep it on her release on August 13 because her prison mates had insisted on signing their names to the garment. The afternoon of the release from prison, Dorothy and her prison mates were brought to join a great crowd of farmworkers and supporters gathered in front of the Fresno courthouse. Banners proclaiming "La Causa" and "La Huelga" and depicting the Virgin of Guadalupe were waving everywhere like giant, outlandishly colored flowers. Priests who had come from many parts of the country to make common cause with the United Farm Workers concelebrated an open-air Mass.

Comforting the Afflicted, Afflicting the Comfortable

To celebrate Dorothy's seventy-fifth birthday, Curtis Books brought out three of her books in economy-type paperbacks. *The Long Loneliness, Loaves and Fishes,* and *On Pilgrimage: The Sixties* were soon on sale in drugstores, grocery store chains, and airports. Dorothy was overjoyed, and remarked that this is what Peter Maurin would have wanted — to bring the story to ordinary men and women as they went about their lives. *On Pilgrimage* was a compendium of her "On Pilgrimage" columns in the *Catholic Worker* and included examples of her unforgettable appeals for help with the work. The last line of *On Pilgrimage* was, "Love is the measure by which we will be judged."[13]

The awarding of the Laetare Medal by Notre Dame University cited Dorothy as one who "comforted the afflicted and afflicted the comfortable." The magazine of the Jesuit Order, *America,* published in New York City, devoted an entire issue to Dorothy and the Catholic Worker movement. From Calcutta came a letter from Mother Teresa, "So much love — so much sacrifice for Him alone."

Maryhouse

The largest, the most encompassing, and the last work of mercy in which Dorothy engaged was the opening of a home for homeless women. She wrote about the tragic plight of shelterless women in New York City — before the opening of Maryhouse, the city had only forty places in public shelters for women. For homeless men, there were many services and shelters conducted by the city. In response to her appeal, a monastery in upper New York State sold some of its acreage and sent her the amount of one hundred thousand dollars. Just two blocks from St. Joseph House, where the Catholic Worker ran a soup line and shelter, stood the Third Street Music School. The operators of the school were anxious to move farther uptown in Manhattan. Their building, a combination of three buildings, was for sale. They agreed to sell it to the Catholic Worker for the money at hand — a small sum for so large a property.

To Frank Donovan, as well as Catholic Worker staff, went the task of overseeing the conversion of the property into a home. A sprinkler system had to be installed as well as costly fire doors. A second appeal by Dorothy brought in the needed funds. Then came the hand of bureaucracy from what Dorothy called "Holy Mother the City." The practice rooms on the second, third, and fourth floors were long, narrow, and high-ceilinged. They were decreed as not large enough for human occupancy. The breaking down of walls and enlarging of rooms would call for an enormous outlay of funds and would be time-consuming. The women finding shelter in the cellars of abandoned buildings, under stairwells, and in vestibules would have to wait until the city granted a C.O., a certificate of occupancy. Compassion won out over bureaucracy. The certificate of occupancy was granted so that a group of the poorest, most destitute women of the city could sleep in warm, clean rooms — albeit of an unusual shape. They brought with them the scars of their former lives and found that their foibles were respected and their quirks endured by young people who served them for a motive other than salary.

The home was called Maryhouse, and it was there that Dorothy spent the last years of her life. On the main floor was a chapel where the Sacrament was reserved with permission of the archdiocese of New York. The building was admirably suited to the needs of the Catholic Worker movement, with an auditorium for large meetings, a dining room for small meetings, and several rooms for the production and circulation of the paper.

For many years, Third Street was well known for a graffito as you entered from Second Avenue. On a yellow background were two sentences in black, foot-high lettering:

> "Mr. Gandhi, What do you think of Western civilization?"
> Mr. Gandhi: "I think it would be a good idea."

Loving Disagreement

Dorothy Day at seventy-nine years of age stood before an audience of over eight thousand persons at the International Eucharistic Congress of 1976 in Philadelphia. The congress was timed to be part of the American Catholic celebration of the founding of the United States of America. The theme of the gathering, to which some women, including Mother Teresa, had already addressed themselves, was "Women and the Eucharist." The date was August 6.

Dorothy's tall frame was slightly stooped and her blue-checked cotton dress hung loosely upon it. Her white, braided hair was wrapped in a kerchief of blue cotton. On the Feast of the Transfiguration of the Lord and the thirty-first anniversary of the nuclear bombing of Hiroshima, a Mass was in progress at the nearby Cathedral of Sts. Peter and Paul.

That morning, as Dorothy Day, Mother Teresa, and I made our way together from the convent that had given us hospitality, Dorothy confided to me that she dreaded the talk. She felt in conscience that she had to chide publicly the organizers of the Eucharistic congress. The Mass at the Cathedral of Sts. Peter and Paul, conducted simultaneously with the gathering on "Women and the Eucharist," was a Mass to honor the military. There was no mention of Hiroshima in the congress program. The armed forces — navy, air, and land forces in all their regalia — were marching to the cathedral as we made our way to the meeting hall.

During the preceding days, she had been deeply worried when she ascertained that the bombing of Hiroshima was not to be acknowledged in the program. What was her duty in this dilemma? She had composed her talk with special care. It was the only occasion in all the years that I knew her that Dorothy came to a meeting with a prepared text. On the lectern, she placed the pages, which had been carefully typed by Jane Sammon of the New York Catholic Worker community.

"It is almost easier to stand before a judge and go to jail than to come before you," she said. "I am usually very diffuse," she continued, "but today, because of the seriousness of this day, I wrote out my paper." Before the words of censure, Dorothy Day declared her gratitude to and love of the Catholic Church. It was the church, she read, that taught her that before approaching the Eucharist, there must be reconciliation with one's fellows and penance for wrongs committed. "And here we are on August 6th," she said, as though the words were being wrung from her, "the day the first atomic bomb was dropped, which ended World War II. There had been holocausts before — massacres after the First World War of the Armenians, all but forgotten now, and the holocaust of the Jews, God's chosen people.... It is a fearful thought that unless we do penance, we will perish. Our Creator gave us life, and the Eucharist to sustain our life. But we have given the world instruments of death of inconceivable magnitude. Today we are celebrating — how strange a word — a Mass for the military, the armed forces." When she asked her audience before her, "Why not a Mass for the military on some other day?" there was a burst of applause. "I plead," she asked the participants, "that we regard the military Mass, and all the Masses today, as an act of penance, begging God to forgive us."[14]

An ovation lasting several minutes broke out after Dorothy Day finished her speech. Mother Teresa and all on the platform joined with the audience in the acclamation.

To the end, Dorothy maintained her position of "loving disagreement," to use a Quaker term, with the church to which she was totally faithful. It was somehow fitting that her last appearance would give her the opportunity to express her "loving disagreement" and emphasize the need of all Christians in every generation to "repent." Calling her church to task when she saw its course as a scandal to the gospel, she still saw it as "Christ made visible." "Though she is a harlot at times," she would say to us, "she is our mother,"[15] and all her disagreements, especially on pacifism, arose out of an overpowering and transparent love.

The Worst Has Already Happened

One evening not long after the August 6, 1976, talk, Dorothy confessed to a feeling of weakness. I was at Maryhouse as usual to lead a Friday evening public meeting. Dr. Marion Moses, of the United Farm Workers, was visiting the Catholic Worker. It was a developing heart attack. Dorothy told us she was planning to travel to the Tivoli farm the next morning for the wedding of her god-child, Dorothy Corbin. We both urged Dorothy to stay in her room at Maryhouse and rest. She maintained that she would be feeling better in the morning. "She is my god-child," she told us. "I've got to be at her wedding." That night her condition became worse, and Dr. Moses was called back. They went by ambulance to Beth-Israel Hospital near Maryhouse. When I visited the recuperating Dorothy at Beth-Israel Hospital, she smiled as she repeated the sentence that she and her sister Della used to say in their youthful days on meeting an event over which they had no control: "If I had known what was about to befall me, I would have rued the day." Dorothy's activities became more and more limited. There came a time when Dorothy's failing heart prevented her from coming from her second-floor room to the Friday evening meetings. The last meeting she attended was in the auditorium when Stanley Vishnewski presented a history of the Catholic Worker in slides. Stanley was his customary jocular self, yet there were moments of poignancy when faces appeared of Catholic Workers who had died in the preceding decades.

When a priest came to celebrate Mass in the Maryhouse chapel, Dorothy insisted on coming down until the doctors expressly forbade it. In her last days, she was grateful that Holy Communion could be brought to her from the Sacrament reserved in the chapel.

When she awoke in the morning, she turned to the Psalms of David, from which she drew unending nourishment. They fulfilled for her what they achieved for St. Ambrose. "History instructs," said the saint, "the law teaches, prophecy foretells, correction improves, and morality persuades — the Psalms do all these things and are a medicine for mankind's healing."[16] The missal with the gospel reading for the day and the short breviary were at her bedside, as were the scriptures and the lives of any number of the saints of the church. When she handed me her New Testament, I was struck by the number of passages set apart by a golden-orange marker. She saw Christ always near, in her prayers or in any child of God who came her way. She told me she was not afraid of death but was apprehensive about the moment of dying.

Dorothy's room became her cell of prayer, but a cell open to the Catholic Workers who needed advice on the paper, or who just wanted to have a visit. Some of the guests for whom Maryhouse was founded also knocked on her door. Dorothy's love of beauty never deserted her. During the spring months, her delight was centered on trellises of morning glories, white, pink, and lavender, which climbed from the railing of Maryhouse to her windows. She watched the children playing and was happy at the presence of Korean children who joined the ethnic mix playing on the opposite sidewalk.

While Dorothy looked unblinkingly at the boundless evils of the day, at the wars, the militarization of the state, the violence and injustice springing up on all continents, she did not succumb to pessimism. She often recalled the words of Dame Julian of Norwich, the fifteenth-century mystic who said that the worst had already happened and had already been repaired, referring to the primal sin and the atonement for it by the crucifixion of Christ. "All shall be well," Dorothy would say in the words of Dame Julian. "All, all shall be well."

One of the books read to her as she sat in the corner of her room with an oxygen tank at the back of her chair was *Resurrection,* by Tolstoy. She kept in contact with the world through television news. For dinner, Dorothy liked company, and I joined her frequently to eat in her room. We chatted about happenings in the world, and we pored over maps, tracing the journeys we had taken together. We reminisced about the people who had come to the Catholic Worker farm for the Pax Tivoli conferences: James Douglass, Gordon Zahn, Barbara Reynolds, Marion Moses, Cecil Gill, Tom Stonier, Seymour Melman, and Archbishop Thomas Roberts. It was Roberts who inveighed against the "blank check" given to governments by the church in the prosecution of wars. He warned that it would bounce when presented at heaven's gate. She wanted to know about Pax Christi, of which she was a sponsor. Was it growing? Did it spread gospel nonviolence?

The Thanksgiving holiday of 1980 brought Tamar, her daughter, to Maryhouse. I had spent Thanksgiving with Tamar's friend, my sister Mabel, and Mabel's family in Albany, New York. I returned to the city on the afternoon of November 29, and immediately telephoned Dorothy. She talked to me of the sufferings of the earthquake victims of southern Italy. She had seen on the television screen how the survivors were trying to keep themselves alive in the snows of the mountainous region. Her faint voice became stronger with compassion when she asked about what was being done for them. I explained that medicines, food, and large supplies of blankets were going by air from Catholic Relief Services. She was relieved at the news and said that blankets could be used to make tents for shelter.

Three hours later, at 5:30 P.M., as the old ecclesiastical year was dying, and the vigil Mass for the First Sunday of Advent was beginning at Nativity Church, around the corner from Maryhouse, Dorothy Day passed quietly out of this life. Tamar was with her, and around her were the poor who had been given shelter and the men and women who served them freely in the name of Christ.[17]

The Task and Duty of Peace

Peter Maurin's dream was that Dorothy Day would become a modern Catherine of Siena, inspiring the church to turn around in its focus to establish houses of hospitality in every diocese. This would be only a beginning — further steps would include a return to the land through agronomic universities where workers and scholars would communicate.

Dorothy did turn the church around, but in a way not envisioned by Peter Maurin. She found a church where just war thinking reigned, and she helped dispel the great myth by which Christians can kill other human beings in good conscience. She awakened many consciences to live by an alternate way of responding to violence and evil, the way of gospel nonviolence. The spirituality she espoused and lived was that of the Beatitudes. Lay Christians need the Beatitudes as a compass for the realities of daily living.

St. Catherine's conversion, though she was already in the church, came when she was about twenty years of age. She realized that her personal love of God could not be divorced from service to others, to the neediest and plague-stricken, and to the church itself. Catherine took no vows in a regular religious order, but as a laywoman in the Third Order of St. Dominic became a peacemaker in fact and in her writings. She dictated hundreds of letters, counseling monarchs and public figures in peacemaking and in fidelity to duty. Catherine's travels took her to Avignon, where her strong conviction helped encourage Pope Gregory XI to return with his curia to Rome in 1376. Like Dorothy, she had the charism of attracting like-minded followers around her, laypeople as well as men of the cloth.

Dorothy made her own, and repeated to us, the words of the fourteenth-century saint, Catherine of Siena: "I have left myself in the midst of you, so that what you cannot do for me you can do for those around you."

Dorothy's conversion brought her into the church and into gospel peacemaking when she was thirty years of age. Her discipleship of peace eventually took her to Rome to pray for enlightenment for the world's bishops when they were involved with updating church teaching on war and peace. Her peace discipleship has been treated at length because, as we approach the third millennium, the weapons (actually instruments of genocide) against which she had witnessed are still being stocked by the five leading powers — as well as by other powers that do not admit of their possession and by "rogue" nations that are secretly attempting to obtain their own stockpiles. Robert Ellsberg, who served as a member of the New York Catholic Worker community and as managing editor of the *Catholic Worker,* said it most succinctly: "It was not what Dorothy Day wrote that was extraordinary, nor even what she believed, but the fact that there was absolutely no distinction between what she believed, what she wrote, and the manner in which she lived."[18]

In this time of peril for humanity, disciples of peace in the Catholic Church have a task and duty of immeasurable significance — first, to their fellow members who are wedded to the power of weapons, and then to humanity, to save it from itself. If they take this duty to heart, they must share at least some of the types of witness of Dorothy Day, who prayed and fasted, who chose poverty, who endured vilification for insisting on loving the enemy and praying for persecutors, who journeyed for peace, who used the spoken and written word for Christ's peace, who protested and resisted preparations for war-making, who suffered the humiliation of being bundled into police wagons, who defended those conscientiously refusing war, and who repeatedly accepted jail terms without anger or acrimony, turning them in fact into occasions for practicing the works of mercy.

The woman who accepted joyfully what came to her in her discipleship of peace was always aware that it was grounded from the beginning in self-sacrifice, the separation from the one she loved, the father of her child. This act of self-transcendence served as the soil of her life of witness.

All Is Grace

In her seventy-third year, Dorothy set herself to writing a book to be entitled *All Is Grace.* She made notes for it during the remaining years of her life but never completed it. She saw the stark three-word title as a development from the statement, "All things work together for good to those who love God."

The notes were retrieved in the Catholic Worker archives[19] and formed the basis of a book by her biographer, William D. Miller. A few citations from what Dorothy put into her meditation on grace give an insight into the spirituality that undergirded her work and the entire Catholic Worker movement.

Referring to the fact that nothing in human nature could draw people to the daring peace witness and the painful and sometimes repulsive tasks assumed by the Catholic Worker, she concluded, "It has to be done by grace, which is the gratuitous gift of almighty God. First, we have to appreciate it, then beg for grace." Her fidelity to prayer was her path to the transformation called for by the witness of peace, voluntary poverty, and mercy to the needy. Referring to the houses of hospitality, the farming communes, and the daily practice of the works of mercy of the Catholic Worker movement, Dorothy asserted, "It is only by the grace of God that they have sprung up and prospered."[20]

The undergirding of Dorothy's conviction of gospel nonviolence rested on her belief in the unity of the human family. "Nationalism," she wrote, "has been superseded by the dogma of the Mystical Body, which is as old as Christianity. It is the mystery of Christ in us." To limit the Mystical Body to Christians or to Roman Catholics was, to her, heresy. "The Mystical Body is the inseparable oneness of the human race from Adam to the cave man.... If men and women recognized this, there would be no more war."[21]

"The incarnation is now," she wrote. "There is no true brotherhood of man unless we see Christ in our brother. The word was in scripture and in the Bread of the altar, and I was to live by both. I had to live by both or wither away."[22]

Always aware of God's love for his human creatures and his will for their ultimate union with him in salvation, Dorothy was confident in the grace that is universally offered. "The purpose of prayer," she wrote, "is to ask for grace, to let Christ grow in us. The reason we have so few saints is because we have not asked.... Prayer has the power to take you into sanctity."[23] "The grace tie," she reminded us, "goes on into eternity. The blood tie ends with death."[24]

During her lifetime, Dorothy saw the discipleship of peace recognized by the universal church in the validation of conscientious objection to military service, the recognition of the witness of gospel nonviolence, and the condemnation of indiscriminate warfare in *The Church in the Modern World.* She would have rejoiced at the statement made by the American bishops in their peace pastoral *The Challenge of Peace.* In discussing the value of nonviolence, they stated that "the nonviolent witness of such figures as Dorothy Day and Martin Luther King has had a profound impact upon the life of the church in the United States."[25] By the time of the issuance of *The Challenge of Peace,* in 1983, Dorothy had been three years dead. A strong statement in *The Challenge of Peace* would have been balm to Dorothy's heart.

The bishops wrote:

> To set out on the road to discipleship is to dispose oneself for a share of the cross (cf. John 16:20). To be a Christian, according to the New Testament, is not simply to believe with one's mind, but also to become a doer of the word, a wayfarer with and a witness to Jesus. This means, of course, that one can never expect complete success within history and that one must regard as normal...even the path of persecution and the possibility of martyrdom.[26]

"We have not yet witnessed unto blood" was Dorothy's way of dismissing those who praised her or commiserated with her on her prison terms or other indignities.

Dorothy gives strength to those who, like her, must sometimes "live on blind and naked faith," especially in their discipleship of peacemaking. She said:

> In this dark night, or in this desert, I know as others have known, that God sends intimations of immortality. We receive enough light to keep going, to follow the path—and I believe that if the will is right, God will take us by the hair of the head, as he did Habakkuk, who brought food to Daniel in the lion's den, and will restore us to the Way and no matter what our wandering, we can still say, "All is Grace."[27]

On January 29, 1998, a number of persons associated with Dorothy Day in the Catholic Worker movement were invited by Cardinal John O'Connor, archbishop of New York, to "explore the appropriateness of initiating the cause of Dorothy Day for canonization." A cause for canonization needs to be initiated by the hierarch in charge of the diocese where the person under consideration lived and worked. After meetings for prayer, exploration, and discussion, the cardinal decided to take the first step in initiating the process toward beatification and canonization.

On September 29, 1998, Cardinal O'Connor wrote me:

> I am writing to update you on the latest developments regarding the proposed canonization of Dorothy Day.
>
> I have written to the Congregation for the Causes of Saints asking that the process for her canonization be initiated. Included in my submission are the letters submitted by those who attended our meetings in the Spring. I have received an invitation to meet with the Prefect for the Congregation during my next trip to Rome. I may have more information for you following that visit.
>
> Thank you again for your willingness to assist me in this process. I hope all is well with you.

Peace Be with You

"Peace be with you" was the greeting of Jesus to the apostles at his sudden appearance among them. They were in a locked room in Jerusalem (Luke 24:36; John 20:19), and behind them was the stupendous fact of the resurrection. Was this a ghost? To reassure them, Jesus showed them his pierced side and the marks of the nails on his hands and his feet. On this momentous meeting, he breathed on them the power of the Holy Spirit and commissioned them to preach in his name to all the nations, beginning from Jerusalem.

This is the greeting that priests give from the Table of the Lord throughout the earth. What more beautiful greeting can believers carry outside church walls to all we meet in daily life?

Wishing peace is wishing the highest good that life can offer. The traditional greeting in Hebrew is *shalom* (peace). *Salaam,* the word for peace, is the salutation in Arabic. In Korean, the age-old greeting is also concerned with peace. The peacemaker can wish it to those who are like-minded or not, to all children of God anywhere, everywhere, in season and out.

Whatever the turmoil, whatever the divisions among humankind, whatever the violence, the followers of Jesus who have accepted his commission can refuse to be moved from his transcendent message of peace. May there come a time when the church, as a peace church without any ties to violence, may greet the human family with the words, "Peace be with you."

Notes

Chapter 1: Will You Restore the Kingdom of Israel?

1. Studs Terkel, *The Good War: An Oral History of World War Two* (New York: Pantheon Books, 1984).

2. John L. McKenzie, *The Civilization of Christianity* (Chicago: Thomas More Press, 1986), 150–51.

3. U.S. bishops, *The Challenge of Peace: God's Promise and Our Response,* in *Catholic Social Thought: The Documentary Heritage,* ed. David J. O'Brien and Thomas A. Shannon (Maryknoll, N.Y.: Orbis Books, 1992), 40.

4. First and Second Maccabees form part of the Old Testament in the Catholic Bible but are not included in the Protestant or Hebrew Bibles. There are questions about some interpolations but not about the outline of the events of the Maccabean uprising.

5. The prohibition against unclean foods helped maintain the separation of the people of Israel: "For you are a people sacred to the Lord, your God, who has chosen you from all nations on the face of the earth to be a people particularly his own" (Deut. 14:12). Among the unclean foods forbidden to the children of Israel was the pig. "And the pig, which indeed has hoofs and is cloven footed, but does not chew the end, is therefore unclean for you. Their flesh you shall not eat, and their dead bodies you shall not touch" (Deut. 14:8).

6. Barabbas, meaning "son of the father," was probably a pseudonym utilized in his capacity as a revolutionary.

7. John M. Oesterreicher, *The New Encounter between Christians and Jews* (New York: Philosophical Libraries, 1985).

8. Josephus, *The War of the Jews,* bk. 2, chap. 5, trans. William Whiston, in *Complete Works of Josephus* (Grand Rapids, Mich.: Kregel Publications, 1978), 574.

9. Oesterreicher, *New Encounter,* 88–89.

10. Cited in ibid., 9.

11. Eusebius, *The History of the Church from Christ to Constantine,* trans. G. A. Williamson (New York: Dorset Press, 1965), 111.

12. Josephus, *Complete Works,* 1–2.

13. Josephus, *War of the Jews,* bk. 5, chap. 12, p. 568.

14. Ibid., bk. 6, chap. 4, p. 580.

15. Ibid., bk. 6, chap. 5, p. 581.

16. Ibid.

17. Eusebius, *History of the Church,* 157–58.

18. In *The Exiled and the Redeemed* (Philadelphia: Jewish Publication Society of America, 1957), Itzhak Ben-Zvi, the second president of Israel, recounts the in-gathering of Jewish groups around the world into Israel. The "redeemed" were those who successfully entered Israel from their exile.

Chapter 2: The New Way

1. John L. McKenzie, *The Power and the Wisdom* (Garden City, N.Y.: Doubleday, 1974), 91.
2. Ibid., 9.
3. After the Roman victory in the Jewish War, a young Jewish leader was part of the booty brought back from Jerusalem. At the end of the parade, he was executed in the temple of Jupiter while the victorious officers feasted.
4. See Ignatius's letters in *The Early Christian Writings*, trans. Maxwell Staniforth (New York: Dorset Press, 1986), 78–79.
5. Ibid., 82.
6. Ibid., 112.
7. Ibid., 104.
8. Eusebius, *The History of the Early Church from Christ to Constantine*, trans. G. A. Williamson (New York: Dorset Press, 1965), 227–28.
9. Ibid.
10. Cited in *Early Christian Writings*, 155.
11. Ibid., 158–59.
12. Ibid., 161.
13. Ibid., 162.
14. John Eppstein, *The Catholic Tradition of the Law of Nations* (London: Burns, Oates and Washbourne, 1935), 32.
15. Ibid., 33.
16. Ibid.
17. Maisie Ward, *Saints Who Made History* (New York: Sheed and Ward, 1959), 29.
18. Justin Martyr, *The First Apology for Christians*, in *Short Breviary*, ed. Monks of St. John's (Collegeville, Minn.: St. John's Abbey Press, 1975), chaps. 66–67, p. 339.
19. Eusebius, *History of the Early Church*, 247.
20. Cited in Eppstein, *Catholic Tradition*, 41.
21. Ibid., 42.
22. Ibid.
23. Ibid., 43.
24. Eusebius, *History of the Early Church*, 273.
25. John C. Cadoux, *The Early Christian Attitude to War* (London: Headlay Publishers, 1919), 246.
26. Ibid., 113. The earliest evidence that baptized Christians volunteered for the Roman army is testified to in a document dated between 198 and 202 C.E.
27. See Hugo Rahner, *Church and State in Early Christianity*, trans. Donald Davis, SJ (San Francisco: Ignatius Press, 1992), 24.
28. Alban Butler, *Lives of the Saints*, trans. Herbert Thurston, SJ, and Donald Atwater, 4 vols. (London: Burns and Oates, 1956), 1:570–74.
29. The quotations in the section on Marcellus are all from ibid., 4:220–21.
30. Ibid., 4:310–12.
31. Lactantius Firmianus, *Epitome of the Divine Institutes*, ed. E. H. Blakeney (London: SPCK, 1950), 74.
32. Cited in Cadoux, *Early Christian Attitude*, 83.
33. Ibid.
34. In Eppstein, *Catholic Tradition*, 38, 39.
35. Cited in Stanley Windass, *Christianity versus Violence* (London: Sheed and Ward, 1964), 6–7.

36. Ibid., 7.
37. Cited in *Early Christian Writings*, 227.
38. Ibid., 231.
39. Cited in Cadoux, *Early Christian Attitude*, 52.
40. See David J. O'Brien and Thomas A. Shannon, eds., *Catholic Social Thought: The Documentary Heritage* (Maryknoll, N.Y.: Orbis Books, 1992), 517.
41. St. Cyprian, *Letters*, in *Short Breviary*, ed. Monks of St. John's (Collegeville, Minn.: St. John's Abbey Press, 1975), pp. 1674–75.
42. Butler, *Lives of the Saints*, 3:564.
43. Ibid., vol. 3, p. 567.
44. St. John Chrysostom, *Homilies on the Gospel of St. Matthew* (Grand Rapids, Mich.: Eerdmans, 1986), sermon 32, p. 221.
45. Ibid.
46. Ibid., p. 337.
47. St. John Chrysostom, *Sermons before His Exile*, in *Short Breviary*, nos. 1–3, p. 1704.
48. St. John Chrysostom, *Homilies*, in *Short Breviary*, ed. Monks of St. John's (Collegeville, Minn.: St. John's Abbey Press, 1975), sermon 33, pp. 220–21.
49. Ibid.

Chapter 3: The Way of Justified Warfare

1. Eusebius, *The History of the Church from Christ to Constantine*, trans. G. A. Williamson (New York: Dorset Press, 1965), 356.
2. Ibid.
3. Hugo Rahner, *Church and State in Early Christianity*, trans. Donald Davis, SJ (San Francisco: Ignatius Press, 1992), 43.
4. Jacob Burckhardt, *The Age of Constantine the Great* (Berkeley: University of California Press, 1949), 271–72.
5. Alban Butler, *Lives of the Saints*, trans. Herbert Thurston, SJ, and Donald Atwater (London: Burns and Oates, 1956), 3:687.
6. The complete title of the work of Ambrose was *De Officiis Ministrorum* (On the duties of ministers).
7. Frederick H. Russell, *The Just War in the Middle Ages* (Cambridge: Cambridge University Press, 1979), 4.
8. Marcus Tullius Cicero, *De Officiis* 1.23, cited in John Eppstein, *The Catholic Tradition of the Law of Nations* (London: Burns, Oates and Washbourne, 1935), 61.
9. Ibid., 59.
10. Eusebius, *History of the Church*, 402.
11. Ibid., 403.
12. Burckhardt, *Age of Constantine the Great*, 294.
13. Eusebius, *History of the Church*, 380–82.
14. Ibid., 413–14.
15. Eppstein, *Catholic Tradition*, 49.
16. Ibid., 51.
17. Donald Senior, *Jesus: A Gospel Portrait* (Mahwah, N.J.: Paulist Press, 1992), 155.
18. See Butler, *Lives of the Saints*, 4:512.
19. Ibid., 4:509.
20. Quoted in Eppstein, *Catholic Tradition*, 59.
21. Ibid., 58.

22. Ibid.

23. Louis J. Swift, *The Early Church Fathers on War and Military Service* (Wilmington, Del.: Michael Glazer, 1983), 101.

24. Ibid., 108.

25. Ibid., 109.

26. Ibid.

27. Ibid., 100.

28. Augustine, *The Confessions of St. Augustine,* trans. John K. Ryan (New York: Image Books, Doubleday, 1960), bk. 3, p. 77.

29. Peter Brown, *Augustine of Hippo* (New York: Dorset Press, 1986), 124.

30. Ibid., 138.

31. Pope John Paul II, "Apostolic Letter on Augustine of Hippo," *Origins* 16, no. 16 (October 2, 1986): 283–93; quotations in the following paragraph are all from this source.

32. Cited in Eppstein, *Catholic Tradition,* 74.

33. Ibid.

34. Cited in Swift, *Early Church Fathers,* 130.

35. Ibid.

36. Ibid., 128–29.

37. J. Bryan Hehir, "The Just War Ethic," in *War or Peace? The Search for New Answers,* ed. Thomas A. Shannon (Maryknoll N.Y.: Orbis Books, 1986), 16.

38. Cited in Eppstein, *Catholic Tradition,* 69.

39. Ibid.

40. Ibid.

41. Ibid., 70.

42. Ibid., 71.

43. Ibid., 70.

44. Swift, *Early Church Fathers,* 140.

45. Augustine, *The City of God,* in *Augustine of Hippo: Selected Writings,* ed. Mary T. Clark (New York: Paulist Press, 1984), bk. 19, chap. 11, p. 452.

46. Ibid., bk. 19, chap. 12, p. 455.

47. Swift, *Early Church Fathers,* 116.

48. Augustine, *The Lord's Sermon on the Mount,* trans. John J. Jepson (Westminster, Md.: Newman Press, 1948), 78.

49. William R. Stevenson Jr., *Christian Love and Just War* (Macon, Ga.: Mercer University Press, 1987), 104–8.

50. Swift, *Early Church Fathers,* 148.

51. St. Augustine, *The Lord's Sermon on the Mount,* 75.

52. Eppstein, *Catholic Tradition,* 77.

53. Ibid.

54. Ibid.

55. Brown, *Augustine of Hippo,* 337.

56. The text of the letter is in Eppstein, *Catholic Tradition,* 77–78.

57. Brown, *Augustine of Hippo,* 424.

58. The text of the letter is in Eppstein, *Catholic Tradition,* 79.

59. Paul Peachey, "Church and Nation in Western History," in *Biblical Realism Confronts the Nation,* ed. Paul Peachey (Scottsdale, Pa.: Fellowship Publications, Herald Press, 1963), 25.

60. Eppstein, *Catholic Tradition,* 67.

Chapter 4: Is It God's Will?

1. Peter Brown, *Augustine of Hippo* (New York: Dorset Press, 1986), 428. Brown relates that toward the end of his life, Augustine gathered his writings in the cupboards that served as bookcases. There were ninety-three of his works, made up of 232 little books, as well as letters and anthologies of his sermons taken down by stenographers. A modern publisher planning to print the complete works of Augustine has already published ten volumes, each over 350 pages. The final total is expected to reach about forty volumes, each of similar length.

2. Before this date excerpts from Augustine's texts on war had been gathered by St. Anselm of Lucca, in 1083, and by Ivo of Chartres, in 1094, but neither compendium gained wide currency.

3. Frederick H. Russell, *The Just War in the Middle Ages* (Cambridge: Cambridge University Press, 1979), 56.

4. Ibid., 85.

5. Cited in John Eppstein, *The Catholic Tradition of the Law of Nations* (London: Burns, Oates and Washbourne, 1935), 81.

6. Ibid., 81–82.

7. Russell, *Just War*, 56.

8. Roland H. Bainton, *Christian Attitudes toward War and Peace* (New York: Abingdon Press, 1960), 98.

9. In many cases popes faced enemy invaders with offers of peace. Notable was that of Pope St. Leo, who faced Attila, called the Hun, and Genseric, king of the Vandals.

10. During the dark period of local wars, a countermovement of peace grew up through the work of the monastic orders. Benedict of Nurcia (d. 547) founded the monastery of Monte Cassino. Monks put down roots in western Europe, bringing the civilizing example of Pax et Labora. One monastery, that of Cluny, had close to a thousand priories, or branches. A book detailing the monasteries' contributions to civilization through their work and the copying and preservation of manuscripts has been written by Thomas Cahill, *How the Irish Saved Civilization* (New York: Doubleday, 1995).

11. Many efforts were made by the church to mitigate the violence and cruelty of warfare, including the Truce of God and the Peace of God, both limiting the periods when war could be waged. The Second Lateran Council of 1139 banned the cross-bow and the siege machine, considered the ultimate weapons of the time, with their power to pierce or crush armor.

12. Eppstein, *Catholic Tradition*, 82.

13. Cited in Henry Treece, *The Crusades* (London: Bodley Head, 1962), 97–98. A less organized Crusade, not at all the one hoped for by Pope Urban, had sprung into action before the formal Crusade was ready to set forth. French and German peasantry, spurred on by the inflammatory speeches of Peter the Hermit of Amiens, Walter the Penniless, also from France, and Gottschalk, a monk of Germany, took to the roads, bound for Constantinople and Jerusalem. In such cities as Cologne and Speyer, they ran amok, finding "infidels" in their midst, the Jews. Bishops tried to save the threatened Jews; the archbishop of Cologne opened his palace to the Jews; the bishop of Speyer took the Jews under his protection. Disaster followed disaster for the fanatic peasantry. At Constantinople, Emperor Alexius ordered them to continue to Nicomedia, across the Bosporus Strait, where they were besieged by Turks. Those who did not perish from hunger or thirst were slaughtered or sold into slavery. Thousands of skeletons of the hapless peasants littered the landscape around Nicomedia. A victory monument erected by the Turks consisted

of a mountain of bones. None of these unfortunate people ever caught a glimpse of holy Jerusalem.

14. See Albert Marrin, ed., *War and the Christian Conscience* (Chicago: Henry Regnery, 1971).

15. Count Raymond saw to it that the emir and his family were protected and lodged safely in Askalon.

16. Majid Khadduri, *War and Peace in the Law of Islam* (Baltimore: Johns Hopkins University Press, 1955), 56.

17. Ibid., 56–57 and passim.

18. Ibid., 18.

19. Bainton, *Christian Attitudes,* 118.

20. George B. Flahiff, CSB, *Ralph Niger: An Introduction to His Life and Works,* Medieval Studies 2 (Toronto: Pontifical Institute of Medieval Studies, 1940), 104–25.

21. George B. Flahiff, CSB, ed., *Deus Non Vult: A Critic of the Third Crusade,* Medieval Studies 3 (Toronto: Pontifical Institute of Medieval Studies, 1941), 162–88; the material that follows on Niger is based on this work, and all citations are from it.

22. Two church councils of 1188 dealing with the Third Crusade forbade women from accompanying the Crusaders, except for laundresses. These were to be women whose character was above reproach.

23. Flahiff, *Deus Non Vult,* 178–79.

24. St. Bonaventure, *The Life of St. Francis of Assisi* (Rockford, Ill.: Tan Books, 1988), 14 and passim.

25. *The Words of St. Francis,* comp. and ed. James Meyer (Chicago: Franciscan Herald Press, 1952), 243.

26. Ibid., 320.

27. Ibid., 274.

28. Ibid.

29. Ibid., 169, 172.

30. *From Gospel to Life: The Rule of the Secular Franciscan Order* (Chicago: Franciscan Herald Press, 1979), 22, 24.

Chapter 5: St. Thomas and the Angels

1. St. Thomas Aquinas, "Treatise on the Angels," in *Summa Theologica,* 5 vols., trans. English Dominican Province (reprint; Westminster, Md.: Christian Classics, 1981), qq. 50–64, pp. 259–324.

2. Ibid., q. 40, aa. 1–4.

3. Ibid., q. 40.

4. Ibid.

5. Ibid.

6. Ibid.

7. Ibid.

8. Ibid.

9. Ibid.

10. Ibid.

11. Ibid.

12. Ibid.

13. Joan D. Tooke, *Just War in Aquinas and Gratian* (London: William Cloves and Sons, 1965), 170.

14. Aquinas, *Summa Theologica,* q. 64, aa. 2–4.

15. Ibid., q. 64, a. 5.

16. A. G. Sertillanges, in comment accompanying *Summa Theologica.*

17. Albert Marrin, *War and the Christian Conscience* (Chicago: Henry Regnery, 1971), 80, 81, 82.

18. Frederick H. Russell, *The Just War in the Middle Ages* (Cambridge: Cambridge University Press, 1979), 179.

19. Ibid., 297.

20. See John Eppstein, *The Catholic Tradition of the Law of Nations* (London: Burns, Oates and Washbourne, 1935), 106.

21. See ibid., 99–107; subsequent references to this work are from this source.

22. See ibid., 110.

23. Erasmus, *In Praise of Folly,* trans. Betty Radice, with introduction by A. H. T. Levy (New York: Penguin Books, 1971), 3.

24. Roland H. Bainton, *Erasmus of Christendom* (New York: Charles Scribner's Sons, 1969), 106–8.

25. Though Erasmus was so highly regarded that he was considered for a cardinalate, an eminence he refused, his works had periods of rejection. Less than a quarter of a century after his death, Pope Paul IV put his writings on the Index. The Council of Trent, however, allowed expurgated editions.

26. Erasmus, letter to Abbot Anthony a Bergis, in *A Peace Reader,* ed. Joseph Fahey and Richard Armstrong (Mahwah, N.J.: Paulist Press, 1992), 181.

27. Ibid.

28. Erasmus, *In Praise of Folly,* 111.

29. Ibid., 102.

30. Erasmus, *The Education of a Christian Prince,* trans. Lester K. Born (New York: Columbia University Press, 1936), 257 and passim; subsequent references to this work are from this edition.

31. See John P. Dolan, *The Essential Erasmus* (New York: Meridian Books, New American Library, 1964), 177 and passim.

32. Ibid., 191.

33. Ibid., 192.

34. Ibid., 202–3.

35. Ibid.

36. James McConica, *Erasmus* (New York: Oxford University Press, 1991), 83.

37. Ibid., 99.

Chapter 6: Just War Challenged

1. Frederick H. Russell, *The Just War in the Middle Ages* (Cambridge: Cambridge University Press, 1979), 308.

2. John Howard Yoder, *When War Is Unjust,* 2d ed. (Maryknoll, N.Y.: Orbis Books, 1996).

3. Bertolt Brecht, *Mother Courage,* trans. Eric Bentley (New York: Grove Press, 1963), 14, 15.

4. Roland H. Bainton, *The Reformation of the Sixteenth Century* (Boston: Beacon Press, 1985), 152.

5. Yoder, *When War Is Unjust,* 21.

6. Russell, *Just War in the Middle Ages,* 3–4.

7. Stanley Windass, *Christianity versus Violence* (London: Sheed and Ward, 1964), 75.

8. "President Bush Defends War as Just," *Origins* 20, no. 35 (February 7, 1991): 571.

9. Ronald J. Musto, *The Catholic Peace Tradition* (Maryknoll, N.Y.: Orbis Books, 1986), 131. An adage cited from M. M. Philips, *Erasmus and the Northern Renaissance* (London: English Universities Press, 1967).

10. See Peter Ackroyd, *William Blake* (New York: Alfred A. Knopf, 1996), 277.

11. J. Bryan Hehir, "The Just War Ethic," in *War or Peace? The Search for New Answers,* ed. Thomas A. Shannon (Maryknoll, N.Y.: Orbis Books, 1986), 15–39.

12. Richard J. Regan, *Just War: Principles and Cases* (Washington, D.C.: Catholic University of America Press, 1996), 7.

13. "Pacifism," s.v., *New Catholic Encyclopedia.*

14. *The Challenge of Peace: God's Promise and Our Response,* in *Catholic Social Thought: The Documentary Heritage,* ed. David J. O'Brien and Thomas A. Shannon (Maryknoll, N.Y.: Orbis Books, 1992), par. 111, p. 517.

15. Ibid., par. 333, p. 562.

16. Ibid., par. 117, p. 518.

17. Ibid., par. 120, p. 518.

18. Ibid., par. 25, p. 497.

19. Ibid., par. 81, p. 511.

20. Ibid., par. 3, p. 492.

21. Introduction to ibid., p. 490.

22. Ibid., par. 173, p. 530.

23. Ibid.

24. Ibid.

25. Ibid., par. 284, p. 553.

26. Ibid., par. 54, p. 503.

27. Ibid., par. 48, p. 502.

28. Ibid., par. 276, pp. 551–52.

29. Ibid., par. 331, p. 562.

30. Ibid., par. 138, p. 522.

Chapter 7: The Works of War

1. *The Day after Trinity: J. Robert Oppenheimer and the Atomic Bomb,* text of documentary (Santa Monica, Calif.: Pyramid Films, 1981), 1.

2. Rachelle Linner, *City of Silence: Listening to Hiroshima* (Maryknoll, N.Y.: Orbis Books, 1995), 13.

3. Victor Frankl, *Man's Search for Meaning: An Introduction to Logotherapy* (Boston: Beacon Press, 1992), 154.

4. "Refusing to Learn to Love the Bomb: Nations Take Their Case to Court," *New York Times,* January 14, 1996, 7.

5. John C. Ford, SJ, *The Morality of Obliteration Bombing* (Woodstock, Md.: Theological Studies, 1944), 267–309.

6. Patricia McNeal, *Harder Than War* (New Brunswick, N.J.: Rutgers University Press, 1992), 67–69.

7. *Day after Trinity,* 16.

8. Ibid., 17.

9. Ibid.

10. Ibid., 29.

11. Ibid., 24.

12. Ibid., 21.
13. Ibid., 24.
14. Ibid.
15. Ibid., 22.
16. Ibid., 23.
17. Ibid., 26.
18. Robert O'Connell, *Of Arms and Men: A History of War, Weapons and Aggression* (New York: Oxford University Press, 1989), 95.
19. See Lynn Montross, *War through the Ages* (New York: Harper Brothers Publishers, 1946); Stanley A. Coblentz, *From Arrow to Bomb* (New York: Beechhurst Press, 1953).
20. Dorothy Day, *Catholic Worker* (New York issue, September 1945), reprinted in Robert Ellsberg, ed., *Dorothy Day: Selected Writings* (Maryknoll, N.Y.: Orbis Books, 1983), 266–67.
21. Thomas Stonier, *Nuclear Disaster* (Cleveland: Meridian Books, 1963), 43 and passim.
22. As cited in Eileen Egan, ed., *The War That Is Forbidden: Peace beyond Vatican II* (New York: Pax Publications, 1968), 48.
23. Roger Ruston, OP, and Angela West, *Preparing for Armageddon* (London: Pax Christi, St. Francis of Assisi Center, 1985), 16.
24. Pope Pius XII, *Justitia et Pax* (Rome: Pontifical Commission, 1982), 174.
25. Francis Stratmann, *War and Christianity Today*, cited in Donald Keys, *God and the H Bomb* (New York: MacFadden Books, 1962).
26. John C. Ford, SJ, "The Hydrogen Bombing of Cities," in *Morality and Modern Warfare,* ed. William Nagle (Baltimore: Helicon Press, 1960), 98.
27. Ibid., 103.
28. Pope John XXIII, *Pacem in Terris,* in *Catholic Social Thought: The Documentary Heritage,* ed. David J. O'Brien and Thomas A. Shannon (Maryknoll, N.Y.: Orbis Books, 1992), 131.
29. Pope Paul VI, *Address to Pilgrims* (New York: Pax Publications, 1965).
30. Robert Heyer, ed., *Nuclear Disarmament: Key Statements of Popes, Bishops, Councils and Churches* (Ramsey, N.J.: Paulist Press, 1982), 26.
31. Ibid., 40.
32. *The Catholic C.O.: The Right to Refuse to Kill* (Erie, Pa.: Pax Christi Publications, 1972), 41.
33. Pax Christi press release, October 8, 1997; in David Robinson, ed., *Catholic Peace Voice* (Erie, Pa. 16502-1343) (fall 1998).
34. Ibid.
35. Ibid.
36. Graham T. Allison et al., *Avoiding Nuclear Anarchy: Containing the Threat of Loose Russian Nuclear Weapons and Missile Material* (Cambridge, Mass.: MIT Press, 1996), passim.
37. Jim Wallis, ed., *Christian Voices from the New Abolitionist Movement* (New York: Harper and Row, 1983), 5.
38. Ibid., 75.
39. Conversation with the author.
40. Wallis, *Christian Voices,* 5.
41. Ibid., 40.
42. Ibid., 41.
43. Ibid., 105–6, 107.
44. Ibid., 30, 31.

45. See Greg Ruggiero and Stuart Sahulka, eds., *Critical Mass: Voices for a Nuclear-Free Future* (Westfield, N.J.: Open Media, 1996), xiv, xv, xxi, 56–76.

46. David Gracie, *Report on Abolition 2000 Conference: Polynesia,* January 21–27, 1997.

47. *Act Now for Nuclear Abolition,* statement of Pax Christi International and World Council of Churches (Geneva, March 1998).

48. Ibid.

49. See Pax Christi bishops in the United States, *The Morality of Deterrence: An Evaluation by the Pax Christi Bishops in the United States* (Erie, Pa.: Pax Christi Publications, 1998); the citations from the bishops in the following paragraphs are from this source.

50. Jonathan Schell, *The Gift of Time: The Case for Abolishing Nuclear Weapons* (New York: Henry Holt, 1998); the citations in this and the following paragraph are from pp. 207–8.

51. Emmanuel Charles McCarthy, *The Nonviolent Trinity: Trinity-Site New Mexico, July 16, 1995* (Baxter, Minn.: Center for Christian Nonviolence, 1995), 3.

Chapter 8: The Works of Mercy and the Works of War

1. War is seen as punishment for sin in one biblical episode. After Nathan said to David, "You are the man" (i.e., the man guilty of taking the wife of Uriah and causing Uriah's death), an evil is brought on David's house. "Now, therefore, the sword shall never depart from your house" (2 Sam. 12:10).

2. Richard A. Falk, Gabriel Kolko, and Robert Lifton, eds., *Crimes of War* (New York: Random House, 1971), 342.

3. This and other details regarding the Persian Gulf War are based on a Greenpeace publication on the conflict: William M. Arkin, Damian Durrant, and Marianne Cherni, *On Impact: Modern Warfare and the Environment: A Case Study of the Gulf War* (London: Greenpeace, 1991), 79.

4. Ibid., 56.

5. Ibid., 148.

6. Ibid., 147–48.

7. Ibid., 149.

8. Ibid., 75.

9. Clergy and Laymen Concerned about Vietnam, *In the Name of America* (Annandale, Va.: Turnpike Press, 1968), 284.

10. General Colin Powell, "On Impact," interview in *USA Today,* March 11, 1991, 29.

11. Ibid., 92.

12. Report for the UN Food and Agriculture Organization by researchers from the Harvard School of Public Health; see Barbara Grosette, "Iraq Sanctions Kill Children, UN Reports," *New York Times,* November 30, 1995.

13. Ibid.

14. The petition was distributed by Peace Media Service, Alkmaar, Holland, June 1994. The originator of the petition was an Icelandic composer, Elias Davidson, who included other examples of economic sanctions in asking his government to end any participation in measures of collective punishment.

15. "At the dawn of the air age in 1909, an Italian, Giulio Douhet, stated that the 'sky is about to become another battlefield.' He pointed out that 'any distinction between belligerents and non-belligerents is no longer admissible today either in fact or

theory'" (Lynn Montross, *War through the Ages* [New York: Harper Brothers, 1946], 691, 776–77).

16. Peter Johnson, *The Withered Garland* (London: New European Publications, 1995); the quotations in the following paragraphs are all from this text.

17. William Shakespeare, *Antony and Cleopatra,* 4.3.

18. Michael S. Teitelbaum, "Right versus Right: Immigration and Refugee Policy in the United States," *Foreign Affairs* (fall 1980).

19. A courageous statement was issued in Switzerland on November 26, 1992, by Monsignor Winko Puljic, archbishop and metropolitan of Sarajevo, Patriarch Pavle of the Serbian Orthodox Church, and Al-Haji Ja'kub efendi Setimoski Rais Ulema, of the Islamic communities. "We have met in common faith in God, the Creator and Benefactor of all people and all nations.... We unanimously and in total unison launch this appeal for peace, this cry to God and to men, this cry of suffering and hope from the bottom of our souls.... We do not beg or implore, but in God's name and justice, in the name of humanity and survival of everyone, we demand the immediate, unconditional and irrevocable end of the war, the reestablishment of peace and the renewal of dialogue, as the only method of solving existing national and political problems.... We address ourselves to all international organizations, to the whole international community, to all countries and all people of goodwill with the request to use all their influence and all morally justifiable means in order to make further appeals like this one unnecessary" ("Round Tables of Reconciliation," in *The Contribution by Religious to the Culture of Peace,* ed. Ivanka Vann Jakic [Nyack, N.Y.: Fellowship of Reconciliation, 1993]).

20. Eileen Egan, *Such a Vision of the Street: The Spirit and Work of Mother Teresa* (New York: Image Books, 1986), 57.

21. Kenneth Hackett, *Report to Catholic Relief Services: Overseas Aid and Development Agency* (Baltimore, Md., June 1995).

22. Peter Hebblethwaite, "In Rwanda Blood Is Thicker Than Water," *National Catholic Reporter,* June 3, 1994, 11.

23. Egan, *Such a Vision.*

Chapter 9: Toward a Theology of Peace

1. John L. McKenzie, SJ, "Gospel of Matthew," in *Dictionary of the Bible* (New York: Macmillan, 1965), 766.

2. A monk of the Eastern church, *Invocation of the Name of Jesus* (London: St. Basil's House, 1960), 19.

3. See "The Beatitudes," in *Encyclopedic Dictionary of the Bible* (New York: McGraw Hill, 1963), 215–17.

4. Domingo Lain, *Latin American Documentation* (LADOC 23 [1975]), based on "He Followed Christ by Taking Up a Rifle," *Diálogo Social* (Panamanian Monthly) (June 1974).

5. Gordon Zahn, "The Bondage of Liberation," in *Vocation of Peace* (Baltimore: Fortkamp, 1992), 81, 82.

6. Gustavo Gutiérrez, *A Theology of Liberation* (Maryknoll, N.Y.: Orbis Books, 1973), 307.

7. John L. McKenzie, *How Relevant Is the Bible?* (Chicago: Thomas More Press, 1981), 15.

8. McKenzie, "Gospel of Matthew," 766.

9. Alfred Bour, preface to Jean Goss and Hildegard Goss-Mayr, *The Gospel and the Struggle for Peace,* trans. Dave Parry (Alkmaar, Holland: International Fellowship of Reconciliation, 1990), iv.

10. Adolfo Pérez Esquivel, introduction to *Christ in a Poncho,* trans. Robert R. Barr (Maryknoll, N.Y.: Orbis Books 1985), 1–19.

11. Ibid., 137.

12. Ibid., 11.

13. Jon Sobrino, *The Principle of Mercy* (Maryknoll, N.Y.: Orbis Books, 1994), 1–11. This and the following citations are from this source.

14. McKenzie, "Gospel of Matthew," 556.

15. Walter Wink, *Jesus' Third Way: Violence and Nonviolence in South Africa* (Philadelphia: New Society Publishers and Fellowship of Reconciliation, 1987). See also Walter Wink, *Engaging the Powers* (Minneapolis: Fortress Press, 1992), 175–93.

16. Donald Senior, *Jesus: A Gospel Portrait* (New York: Paulist Press, 1992), 93.

17. Bishop Procopius of Gaza, *Commentary on the Proverbs,* in *Short Breviary,* ed. Monks of St. John's (Collegeville, Minn.: St. John's Abbey Press, 1975), 189.

18. St. John Chrysostom, *Homilies on the Gospel of St. Matthew: Nicean and Post-Nicean Fathers* (Grand Rapids, Mich.: Eerdmans, 1986), 495.

19. Cyril of Alexandria, *Commentary on John,* in *Short Breviary,* ed. Monks of St. John's (Collegeville, Minn.: St. John's Abbey Press, 1975), bk. 11, pp. 380–81.

20. The Real Presence is described as transubstantiation in the Catholic tradition; the very substance of the bread and wine is changed and becomes the body and blood of the glorified Lord. The consecrated wine becomes "the blood of the new and everlasting covenant." While the Godhead is present everywhere, there is a uniqueness in the way Christ becomes present in the Supper of the Lord. Other faith groups have various approaches, including consubstantiation. The bread and wine are already the body and blood of our Lord but remain bread and wine. The being or essence of the gifts has not been changed. The receptionist view is a spiritual understanding of presence in which Christ becomes present to the individual as he or she eats and drinks the blessed bread and wine. For some, the Supper of the Lord is a memorial without any special presence of Christ in connection with the service.

21. St. Gregory of Nyssa, *The Perfect Christian,* in *Short Breviary,* ed. Monks of St. John's (Collegeville, Minn.: St. John's Abbey Press, 1975), 554.

Chapter 10: Crime against God and Man

1. Vatican I, the twentieth ecumenical council in the church's history, was convened by Pope Pius IX in 1869. It is remembered chiefly for the intense debate on the nature and extent of papal infallibility. A definition was arrived at in July 1870. Disputes arose on other matters. After the invasion of the Papal States by Italy, Pope Pius IX suspended the council indefinitely, and it was never reconvened.

2. See Eileen Egan, ed., *The War That Is Forbidden: Peace beyond Vatican II* (New York: Pax Publications, 1968), 80.

3. See ibid., 66.

4. It should be noted that the observer-delegates from the Orthodox Church paid special respect to the patriarch by rising and removing their clerical headdress when he spoke. They explained that one stands bareheaded when a patriarch speaks, just as one does at the readings of the word of God. Thomas E. Bird, in writing of the patriarch, pointed out that His Beatitude Maximos IV Saygh had much in common with the Orthodox Church, since as a member of the Melkite community, he led his followers in the celebration of the liturgy in the Byzantine rite. "The rite of Byzantium," Bird points out, "encompasses nearly 8,200,000 Catholics; numerically, it is second only to the Roman

rite. Moreover, the Byzantine rite is the rite of more than 200,000,000 Christians who belong to the Orthodox Church."

5. Egan, ed., *War That Is Forbidden,* 67.

6. Ibid.

7. Pope John XXIII, *Pacem in Terris,* in *Catholic Social Thought: The Documentary Heritage,* ed. David J. O'Brien and Thomas A. Shannon (Maryknoll, N.Y.: Orbis Books, 1992), 131–62.

8. E. M. Borgese, "Pacem in Terris," *Nation* (April 1963).

9. In Patricia McNeal, *Harder Than War* (New Brunswick, N.J.: Rutgers University Press, 1992), 97.

10. The donor was Mrs. Valerie Delacorte, whose husband allowed Pax members to use a part of his publishing office in Manhattan to address the envelopes to the world's bishops. Mrs. Delacorte, a native of Hungary, knew war at firsthand, having been in Budapest when the Soviet army took the city.

11. *Catholic Worker* (New York issue, July 1965).

12. McNeal, *Harder Than War,* 96.

13. Cited in James Finn, *Protest, Pacifism, and Politics* (New York: Vintage Books, 1968), 63.

14. Egan, ed., *War That Is Forbidden,* 83.

15. Ibid., 72–73.

16. Ibid., 82.

17. Ibid., 70–71.

18. Ibid., 77.

19. Xavier Rynne, *Vatican Council II* (New York: Farrar, Straus and Giroux, 1968), 566.

20. See O'Brien and Shannon, *Catholic Social Thought,* 222.

Chapter 11: Violence Forestalled: The Philippines and the Iron Curtain

1. Niall O'Brien, *Island of Tears, Island of Hope* (Maryknoll, N.Y.: Orbis Books, 1993), 116–17. According to Niall O'Brien, there had been official efforts to block the pastoral letter, at least to delay its publication, until after the inauguration of Marcos.

2. Jean Goss and Hildegard Goss-Mayr, *The Gospel and the Struggle for Peace,* trans. Dave Parry (Alkmaar, Holland: International Fellowship of Reconciliation, 1990), 4.

3. Ibid., 2.

4. Ibid., 5.

5. Ibid., 4.

6. Ibid., 7.

7. Conversation of the author with Richard Deats. Akkapka, Active Nonviolence, published a leaflet with its credo:

Credo

We are a people of God.
We believe in justice, democracy and peace.
But most of all in the absolute value of the human being.
And in the solidarity of all peoples.

We are opposed to all forms of injustice and oppression
now prevalent in our society:
Any authoritarian form of government

The discrimination against the poor
The gross violation of human rights
The foreign domination over our economic,
political and cultural system.

We espouse a society that fosters
Equality,
Protects the rights,
And holds sacred the dignity of every person.

We commit ourselves to the construction
and preservation of a just Filipino society.

But in all our deeds, we vow:
Never to kill
Never to hurt
To convert our oppressors to the truth
And to remain united in our struggle.

And that this Credo may become
A way of life,
We humbly call on God
To favor us with his help.
We ask each sister and brother
To tell us when we fail to be true
To this Credo.

The credo reflected the pacifism of the Fellowship of Reconciliation.

The Fellowship of Reconciliation took its beginnings from the pledge of two Christians, an Englishman and a German, to refuse to bow to the enmity that was descending on their nations at the outbreak of World War I. The growth of the organization between the two world wars responded to the revulsion of Christians against the bloody face of twentieth-century war, its aftermath of poverty and disillusion, and the hunger for a reassertion of the message of Jesus as above all claims of nations. Fellowships were founded in various countries, including England, Germany, Austria, and the United States. Kaspar Mayr, a resident of Vienna, became one of the founders of the International Fellowship of Reconciliation, binding all Fellowships of Reconciliation in the common ground of pacifism in the spirit of Jesus. In addition, fourteen Peace Fellowships were founded to serve particular groups, including the Episcopal Peace Fellowship, the Brethren Peace Fellowship, the Catholic Peace Fellowship, the Presbyterian Peace Fellowship, and the fellowship associated with Buddhist, Jewish, and Muslim groups.

8. Niall O'Brien, *Revolution from the Heart* (Maryknoll, N.Y.: Orbis Books, 1990).

9. See n. 1, above.

10. Morina Hillarey Mercado, in "Preface and Scenario," *People Power: An Eyewitness History of the Philippine Revolution of 1986,* ed. Francisco S. Tarad (Manila: James B. Reuter, SJ, Foundation, 1987), 77–78. This and other citations on the events in the Philippines come from this source.

11. Niels Nielsen, *Revolutions in Eastern Europe* (Maryknoll, N.Y.: Orbis Books, 1991), 7.

12. David Cartright, *Report on Nonviolent Mass Action at Albert Einstein Institution, Boston, Mass., February 8–10, 1990* (newsletter of the Institute for International Peace Studies, Notre Dame University), 11.

13. Ibid.

14. Lech Walesa, *A Way of Hope: An Autobiography* (New York: Henry Holt and Company, 1987).

15. Nielsen, *Revolutions in Eastern Europe.*

16. Cited in John Feffer, *Shock Waves: Eastern Europe after the Revolution* (Montreal: Black Rose Books, 1998), 179.

17. Nielsen quotes one Orthodox monitoring group that found that seventeen metropolitans and sixty-some priests died in prison. In all some four thousand priests of all faith groups were legally dead, as far as public witness was concerned. They were in Romania's prisons. See Nielsen, *Revolutions in Eastern Europe,* 110.

18. Mary Evelyn Jegen, SND, carried these words on a poster for seventeen days in 1990 during the Christmas season when there was hope that President Bush could be deflected from declaring war on Iraq. She walked with the poster in front of the White House in Washington despite rain and snowfall.

19. Pope John Paul II, "Address to the General Assembly of the United Nations," fiftieth anniversary session, October 5, 1995 (text supplied by the Observer Missions to the United Nations).

Chapter 12: Prophets of Nonviolence

1. M. K. Gandhi, *The Science of Satyagraha,* ed. A. Hingorani (Bombay: Bharatiya Vidya Bhavan, 1970), 1.

2. Ibid., 2.

3. M. K. Gandhi, *What Jesus Means to Me* (Ahmedabad: Navajivan Publishers, 1959), 12.

4. Ahimsa, noninjury to all living things, has deep roots in Hinduism.

5. Gandhi, *What Jesus Means,* 16–17.

6. See S. K. George, *Gandhi's Challenge to Christianity* (Ahmedabad: Navajivan Publishers, 1960), 93.

7. Gandhi, *Science of Satyagraha,* 40.

8. Krishnalal Shridharani, *War without Violence* (Bombay: Bharatiya Vidya Bhavan, 1962), 151–52.

9. Ibid., 10.

10. Joan V. Bondurant, *Conquest of Violence: The Gandhian Philosophy of Conflict* (Princeton, N.J.: Princeton University Press, 1958), 46–52.

11. M. K. Gandhi, *Caste Must Go and the Sin of Untouchability* (Ahmedabad: Navajivan Publishers, 1964), 65.

12. John J. Ansbro, *Martin Luther King, Jr.: The Making of a Mind* (Maryknoll, N.Y.: Orbis Books, 1986), vi.

13. Patricia McNeal, *Harder Than War* (New Brunswick, N.J.: Rutgers University Press, 1992), 149.

14. Cited in Fred A. Wilcox, *Uncommon Martyrs* (New York: Addison-Wesley, 1991), 35.

15. Daniel Berrigan, SJ, *Whereon to Stand: The Acts of the Apostles and Ourselves* (Baltimore: Fortkamp, 1991), 286.

16. Ibid.

17. McNeal, *Harder Than War,* 210.

18. *The Seven Storey Mountain* owes its title to Dante's *Purgatorio,* which describes the seven heights that lead one out of purgatory.

19. Michael Mott, *The Seven Mountains of Thomas Merton* (Boston: Houghton Mifflin, 1984), 169.

20. Thomas Merton, *Cold War Letters* (Abbey of Gethsemani, Ky., 1972, private printing), no. 11.

21. Cited in William H. Shannon, *The Hidden Ground of Love* (New York: Harcourt Brace, 1993), 136–37.

22. *Catholic Worker* (October 1961).

23. Thomas Merton, "The Prison Meditations of Fr. Delp," mimeographed version; reprinted in *Passion for Peace: The Social Essays of Thomas Merton* (New York: Crossroad, 1995), 148.

24. Letter from Thomas Merton to author, June 10, 1962.

25. Ibid.

26. Letter from Thomas Merton to author, July 16, 1968.

27. Thomas Merton, "Peace and Revolution: A Footnote from *Ulysses*," *Peace* (1968): 5; reprinted in Gordon C. Zahn, ed., *The Nonviolent Alternative* (New York: Farrar, Straus and Giroux, 1980), 35.

28. Ibid.

29. Gordon Zahn, *Vocation of Peace* (Baltimore: Fortkamp, 1992), 142.

30. Merton, "Peace and Revolution," 35.

Chapter 13: The Right to Refuse to Kill

1. William Shakespeare, *Henry V*, 4.1.

2. As cited in Richard D. Challoner, *French Theory of the Nation in Arms, 1866–1939* (New York: Columbia University Press, 1995), 3.

3. See Lynn Montross, *War through the Ages* (New York: Harper Brothers, 1946), 452.

4. Ibid., 4.

5. From the eleventh edition of the *Encyclopaedia Britannica* (1910), as cited by E. Raymond Wilson, *Background Information on Military Conscription* (Washington, D.C.: Friends Committee on National Legislation, July 1966).

6. Cited in John Eppstein, *The Catholic Tradition of the Law of Nations* (London: Burns, Oates and Washbourne, 1935), 213.

7. Ibid., 216.

8. Roland H. Bainton, *The Reformation of the Sixteenth Century* (Boston: Beacon Press, 1985), 152. Among the thirty-nine articles of the Church of England was one which stated (in 1571) that it was lawful for Christian men to carry weapons and serve in wars. In the Augsburg Confession (1530), the Lutheran Church taught that Christians might take part in governance, act as soldiers, and engage in just war. The Augsburg Confession condemned the Anabaptists, who are forbidden to take part either in governance or in war. The Westminster Confession of Faith of the Presbyterians (1647) asserted that it is licit for Christians to take part in war in just and necessary situations.

9. Bainton, *Reformation*, 152.

10. David A. Martin, *Pacifism: An Historical and Sociological Study* (New York: Schocken Books, 1966), 43.

11. Bainton, *Reformation*, 143–44. The theocracy led by Ulrich Zwingli in Zurich was part of the revolutionary wing of the Anabaptists, which, in common with the times, believed in executing dissenters. Zwingli had been a Catholic priest who early in his life had been influenced by Erasmus (in particular by his Greek New Testament).

12. *The Great Chronicle of the Hutterian Brethren* (Farmington, Pa.: Plough, 1995), 3.

13. Bainton, *Reformation*, 104.

14. Basil Penington, introduction to Eberhard Arnold and Thomas Merton, *Why We Live in Community* (Farmington, Pa.: Plough, 1995), 3.

15. See Bainton, *Reformation,* 107.

16. See Peter Brock, *Pacifism in the United States from the Colonial Era to the First World War* (Princeton, N.J.: Princeton University Press, 1968), 260. In general, the Mennonite conscientious objectors belonging to larger communities were subjected to monetary fines. The more isolated groups of Mennonites found that their unwillingness to join in the war inflamed their neighbors, who despoiled them of all their possessions, including provisions, furniture, and animals.

17. Dale W. Brown, *Biblical Pacifism: A Peace Church Perspective* (Elgin, Ill.: Brethren Press, 1986). I am indebted to Dale W. Brown for information on the Church of the Brethren.

18. Cited in Brock, *Pacifism,* 263.

19. See Brown, *Biblical Pacifism,* 163–64.

20. See ibid., 18–19.

21. Cited in Leonard Kenworthy, *The American Friends Peace Testimony* (Dublin, Ind.: Prinit Press, 1975), 5.

22. Cited in Brown, *Biblical Pacifism,* 185.

23. Ibid.

24. Ibid., 45.

25. National Advisory Committee on Selective Service, statement, Washington, D.C., 1966.

26. Thomas Merton, letter to the author, 1966.

27. Don Edwards, "Pax Rights of Conscience," *Commonweal* (June 21, 1968): 408–9.

28. U.S. Catholic Bishops, *Human Rights in Our Day,* in *Catholic Social Thought: The Documentary Heritage,* ed. David J. O'Brien and Thomas A. Shannon (Maryknoll, N.Y.: Orbis Books, 1992).

29. *New York Times,* November 21, 1968.

30. Pax statement, UN Commission on Human Rights, March 17, 1970. The statement was a UN document listed as E/CH.4/NGO 153.

31. Devi Prasad and Tony Smythe, eds., *Conscription: A World Survey* (London: War Resisters International, 1968), 131–37.

32. U.S. Catholic bishops, *Resolution on South East Asia* (statement dated November 19, 1971), in *Peace Magazine: Catholic Conscience and the Draft* (New York) (1972): 24.

33. Ibid., 25.

34. Gordon Zahn, *In Solitary Witness: The Life and Death of Franz Jagerstatter* (Springfield, Ill.: Templegate Publishers, 1964), 233.

Chapter 14: Peace of Christ

1. Cited in Francis Stratmann, OP, *War and Christianity Today,* trans. John Doebele (London: Blackfriars, 1956), 35, 36.

2. Brian Wicker, ed., *Studying War No More? From Just War to Just Peace* (Grand Rapids, Mich.: Eerdmans, 1994).

3. See William H. Shannon, "Christian Conscience and Modern Warfare," in *Studying War No More?* 142 and passim.

4. See Wicker, *Studying War,* 151.

5. Pax Christi (Metro N.Y.), *The Way of the Cross: The Way to Peace* (New York: Pax Christi, 1998).

6. "Just Peace," *Journal of Pax Christi* 1 (London) (March 1997).

Chapter 15: The Lord's Prayer and the Reign of God

1. The words "For yours is the kingdom, the power and the glory" came from the early liturgy. They appear in the *Didache* (The Teaching), used by the early church between 100 and 150 C.E.

2. The tapes and publications of Father Emmanuel Charles McCarthy are available from the Center for Christian Nonviolence, 293 Kenwood Court, Baxter, MN 56425. Included are the following six works: *Who Is Your King? Who Is Your God? The Nonviolent Eucharist; August Ninth; First Epistle to the Twentieth Century;* and *Stations of the Nonviolent Cross.*

3. Father Emmanuel Charles McCarthy, "'I Was Told It Was Necessary': Interview with Father George Zabelka," *Sojourners* (August 1980): 12–15; subsequent quotations in this section are from this text.

4. Cardinal Jean-Marie Lustiger, *The Lord's Prayer* (Huntington, Ind., 1988), 100.

5. St. Cyprian of Carthage, *The Lord's Prayer* (Westminster, Md.: Christian Classics, 1983), 52, 53.

6. St. Augustine, "Sermon 329," in *Short Breviary,* ed. Monks of St. John's (Collegeville, Minn.: St. John's Abbey Press, 1975), 1723.

7. Lustiger, *Lord's Prayer,* 106.

8. Ibid., 117.

9. Bogside refers to the swamp and the bogs where the local Catholic population was driven to make place for the settlers introduced under British rule. In areas ripe for conflict, history lies heavy and has monuments and relics that are a constant provocation to wreak vengeance on the present.

10. The text of the "Declaration of the Peace People":

> We have a simple message to the world from this movement for peace. We want to live and love and build a just and peaceful society.
>
> We want for our children as we want for ourselves, our lives at home, at work, at play, to be lives of joy and peace.
>
> We recognize that to build such a life demands of us all, dedication, hard work and courage.
>
> We recognize that there are many problems in our society which are a source of conflict and violence.
>
> We recognize that every bullet fired and every exploding bomb makes that work more difficult.
>
> We reject the use of the bomb, the bullet and all the techniques of violence.
>
> We dedicate ourselves to working with our neighbours, near and far, day in and day out, to building that peaceful society in which the tragedies we have known are a bad memory and a continuing warning.

(As published in *Citizen,* a publication of the Peace People: Belfast, Northern Ireland, 1996).

11. Lawrence Martin Jenco, OSA, *Bound to Forgive: The Pilgrimage to Reconciliation of a Beirut Hostage* (Notre Dame, Ind.: Ave Maria Press, 1995).

12. Ibid., 73.

13. Ibid., 14.

14. Ibid.

15. Celine Morgan, OP, "The Lord's Prayer (Our Father)," in *The Modern Catholic Encyclopedia* (Collegeville, Minn.: Liturgical Press, 1996), 527.

Chapter 16: Dorothy Day: Christ's Way or the World's Way (1927–59)

1. See Robert Ellsberg, ed., *Dorothy Day: Selected Writings* (Maryknoll, N.Y.: Orbis Books, 1983), 261.
2. See Paul Peachey, "Church and Nation in Western History," in *Biblical Realism Confronts the Nation,* ed. Paul Peachey (Scottsdale, Pa.: Fellowship Publications, Herald Press, 1963).
3. Ellsberg, *Dorothy Day,* 262.
4. Ibid., 263.
5. From a notebook of Day's cited in William D. Miller, *All Is Grace* (Garden City, N.Y.: Doubleday, 1987), 172.
6. Dorothy Day, *The Long Loneliness* (New York: Harper and Row, 1981), 34.
7. See Ellsberg, *Dorothy Day,* 13.
8. William D. Miller, *Dorothy Day: A Biography* (San Francisco: Harper and Row, 1982), 84.
9. Ellsberg, *Dorothy Day,* 84.
10. Day, *Long Loneliness,* 87.
11. Ellsberg, *Dorothy Day,* 9.
12. Ibid., 33.
13. Dorothy Day, *On Pilgrimage: The Sixties* (New York: Curtis Books, 1972), 383.
14. Ellsberg, *Dorothy Day,* 39.
15. Day, *Long Loneliness,* 166.
16. Dorothy Day, introduction to Peter Maurin, *Easy Essays* (Chicago: Franciscan Herald Press, 1977), n.p.
17. Ellsberg, *Dorothy Day,* 42.
18. See Peter Maurin, "Dorothy Day, Postscript to," in *Easy Essays* (Fresno, Calif.: Academy Guild Press, 1961).
19. The preamble to the constitution of the Industrial Workers of the World stated: "By organizing industrially we are forming the structure of a new society within the shell of the old." The preamble also called for the "abolition of the wage system." See Staughton Lynd and Alice Lynd, *Nonviolence in America* (Maryknoll, N.Y.: Orbis Books, 1995).
20. Leo XIII, *The Condition of Labor,* in *Catholic Social Thought: The Documentary Heritage,* ed. David J. O'Brien and Thomas A. Shannon (Maryknoll, N.Y.: Orbis Books, 1992), 15.
21. Pius XI, *After Forty Years,* in *Catholic Social Thought,* 52.
22. Emmanuel Mounier, *Personalism* (Notre Dame, Ind.: University of Notre Dame Press, 1952), xiii.
23. Maurin, *Easy Essays* (1977), 3. Peter, in accord with earlier teachings of the church, rejected usury, the breeding of money by money.
24. Ellsberg, *Dorothy Day,* 51.
25. Conversation with author.
26. Mel Piehl, *Breaking Bread: The Catholic Worker and the Origin of Catholic Radicalism in America* (Philadelphia: Temple University Press, 1982), 198.
27. Ronald Musto, *The Catholic Peace Tradition* (Maryknoll, N.Y.: Orbis Books, 1967), 9.

28. Dorothy Day, *Catholic Worker* (May 1936): 8.
29. Ellsberg, *Dorothy Day,* 78.
30. Ibid.
31. Dorothy Day, "Aims and Purposes," *Catholic Worker* (January 1939): 7.
32. U.S. Congress, hearings of the House Committee on Military Affairs (July–August 1940).
33. Conversation with author.
34. The account of one Catholic conscientious objector of World War I is told in Torin R. T. Finney, *Unsung Hero of the Great War: The Life and Witness of Ben Salmon* (New York: Paulist Press, 1989). Salmon, a devout Catholic, based his objection on the command not to kill and on the plea of St. Paul "to overcome evil with good." Arrested for refusing induction, he refused any work that would aid the military. For this stance of civil disobedience he was placed in solitary confinement and for a time in a mental hospital. When he was still in prison one year after World War I, Salmon, of Irish origin, took the lead of Terence McSwiney, lord mayor of Cork, and began a hunger strike. This called attention to his plight and that of other conscientious objectors. On November 24, 1920, the day before Thanksgiving, the conscientious objectors were released from prison. Salmon directed his attacks at the validity of the just war theory and became a pioneer in Catholic pacifism in the United States.
35. Letter from John Cogley to Dorothy Day, November 1941, archives CW (Marquette University).
36. Patricia McNeal, *Harder Than War* (New Brunswick, N.J.: Rutgers University Press, 1992), 55. McNeal cites figures supplied by the National Service Board for Religious Objectors (NSBRO).
37. Gordon Zahn, *Another Part of the War: The Camp Simon Story* (Amherst: University of Massachusetts Press, 1979).
38. Arthur Sheehan, ed., *The Catholic CO* (October 1942).
39. Ibid.
40. Day, *On Pilgrimage,* 254.
41. Miller, *All Is Grace,* 48.
42. Conversation with the author.
43. Dorothy Day, "We Go on Record," *Catholic Worker* (September 1945): 1.
44. Arthur Brown, "War Resisters League," in *What Happened on June 15?* (pamphlet) (New York: Provisional Defense Committee, 1955), 1.
45. Ibid., 2.
46. Dorothy Day, *Loaves and Fishes* (New York: Harper and Row, 1981), 161.
47. Miller, *All Is Grace,* 171–72.
48. Conversation of Judith Malina with author.
49. Day, *Loaves and Fishes,* 160.
50. Berrigan, foreword to Day, *Long Loneliness,* xxiii.

Chapter 17: Dorothy Day: Love Is the Measure (1960–80)

1. See Robert Ellsberg, ed., *Dorothy Day: Selected Writings* (Maryknoll, N.Y.: Orbis Books, 1983), 308.
2. Ibid.
3. Ibid.
4. Ibid., 165.
5. Ibid.

6. Pope John XXIII, *Peace on Earth,* in *Catholic Social Thought: The Documentary Heritage,* ed. David J. O'Brien and Thomas A. Shannon (Maryknoll, N.Y.: Orbis Books, 1992), 158.

7. *Fear in Our Time* (Catholic Worker Archives, Marquette University, Milwaukee, Wisconsin).

8. Dorothy Day, *On Pilgrimage: The Sixties* (New York: Curtis Books, 1972), 250.

9. Ibid.

10. Ibid., 252.

11. Ibid., 253.

12. As of 1996, there were two Catholic Worker houses in Australia. One publishes a newspaper entitled *Mutual Aid.*

13. Day, *On Pilgrimage,* 383; idem, "Bread for the Hungry," *Catholic Worker* (September 1976): 1, 5.

14. Ellsberg, *Dorothy Day,* 339.

15. Day, *On Pilgrimage,* 383.

16. St. Ambrose, *Commentary on the Psalms,* in *Short Breviary,* ed. Monks of St. John's (Collegeville, Minn.: St. John's Abbey Press, 1975), 431.

17. The funeral Mass for Dorothy was held at Nativity Church on Second Avenue near Second Street on New York's Lower East Side. As many people were crowded on the sidewalk outside the church as were able to find places in the church. To the wake at Maryhouse came a stream of people from every part of the nation and from Canada. They were of many religions. Some came to pray and some only to stand at the bier of the poor woman lying in an unpainted pine coffin. Round her hair was the blue cotton scarf she had worn in life and on her body her usual cheap cotton dress.

At the public memorial Mass held two months later in St. Patrick's Cathedral, Cardinal Terence Cooke concelebrated with twenty-five priests. The cardinal pointed out in his homily that Dorothy Day had been called upon to make "a voluntary and complete surrender of human life to divine love." The drama which followed Dorothy's life did not desert her in death. Facing the cardinal in the first pew was the "human love" in person, Forster Batterham, eighty-six, alert, straight-backed, wearing thick-lensed glasses. The child born to him and Dorothy, Tamar Teresa, sat next to him with three of her nine children. When I presented Tamar to Cardinal Cooke after the Mass, the cardinal told us that he objected to some statements in the homily belittling the process of formal beatification of holy people. "The saintly person's example must be enshrined for future centuries, for people living three hundred and five hundred years from now," the cardinal said. Formal declaration of a person as a saint, he explained, makes it more likely that that person's example and inspiration will be cherished by future generations.

I had a thought after the Mass, irreverent perhaps. Could not the paramour of St. Augustine have outlived him, since he died at seventy-six? Might she not, though she was cast aside as a human love, have attended a memorial service for him in Hippo?

18. Ellsberg, *Dorothy Day,* xv. Dorothy Day went to Rome during Vatican II to fast for the bishops as they addressed themselves to the peace of the world. While no woman was permitted to speak to the bishops at the council, an astonishing mission was assigned to women at the end of the council, on December 8, 1965: "Women of the entire universe, whether Christian or nonbelieving, you to whom life is entrusted at this grave moment in history, it is for you to save the peace of the world." The message to "women" was one of seven special messages issued by Pope Paul VI at the council's close: other messages went to "workers," "youth," "artists," "rulers," "men of thought and science," and "the poor, sick and suffering."

"The hour is coming," asserted the message to women, "in fact has come, when the vocation of woman is being achieved in its fullness, the hour in which woman acquires in the world an influence, an effect, and a power never hitherto achieved. . . . Our technology runs the risk of becoming inhuman. Reconcile men with life and above all, we beseech you, watch carefully over the future of our race. Hold back the hand of man, who, in a moment of folly, might attempt to destroy human civilization" (see Walter Abbott, SJ, ed., *The Documents of Vatican II* [New York: Herder and Herder, 1966], 732–33).

Dorothy Day was a woman who in a grave time embraced as her mission the saving of the peace of the world. Her leadership for peace had a transforming effect on the lives of others who were moved, like her, to stay the hand of violence in a nuclear-fragile world.

19. Extensive archives of the Catholic Worker movement have been lodged in Marquette University libraries (Wisconsin Avenue; PO Box 314; Milwaukee, WI 53201).

20. William D. Miller, *All Is Grace* (Garden City, N.Y.: Doubleday, 1987), 103.

21. Ibid., 146–47.

22. Ibid., 63.

23. Ibid., 109–10.

24. Ibid., 92.

25. U.S. Catholic bishops, *The Challenge of Peace: God's Promise and Our Response,* in O'Brien and Shannon, *Catholic Social Thought,* par. 276, p. 531.

26. Ibid.

27. Miller, *All Is Grace,* 77.

Index

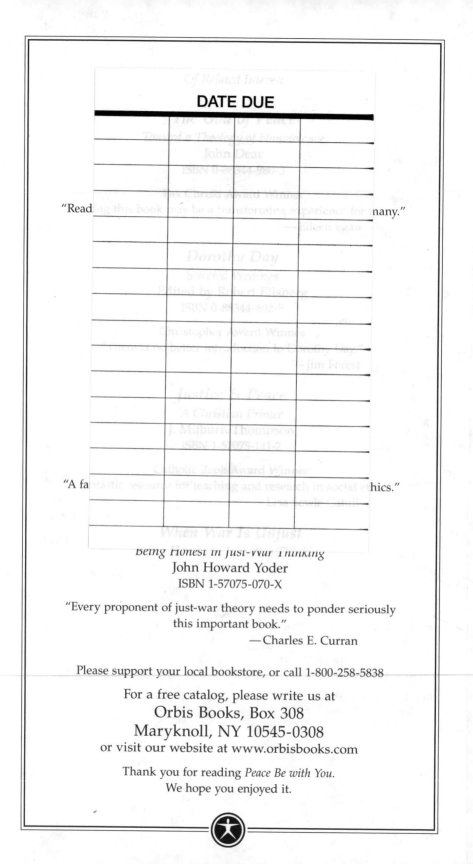

Of Related Interest

The God of Peace

Toward a Theology of Nonviolence

John Dear

ISBN 0-88344-980-3

Pax Christi Award Winner

"Reading this book may be a transforming experience for many."

—Eileen Egan

Dorothy Day

Selected Writings

Edited by Robert Ellsberg

ISBN 0-88344-802-5

Christopher Award Winner

"There is no better introduction to Dorothy Day."

—Jim Forest

Justice & Peace

A Christian Primer

J. Milburn Thompson

ISBN 1-57075-141-2

Catholic Book Award Winner

"A fantastic resource for teaching and research in social ethics."

—Lisa Sowle Cahill

When War Is Unjust

Being Honest in Just-War Thinking

John Howard Yoder

ISBN 1-57075-070-X

"Every proponent of just-war theory needs to ponder seriously this important book."

—Charles E. Curran

Please support your local bookstore, or call 1-800-258-5838

For a free catalog, please write us at

Orbis Books, Box 308

Maryknoll, NY 10545-0308

or visit our website at www.orbisbooks.com

Thank you for reading *Peace Be with You.*

We hope you enjoyed it.